Handbook of Plant Salinity Tolerance

Handbook of Plant Salinity Tolerance

Editor: Walker Williams

www.callistoreference.com

Callisto Reference,
118-35 Queens Blvd., Suite 400,
Forest Hills, NY 11375, USA

Visit us on the World Wide Web at:
www.callistoreference.com

ISBN: 978-1-64116-819-9 (Hardback)

Trademark Notice: Registered trademark of products or corporate names are used only for explanation and identification without intent to infringe.

Cataloging-in-Publication Data

Handbook of plant salinity tolerance / edited by Walker Williams.
    p. cm.
Includes bibliographical references and index.
ISBN 978-1-64116-819-9
1. Plants--Effect of salts on. 2. Salt-tolerant crops--Physiology. 3. Plant physiology.
4. Plants--Hardiness. I. Williams, Walker.
QK753.S3 S25 2023
581.133 54--dc23

Table of Contents

Permissions

List of Contributors

Index

Preface

Plants are exposed to a number of biotic and abiotic stresses. Soil salinity is one of the major abiotic stresses, which negatively affects plant growth and crop production. Stress due to salinity leads to various changes in the physiological and metabolic processes based on the severity and duration of the stress. In the initial stages, soil salinity leads to osmotic stress, which causes physiological changes such as nutrient imbalance, interruption of membranes, decreased photosynthetic activity, and inhibition of the ability to detoxify the reactive oxygen species. In the later stage, soil salinity leads to ionic toxicity that leads to nutrient imbalance in the cytosol. In order to overcome salinity stress, plants have developed several mechanisms such as ion homeostasis and compartmentalization, ion transport and uptake, generation of nitric oxide (NO), and synthesis of polyamines. This book includes some of the vital pieces of works being conducted across the world on various topics related to plant salinity tolerance. It will help the readers in keeping pace with the rapid changes in this area of study.

This book is a result of research of several months to collate the most relevant data in the field.

When I was approached with the idea of this book and the proposal to edit it, I was overwhelmed. It gave me an opportunity to reach out to all those who share a common interest with me in this field. I had 3 main parameters for editing this text:

1. Accuracy – The data and information provided in this book should be up-to-date and valuable to the readers.

2. Structure – The data must be presented in a structured format for easy understanding and better grasping of the readers.

3. Universal Approach – This book not only targets students but also experts and innovators in the field, thus my aim was to present topics which are of use to all.

Thus, it took me a couple of months to finish the editing of this book.

I would like to make a special mention of my publisher who considered me worthy of this opportunity and also supported me throughout the editing process. I would also like to thank the editing team at the back-end who extended their help whenever required.

Editor

Effect of Salinity Stress on Growth and Metabolomic Profiling of *Cucumis sativus* and *Solanum lycopersicum*

Ibrahim Bayoumi Abdel-Farid [1,2,*], **Marwa Radawy Marghany** [2],
Mohamed Mahmoud Rowezek [1] **and Mohamed Gabr Sheded** [2]

[1] Biology Department, College of Science, Jouf University, Sakaka P.O. Box 2014, Saudi Arabia;
     mmroweazk@ju.edu.sa

[2] Botany Department, Faculty of Science, Aswan University, Aswan 81528, Egypt;
     marwa_88@aswu.edu.eg (M.R.M.); msgber@aswu.edu.eg (M.G.S.)

*   Correspondence: bayoumi2013@aswu.edu.eg

Abstract: Seeds germination and seedlings growth of *Cucumis sativus* and *Solanum lycopersicum* were monitored in in vitro and in vivo experiments after application of different concentrations of NaCl (25, 50, 100 and 200 mM). Photosynthetic pigments content and the biochemical responses of *C. sativus* and *S. lycopersicum* were assessed. Salinity stress slightly delayed the seeds germination rate and significantly reduced the percentage of germination as well as shoot length under the highest salt concentration (200 mM) in cucumber. Furthermore, root length was decreased significantly in all treatments. Whereas, in tomato, a prominent delay in seeds germination rate, the germination percentage and seedlings growth (shoot and root lengths) were significantly influenced under all concentrations of NaCl. Fresh and dry weights were reduced prominently in tomato compared to cucumber. Photosynthetic pigments content was reduced but with pronounced decreasing in tomato compared to cucumber. Secondary metabolites profiling in both plants under stress was varied from tomato to cucumber. The content of saponins, proline and total antioxidant capacity was reduced more prominently in tomato as compared to cucumber. On the other hand, the content of phenolics and flavonoids was increased in both plants with pronounced increase in tomato particularly under the highest level of salinity stress. The metabolomic profiling in stressful plants was significantly influenced by salinity stress and some bioactive secondary metabolites was enhanced in both cucumber and tomato plants. The enhancement of secondary metabolites under salinity stress may explain the tolerance and sensitivity of cucumber and tomato under salinity stress. The metabolomic evaluation combined with multivariate data analysis revealed a similar mechanism of action of plants to mediate stress, with variant level of this response in both plant species. Based on these results, the effect of salinity stress on seeds germination, seedlings growth and metabolomic content of plants was discussed in terms of tolerance and sensitivity of plants to salinity stress.

Keywords: cucumber; metabolomic; PCA; photosynthetic pigments; root length; tomato; shoot length

1. Introduction

One third of irrigated lands all over the world are significantly affected by salinity. Searching for tolerant plants to increase the productivity of these lands is considered a great challenge [1]. In Egypt, the problem has been increasing recently as larger areas of irrigated lands have become free from cultivation due to salinity.

Salinity is a serious problem causing decrease of crop productivity through its negative effect on seeds germinations and seedlings growth. It affects plant growth through disturbing plant osmosis, imbalance nutrition channels and ionic toxicity [2,3]. The alteration in these main mechanisms leads to metabolic and physiological changes and poses a negative impact on seed's germination, seedling's growth as evident from retarded shoot and root length, fresh and dry weight, chlorophyll content and its synthesis [2,4–7].

Salinity stress significantly decreases the seeds germination percentage in a number of plants e.g., in *Hordem vulgare* cultivars affecting dry weight [5], rice genotypes by retarding its leave area [2], cotton (*Gossypum hirsutum*) by effecting its various growth parameters including fresh weight, root and shoot lengths, as well as in *Brassica juncea* or other plants [7–9].

The most prominent effect of salinity stress was noticed on photosynthetic pigments in plants including chlorophyll a,b and carotenoids. In most species, photosynthetic pigments are decreased in plants subjected to salinity stress such as in rice cultivars and rice genotypes [2,10], in Pisum species [6], in *Sesbania grandiflora* [11], in *H. vulgare* cultivars [5], in *Viciafaba* [12], in *Ociumumbasilicum* genotypes [13], in *Helianthusannus* [14] and in beet cultivars [15].

Salinity stress not only affects the photosynthetic pigments and chlorophyll synthesisbut also interferes with plant's metabolic and physiological activities in plants through altering primary and secondary metabolites fluxes. Furthermore, the carbohydrates and protein contents are found to vary in plants undersalinity stress [5,9]. In particular, proline as an osmoregulator metabolite is associated with various stresses such as drought and allelochemicals stresses [16–18]. Proline content is increased in plants subjected to salinity stress such as in *H. vulgare* cultivars [5] and in *O. basilicum* genotypes [13].

Furthermore, the alteration in secondary metabolomics poolsuch as polyphenolics, flavonoids, saponins, anthocyanins and tannins is reported in many plants subjected to salinity stress [7,19]. This metabolomics alteration (positive or negative) under salinity stress is controlled by many factors including plant species, developmental stages, magnitude and duration of stress itself.

In most studies, the precursors of secondary metabolites are channelized, leading to an increase in flavonoids and phenolics contents [20]. Flavonoids and phenolics were significantly increased in *Plantago ovata*, rice genotypes [19,21], wheat cultivars and *Z. mays* under salinity stress [7,22]. In another study, the saponin content was also significantly increased in salt stressed *Acalypha wilkesiana* [23].

Althoughsalinity stress is well evaluated in a number of crops, unfortunately there is limited information regarding the impact of salinity on metabolomic changes and metabolomics pathways of the stressful plants during their growth under salinity stress.

Evaluation of the strategies of the interaction of cucumber and tomato with salinity stress may be achieved through having a complete picture of their metabolites under salinity stress. Gaining information considering their metabolomic pathways under salinity stress may be very valuable for genetic engineering's researchers to grow and breed more tolerant varieties or cultivars of these crops. Moreover, multivariate data analysis will be used for the first time to dissect the effect of salinity stress on morphological performance and metabolomic changes in stressful tomato which may give a clear picture about the effect of salinity on tomato metabolomic under salinity stress.

So, the present study was conducted to evaluate the effect of salinity stress on the growth of cucumber and tomato plants and the objectives will also extend to assess the metabolomic changes of the salt stressed plants.

2. Results

2.1. Effect of Salinity Stress on Seeds Germination Rate of C. sativus and S. lycopersicum

In vitro experiment revealed thatsalt stress slightly delayed the seeds germination in cucumber and the delay was more obvious under the highest salt concentrations (200 mM) (Figure 1A). However, comparatively, the seed's germination rate was more prominently affected by salinity stress (Figure 1B). Concentrations of 25 mM and 50 mM of salt delayed the seeds germination as the emergence of radicle was noticed only in the third day under 25 mM and in the seventh day under 50 mM, while it was emerged in the first day in control seeds.

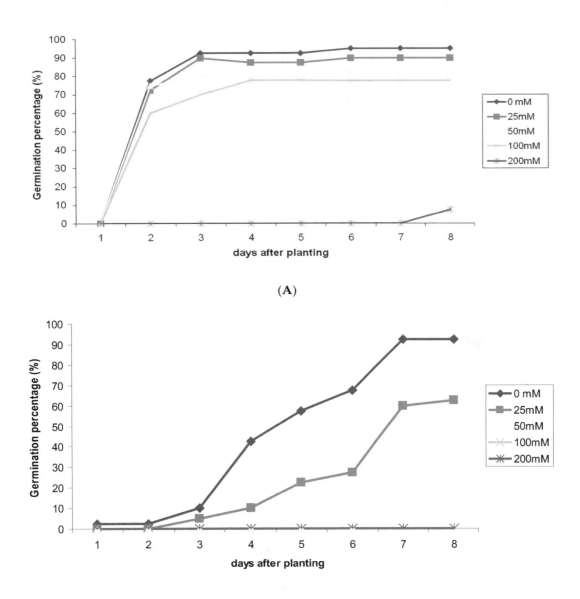

Figure 1. Effect of salinity stress on germination rate in cucumber (**A**) and tomato (**B**).

2.2. Effect of Salinity Stress on Percentage of Seeds Germination of C. sativus and S. lycopersicum

The percentage of seeds germination (8 days post salt application) was recorded for cucumber and tomato (Figure 2A,B). Gradual decreasing in the percentage of seeds germination in both cucumber and tomato is observed with increasing salt concentration (Figure 2A,B). In cucumber, the percentage of seeds germination was significantly reduced only under the highest concentration used (200 mM) (Figure 2A), while the lower concentrations (25–100 mM) did not affect it significantly (Figure 2A). In tomato, the percentage of seeds germination was reduced significantly after application of all doses of salt (i.e., 50–200 mM) (Figure 2B). This behavior is increased with increasing salt concentration (Figure 2B). Surprisingly, the highest concentrations of NaCl (100 and 200 mM) completely inhibited seed germination in tomato (Figure 2B).

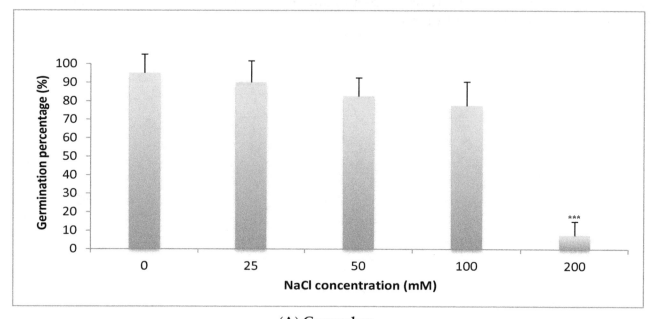

(A) Cucumber

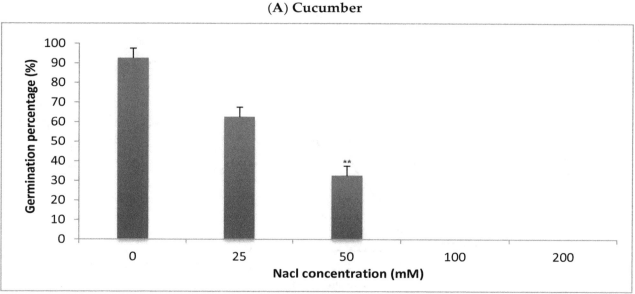

(B) Tomato

Figure 2. Effect of salinity stress on the percentage of seeds germination in cucumber (**A**) and in tomato (**B**). ** = highly significant and *** = very highly significant.

2.3. *Effect of Salinity Stress on Seedlings Growth (Shoot and Root Lengths) in C. sativus and S. lycopersicum*

The shoot length in cucumber was reduced slightly at the lowest concentrations (25 mM and 50 mM) by 3% and 6%, respectively below the control. In contrast, a strong reduction in shoot length was observed under the concentrations of 100 and 200 mM of NaCl (Figure 3A). The shoot length in tomato was significantly decreased under any dose of salt starting from 25 to 200 mM (Figure 3B). The reduction in shoot length was by 32.5% at 25 mM and 53.8% at 50 mM below the control (Figure 3B).

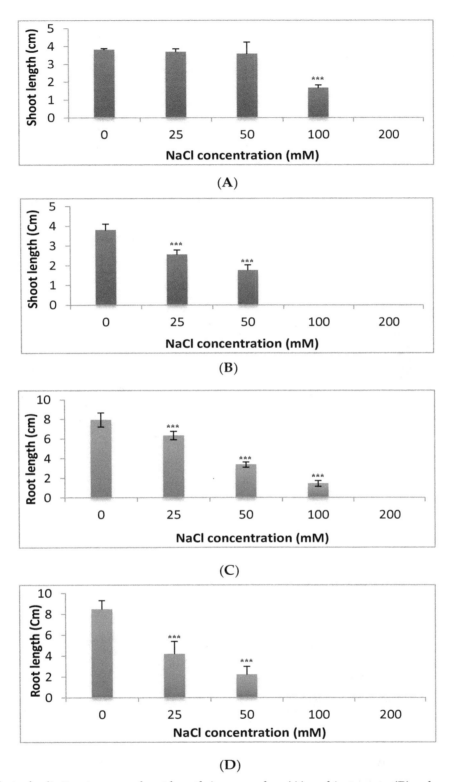

Figure 3. Effect of salinity stress on shoot length in cucumber (**A**) and in tomato (**B**)and on root length in cucumber (**C**) and in tomato (**D**). *** = very highly significant.

Root length was also significantly reduced under all doses of salt in both cucumber and tomato (Figure 3C,D). Salinity stress exerted a dramatic effect on root growth by a significant reduction in root length in cucumber.This effect was consistent with increase in NaCl concentration. Maximum root length (7.92 cm) was observed in seedlings grown under control while root lengths of 6.33, 3.37 and 1.43 cm were recorded under 25, 50 and 100 mMNaCl, respectively.

Similarly, in tomato seedlings, a significant decrease was observed in root length under 25 mM and 50 mMNaCl and the reduction was by 50.6% at 25 mM and 73.9% at 50 mM below the control level (Figure 3D). The results indicated that the root length in both cucumber and tomato was more sensitive than shoot length to salinity stress (Figure 3A–D).

2.4. Effect of Salinity Stress on Fresh and Dry Weights in C. sativus and S. lycopersicum

Fresh weight of both cucumber and tomato was increased under lower levels of salinity stress but was reduced dramatically under higher levels (no growth was recorded under the highest salinity level of both plants) (Figure 4A,B). Dry weight of cucumber was slightly increased at the lower levels of salt stress but decreased prominently under higher level (200 mM) of NaCl application. In case of tomato, the dry weight was reduced dramatically under all levels of salinity stress (Figure 4C,D).

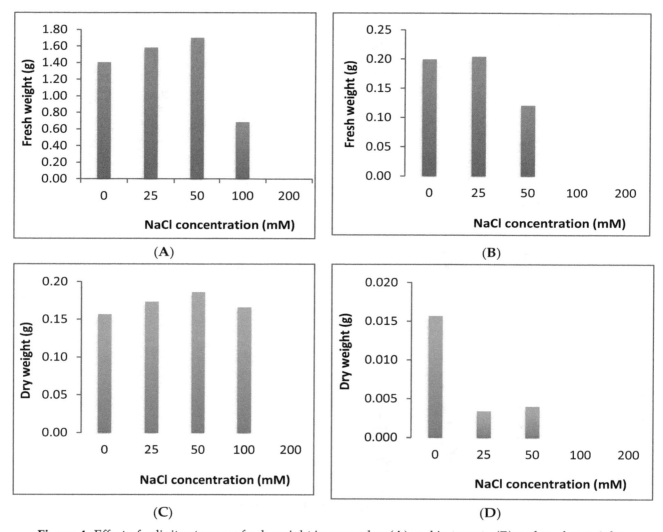

Figure 4. Effect of salinity stress on fresh weight in cucumber (**A**) and in tomato (**B**) and on dry weights of cucumber (**C**) and tomato (**D**).

2.5. Effect of Salinity Stress on Physiological and MetabolomicResponse in C. sativus and S. lycopersicum

2.5.1. Effect of Salinity Stress on the Content of Secondary and Primary Metabolites

Plants grow in pots experiments were analysed for primary and secondary metabolites. The content of flavonoids was reducedunder lower concentrations from NaCl (with a range from 2% to 6% in case of cucumber and 19% to 26% in case of tomato, as below control level).Interestingly, the content of flavonoids was increased in both plants under the highest concentrations used (200 mM in case of cucumber and 100 and 200 mM in case of tomato). The increase was 2% above control in cucumber and 11% and 30% above control in tomato, respectively (Figure 5 and Supplementary Table S1).

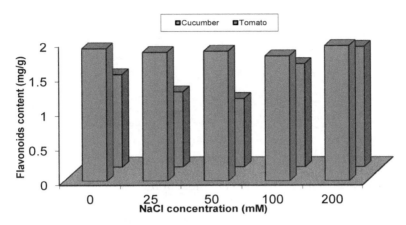

Figure 5. Flavonoids content of cucumber and tomato under NaCl stress.

Effect of salinity on the total phenolics content in cucumber and tomato was presented in Figure 6. Although no significant change was observed in the phenolics content in cucumber under salinity stress, a significant increase in phenolics content was recognized in tomato that is subjected to any dose of NaCl from 50–200 mM. The highest increase wasobserved as 70% above control in tomato that is subjected to 200 mM of NaCl, whereas the lowest increase was 29% above control in plants that are subjected to 100 mM of NaCl (Figure 6).

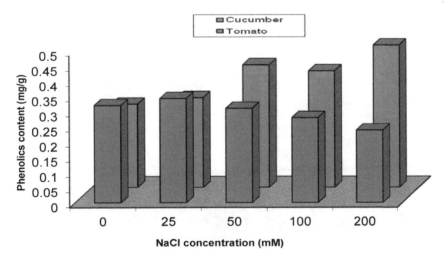

Figure 6. Phenolics content of cucumber and tomato under NaCl stress.

In cucumber, the biosynthesis of saponins was increased significantly ($p < 0.05$) in plants that were subjected to 25 mM of NaCl (4% increase above control) and the content of saponins was reduced significantly with the highest concentration used from NaCl (200 mM) (Figure 7). On the other hand, in case of tomato, the content of saponins was decreased significantly ($p < 0.05$) with lower levels of salinity (25, 50 and 100 mM), but interestingly a significant increase was observed at the highest level (200 mM) of NaCl application (Figure 7). Maximum decrease in saponin content in tomato was recorded in plants exposedto 25 mM NaCl concentration, followed by 50 mM and finally plants under 100 mM of NaCl stress (Figure 7 and Supplementary Table S1).

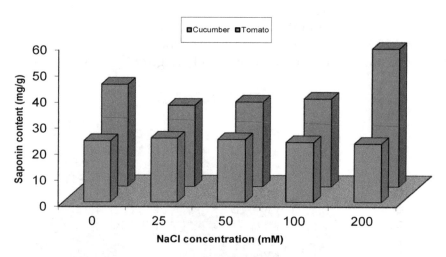

Figure 7. Saponin content in cucumber and tomato under NaCl stress.

The effect of salinity on total antioxidant capacity in cucumber and tomato was presented in Figure 8. In cucumber, no significant change in total antioxidant capacity was noticed under salinity levels, but with increase in the salt concentrations, a slight decreasein the total antioxidant capacitywas observed. In tomato, with increasing salt concentrations, the total antioxidant capacity was significantly reduced ($p < 0.05$) (Figure 8). Pearson's correlation was performed between TAC and the detected secondary metabolites (flavonoids, phenolics and saponins), where TAC was correlated positively with flavonoids content ($r = 0.662$, $p = 0.001$) and negatively with saponins content ($r = -0.0772$, $p = 0.001$).

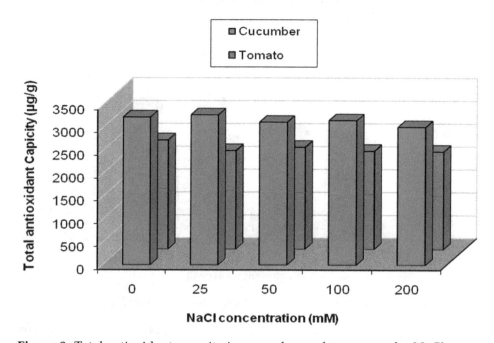

Figure 8. Total antioxidant capacity in cucumber and tomato under NaCl stress.

The change in proline content in cucumber and tomato under salinity stress is presented in Figure 9. In cucumber, even though there was decrease in proline content with different salinity levels, only a significant reduction was observed in plants that were exposed to the highest concentrations of NaCl (100 and 200 mM) (Figure 9 and Supplementary Table S1). While in tomato, the content of proline was decreased at all levels of NaCl, as compared to control plants (Figure 9).

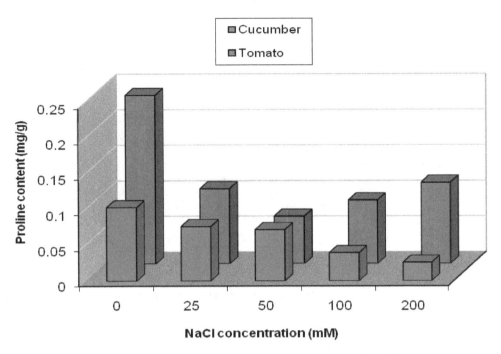

Figure 9. Proline content in cucumber and tomato under salinity stress.

2.5.2. Effect of Salinity Stress on the Content of Photosynthetic Pigments

The content of photosynthetic pigments (chlorophyll a and chlorophyll b) was estimated in the leaves of cucumber and tomato and presented in Figure 10.

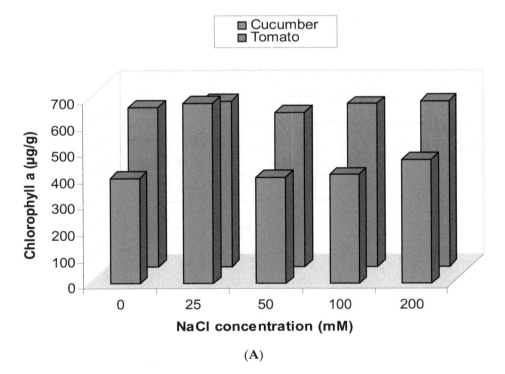

(**A**)

Figure 10. *Cont.*

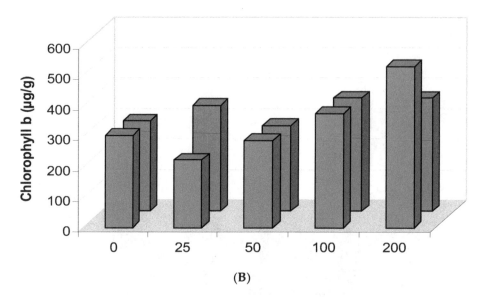

Figure 10. Photosynthetic pigment content in cucumber and tomato under salinity stress. Total chlorophyll a (**A**) and chlorophyll b (**B**).

In cucumber, the content of chlorophyll a was increased in plants under different levels of salinity (25, 50, 100 and 200 mM) ($p > 0.05$). In tomato, a reduction in chlorophyll a content was observed in plants subjected to 50 mM of NaCl and increased under other levels of salt but the increase was not significant (Figure 10A).

In cucumber, chlorophyll b content was increased with the highest salinity levels and the increase was only significant in plants subjected to the highest concentration of NaCl (200 mM) ($p < 0.01$) (Figure 10B). In tomato, a reduction was noticed in chlorophyll b content at low salinity level. At the highest concentrations used from NaCl (50–200 mM), the content of chlorophyll b was increased (Figure 10B).

2.6. Dissecting the Effect of Salinity Stress on Tomato Using Metabolomic and Multivariate Data Analysis

Because of the economical importance of tomato and its use as a model plant, the effect of salinity stress on growth traits and metabolomic profiling of stressful tomato was studied using PCA of multivariate data analysis (MVDA). Growth criteria and metabolomic data combined with PCA showed the change of growth parameters and also metabolomic alteration due to salinity stress in simple figures of score scatter, score loading plots and score biplot (Figure 11A–C). Two distinct groups were obtained;in this case, one included control plants together with plants subjected to lower levels of salt and the second group included plants that are exposed to the highest levels of salt (Figure 11A).

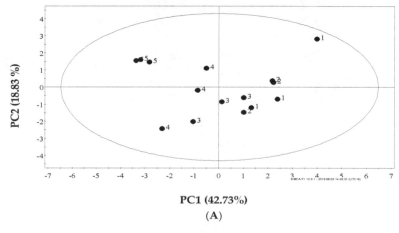

PC1 (42.73%)

(A)

Figure 11. *Cont.*

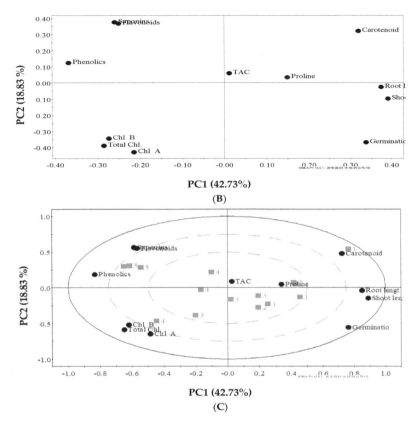

Figure 11. Principal component analysis (PCA) of tomato data under salinity stress. Score scatter plot of PC1 vs. PC2 (**A**), score loading plot of PC1 vs. PC2 (**B**) and score biplot(**C**). 1 = control, 2 = 25 mM, 3 = 50 mM, 4 = 100 mM and 5 = 200 mM of NaCl.

With increasing salinity stress, the content of phenolics, flavonoids, saponins and photosynthetic pigments were increased, whereas percentage of germination, shoot and root lengths, proline, carotenoids and TAC were decreased (Figure 11B,C). Score biplot of PCA confirmed the results of score scatter and loading plots. Stressful plants under the highest concentrations of salt (200 mM) were characterized with higher content of secondary metabolites such as phenolics, flavonoids, saponins and chlorophyll (a, b and total) and lower content of carotenoids, TAC, proline and lower shoot and root length and seeds germination percentage (Figure 11C).

3. Discussion

Salinity stress causes delaying of seeds germination in many crops such as lettuce [24], radish [25], canola [26] and spinach cultivars [27]. The delaying of seeds germination under salinity stress may be used as a clue for its sensitivity [28]. Similarly, germination percentage was significantly declined in tomato seeds even under lower concentrations of salt (i.e., 50 mM), whereas cucumber percentage of germination was significantly reduced only under the highest concentration of salt (200 mM). Consequently, tomato seeds proved to be more sensitive to salinity stress as compared with cucumber seeds. These results are in line with previous studies on different cultivars and genotypes of tomato [29,30], cucumber [31] and also in radish, canola and spinach cultivars [25–27].

Cucumber showed more tolerance to salinity stress than tomato as the significant reduction started only when seeds were subjected to the highest salt concentration (200 mM). These results are in agreement with Passam and Kakouriotis (1994) [31]. They reported a 90% germination loss in seeds subjected to 150 mM. In our study, 20 and 90% germination loss in cucumber seeds was observed under 100 and 200 mM of NaCl concentration, respectively. Different plants showed different sensitivity or tolerance to salinity stress [24]. Generally, the reduction in the percentage of seeds germination under salinity stress particularly with the highest levels may be attributed to the interferenceof salt with the germination starting enzymes. The decrease of seeds germination percentage was attributed also to the drought stress (an indirect effect of salinity) which affects plant metabolism. The effect of salinity stress may cause accumulation of toxic ions and imbalance in nutritional elements in stressful seeds [5].

Shoot length was increased slightly in cucumber until 100 mM, whereas it was significantly reduced in tomato, which is subjected to the lowest concentration used (25 mM). The significant decrease in root length in tomato was more pronounced where an intense reduction in root length was observed with increasing the concentrations of salt. In agreement with the results obtained from the evaluation of shoot length in cucumber, *V. faba* that is exposed to different levels of salinity stress showed significant increase in growth criteria under low and moderate levels, where it showed a significant reduction under high level of salinity stress [12]. A significant reduction was observed in all growth parameters in *H. vulgare* subjected to different levels of salinity stress [5]. Similarly, Jojoba (*Simmondsia chinensis*) under different levels of salinity stress showed a significant reduction in shoot length [32]. Under salinity stress, injury of transpiring leaves due to entering of salt to transpiration steam may consequently affect plant growth [13]. The decrease of water uptake and also toxicity of sodium chloride itself may be also a reason responsible for decreasing plant growth through reduction of shoot length, leaf area and leaf number [2]. Root system is in direct contact with salt and this may affect negatively the enzyme activity and also cell division in root tips causing reduction in root length. The increase in the fresh and dry weights in cucumber under lower level of salinity may be attributed to the increase in shoot length under lower concentrations of NaCl. Whereas the shoot length was intensively reduced in tomato under any levels of salinity stress, this affected negatively the fresh and dry weight in tomato. Decreasing in dry weight of plumule and radicle was reported in *C. officinalis*exposed to different levels of salinity stress [33]. In line with our results considering dry weight in cucumber, *S. chinensis*, subjected to different levels of salinity stress, showed an increase in dry weight till the moderate level of salinity and then decreased significantly under the highest concentrations [32]. Abdul-Qados (2011) [12] reported increase in fresh and dry weight in *V. faba* exposed to different levels of salinity stress even under the highest concentration used (240 mM). Many reports showed the positive effect of salinity stress on fresh and/or dry weight in plants such as the study on *Lactuca sativa* by Andriolo et al. (2005) [34], on *Vigna unguiculata* by Dantus et al. (2005) [35], on *B. vulgaris* and *B. maritime* by Niaz et al. (2005) [36] and on *Atriblex halimus* by Nedjimi et al. (2006) [37]. On thecontrary, some reports showed that the exposure to different salinity stress reduced significantly fresh and/or dry weight in different plants [2,5,8,13,19,25,33,38,39]. Ali et al. (2004) [2] reported a significant decrease in leaf area and leaves number under salinity stress and this reduction in both leaf area and leaves number may also affect the fresh and dry weights of plants under salinity stress. From previous studies, we had information that the germination and growth traits of both cucumber and tomato were negatively affected by the highest levels of salinity stress [29–31]. In these studies, the effect was evaluated only on the level of seeds germination and seedlings growth and there is a lack of information considering the metabolic changes in these stressful plants, which is a crucial step in suggestion of nontraditional solutions to overcome the negative impacts of salinity stress on such economic plants.

Our study has not only evaluated the effect of salinity stress on morphological performance such as seeds germination, seedlings growth and fresh and dry weights of stressful plants but also the changes in photosynthetic pigments, primary and secondary metabolites were monitored in stressful plants showing difference strategies of both cucumber and tomato interacting with salinity stress. Both cucumber and tomato showed different pattern of variation in secondary metabolites content in plants subjected to different levels of salinity stress. Bioactive secondary metabolites such as phenolics and flavonoids might have a hormone-like activity that ultimately elicits an improved tolerance to salinity stress.Phenolics were accumulated in higher rates in tomato subjected to higher levels of salinity stress and this is in agreement with the previous reports. Al Hassan et al. (2015) [40] reported higher phenolics accumulation in cherry tomato under salinity stress. An increase in phenolics content was detected in some rice genotypes under different level of salinity stress [19]. In line with our results, phenolics content also increased under salinity stress in *Chenopodium quinoa* [41] and in *Fagopyrum esculentum* [42]. Contrary to these results, a significant decrease in phenolics content in some plants subjected to different levels of salinity stress such as in *S. chinensis* [32] and in wheat cultivars was observed [22]. Salinity stress had not changed the content of phenolics in wheat and bean cultivars exposed to salinity stress [43]. Flavonoids content was increased in tomato and cucumber under the highest level of salinity stress with prominent increase in tomato. Many previous reports indicated higher accumulation of flavonoids in plants under different levels of salinity stress such as in some rice genotypes [19], in cherry tomato [40] and in *Prosopis strombulifera* [44]. Contrary to these results, *A. wilkesiana*that is exposed to salinity stress showed a significant decrease in flavonoids content [23]. Saponins content was enhanced in tomato subjected to the highest concentration of salt and was reduced significantly under lower concentrations, whereas in cucumber, saponins content was increased under low and moderate levels of salinity. Saponinscontent was accumulated in higher level comparing to control in *A. wilkesiana* subjected to salinity stress [23]. Under low concentration of sodium chloride (15 mM), saponins content was enhanced where it decreased under higher levels of salinity [45]. This is in line with our results in this study on cucumber subjected to salinity stress. In another study, saponins content was decreased in *C. quinoa* subjected to salinity stress [41]. Total antioxidant capacity (TAC) was reduced under salinity stress in tomato and increased in cucumber under low level of salinity. This is inconsistent with other reports which showed higher total antioxidant capacity under salinity stress [22,39,42,44,46]. Plant species differed from each other regarding the TAC due to their content from secondary metabolites accumulated after stress as TAC is correlated with the content of secondary metabolites present in plants either in normal conditions or under stress [47–49].

Generally, proline content showed significant increase after salinity stress [5,13]. In our results, plants subjected to salinity stress either showed a decrease in proline content or the content was not significantly affected. This may be attributed to the replacement of proline by another osmoregulator compound under salinity stress [50,51].

In cucumber and tomato, the content of chlorophyll a and b was increased in plants that are subjected to the highest levels of salinity stress. Induction of chlorophyll under salinity stress as reported in cucumber and tomato was supported by many studies such as that of Misra et al. (1997) and Jamil et al. (2007) [25,52] on their studies on *O. sativa* and *B. vulgaris* under salinity stress, respectively. Under low levels of salinity, the content of chlorophyll a and b was decreased in both plants. Many previous studies support our findings regarding the reduction of chlorophyll in tomato and cucumber under low level of salinity stress [2,5,10,12,15,39,43].

Plants under salinity stress showed different growth and metabolomic responses. This depends on their genetic variation and their content from primary and secondary metabolites. The profiling of secondary metabolites in plants under stress may give a clue concerning tolerance or sensitivity of plants under stress which depend on the severity of salinity stress and also the present of the plant resistance genes [53]. Some sensitive plants respond to stress by accumulating some primary and secondary metabolites as a mechanism to tolerate different types of stresses [54].

4. Materials and Methods

4.1. Germination Experiment

Seeds of cucumber (cv. Beta Alfa) and seeds of tomato were obtained from the National Research Centre (NRC), Giza, Egypt. Homogenized size of *C. sativus* (cucumber) and *S. lycopersicum* (tomato) seeds were soaked in distilled water for 30 min. 70% ethanol was used for seeds surface sterilization for 30 s. After that the seeds were shaken for 10 min in 5% sodium hypochlorite and finally, they were washed 4 times in sterilized distilled water. After sterilization, 10 seeds were transferred into 12 cm sterile Petri dishes containing filter papers. 8 mL of distilled water was used as control. Different concentrations of treatment solution (25, 50, 100, 200 mM NaCl) were used. Distilled water was added when necessary. Four replicates were used for each treatment. Each treatment was replicated

4 times. All previous steps were performed under laminar flow. Petri dishes (control and treated) were numbered and moved to a growth chamber with 23 °C ± 2 and 14/10 h light/dark illumination. When a 2 mm radicle emerged from seed coat and became visible by naked eye, the germinated seeds were counted [55,56]. Counting process was daily recorded at the same time from the day. Percentage of germination, shoot length, root length, fresh weight and dry weight were registered 8 days post germination. The root and shoot lengths were determined using a ruler. Data representing seedlings shoot and root lengths were based on the randomly selected number of seedlings from replicates of each treatment. Fresh weights were determined directly after harvesting, whereas dry weights were calculated after removal of water content at oven at 105 °C for 24 h.

4.2. Pots Experiments

After sterilization of seeds, as described above, 6 seeds of cucumber and 8 seeds of tomato were sown in a plastic pot (15 cm height and 10 cm diameter) that contained 200 g of peat- sand and moss (1.1:5). Pots were transferred to the growth chamber adjusted to 23 °C ± 2 and 14/10 h light/dark illumination. Pots were watered every two days. After 22 days, nutrient solution (Hoagland's solutions) was used twice to irrigate pots. Twenty seven days after plantation, thinning process of seedlings was carried out where each pot contained 3 plants of cucumber and 5 plants of tomato. For salinity treatments, non-salt-treated plants were kept as control and salt-stressed plants were subjected to 25, 50, 100, 200 mM of NaCl solution. Four replicates were used for each treatment. 7 days post saline solutions application, plants were harvested. The harvested samples were washed in distilled water to remove salts and soil remains from the surface tissues. Fresh samples (leaves)were used for the determination of photosynthetic pigments contents (chlorophyll a, chlorophyll b, carotenoids) and proline. The remaining plant samples were dried at room temperature then ground to fine powder which was used in determination of saponins, flavonoids, total phenolics and total antioxidant capacity.

4.3. Plant Analysis

4.3.1. Determination of Total Phenolics (TP)

Folin–Ciocalteau method was used for determination of total phenolics content according to Singleton et al. 1999 [57]. Briefly, 1 mL of the plant extract or different concentrations of standard (gallic acid) were mixed with 1 mL of Folin reagent. Then 1 mL of 10% (w/v) Na_2CO_3 solution was added and mixed. The mixture was allowed to stand for 1 h at room temperature and the absorbance was measured at 700 nm using spectrophotometer. Total phenolics content was expressed (mg) as gallic acid equivalent g^{-1} dry weight.

4.3.2. Determination of Total Flavonoid (TF)

The content of total flavonoids was determined colorimetrically using aluminum chloride according to Zhishen et al. 1999 [58]. Briefly, 1 mL extract or different concentrations of quercetin standard solution was mixed with 0.3 mL of 5% (w/v) $NaNo_2$ solution. After 6 min, 0.3 mL of 10% (w/v) Al $(Cl)_3$ solution was added and the mixture was allowed to stand for a further 6 min before 0.4 mL of 1 MNaOH was added. The mixture was mixed well and placed for 12 min and the absorbance was measured at 510 nm. The results were expressed as mg quercetin equivalent g^{-1} dry weight.

4.3.3. Determination of Total Saponins

The content of total saponins was determined using vanillin solution according to Ebrahimzadeh and Niknam, 1998 [59]. 2.5 mL of vanillin reagent (2 g vanillin/100 mL of sulfuric acid) were added to 1 mL of each extract or different concentrations of saponin standard. The samples and the standard were vortexed for 10 s, then incubated at 60 °C. After 1 h, samples and standards were placed in ice bath for 10 min. The absorbance was measured at 473 nm. The content of total saponins was expressed as mg saponin equivalent g^{-1} dry weight.

4.3.4. Determination of Total Antioxidant Capacity (TAC)

Measurement of total antioxidant capacity was performed according to Prieto et al., (1999) [60]. The reduction of Mo (VI) to Mo (V) by extracts and subsequent formation of green phosphate/Mo (V) complex at acidic pH is the basis of TAC assay. One mL of plant extract was mixed with 1 mL of reagent solution (0.6 M sulfuric acid, 28 mM sodium phosphate and 4 mM ammonium molybdate) and incubated at 90 °C for 90 min. Series of ascorbic acid concentrations was prepared as above. After cooling to room temperature, the absorbance of reaction mixture of samples and standard was measured at 695 nm against blank (containing only reagent and extraction solvent of samples). The total antioxidant activity was expressed as mg ascorbic acid equivalent g^{-1} dry weight.

4.3.5. Determination of Free Proline

Twenty five mg of grinded fresh samples was dissolved in 2 mL of 3% (w/v) aqueous 5-sulfosalicylic acid solution and centrifuged at 7000 rpm for 20 min. One mL of the supernatant was mixed with 2 mL of acidic ninhydrin reagent (2.5 g ninhydrin/100 mL of a solution containing glacial acetic acid, distilled water and orthophosphoric acid 85% at a ratio of 6:3:1) and boiled in a water bath for 1 h and then cooled in an ice bath. Two mL of toluene was added to the mixture and vortexed for 20 s. The colored toluene layer was decanted from the aqueous phase and warmed at room temperature. The absorbance was read at 520 nm using toluene as a blank and the free proline content was determined from a curve constructed with proline standard according to the method of Bates et al., 1973 [61]. The proline concentration was calculated as a fresh weight basis (mg g^{-1} FW).

4.3.6. Determination of Photosynthetic Pigments

The content of photosynthetic pigments was estimated according to the method of Lichtenthaler and Wellburn (1985) and Dere et al. (1998) [62,63]. Fifty mg of fresh samples was extracted using 5 mL of absolute methanol overnight then homogenized and centrifuged for 10 min at 1000 rpm. The supernatant was separated and the absorbance was read at 666, 653 and 470 nm. Chlorophyll a, chlorophyll b, total carotene and total chlorophyll were calculated according to (Dere et al., 1998) [63]. The pigment level was expressed as µg g^{-1} FW.

4.4. Data Analysis

One way analysis of variance (ANOVA) from Minitab version 12.21 was used to assess the significant difference between percentages of germination, shoot length, root length, fresh and dry weight. The significant difference in the content of secondary metabolites (phenolics, flavonoids, saponins, total antioxidant capacity) and photosynthetic pigments (chlorophyll a, chlorophyll b and carotene) in control and in salinity stressful plants was also evaluated using ANOVA from Minitab. Data is a mean with standard deviation of three or four replicates. $p < 0.05$ considered significant, $p < 0.01$ considered highly significant and $p < 0.001$ considered very highly significant. Principal component analysis (PCA) was performed with the SIMCA-P software (v. 12.01, Umetrics, Umeå, Sweden) for reduction of dimensionality among the metabolomic data and morphological data to evaluate the effect of salinity on plant growth and their metabolomic composition.

5. Conclusions

Plants are smart for responding to abiotic stress which is specific to plant species by responding differently or at different levels through metabolomic alterations. Growth traits were significantly reduced in tomato under salinity stress. Cucumber growth traits were not affected under low or moderate salinity stress but root length was significantly reduced. According to the results of seeds germination and seedlings growth and also the metabolic changes in both plants, one can conclude that cucumber is moderately salt tolerant, whereas tomato is moderately salt sensitive. Both plant species showed different strategies for tolerating the salinity stress, probably due to the difference in their metabolic contents. Sensitive plants respond to salinity stress by accumulating some secondary metabolites as a way to increase their tolerance against salinity stress. Searching for some safe and friendly ways for the environment to overcome the negative effect of salinity on the growth and development of these crops is desirable.

Author Contributions: Conceptualization, I.B.A.-F. and M.R.M.; formal analysis, I.B.A.-F. and M.R.M.; methodology and investigation, M.R.M.; data curation, M.R.M. and M.M.R.; software, I.B.A.-F.; validation, I.B.A.-F.; supervison, I.B.A.-F. and M.G.S.; visualization, M.M.R.; writing–original draft preparation, I.B.A.-F.; writing–review & editing, I.B.A.-F., M.R.M., M.M.R. and M.G.S. All authors have read and agreed to the published version of the manuscript.

References

1. Maas, E.V.; Hoffman, G.J. Crop salt tolerance-current assessment. *J. Irrig. Drain. Div.* **1977**, *103*, 115–134.
2. Ali, Y.; Aslam, Z.; Ashraf, M.Y.; Tahir, G.R. Effect of salinity on chlorophyll concentration, leaf area, yield and yield components of rice genotypes grown under saline environment. *Int. J. Environ. Sci. Technol.* **2004**, *1*, 221–225. [CrossRef]
3. Kamran, M.; Parveen, A.; Ahmar, S.; Malik, Z.; Hussain, S.; Chattha, M.S.; Saleem, M.H.; Adil, M.; Heidari, P.; Chen, J.T. An overview of hazardous impacts of soil salinity in crops, tolerance mechanisms, and amelioration through selenium supplementation. *Int. J. Mol. Sci.* **2020**, *21*, 148. [CrossRef] [PubMed]
4. Sonar, B.A.; Desai Nivas, D.; Gaikwad, D.K.; Chanven, P.D. Assessment of salinity-induced antioxidative defense system in Colubrina asiatica Brong. *J. Stress Physiol. Biochem.* **2011**, *7*, 193–200.
5. Movafegh, S.; Jadid, R.R.; Kiabi, S. Effect of salinity stress on chlorophyll content, proline, water soluble carbohydrate, germination, growth and dry weight of three seedling barley (*Hordeum vulgare* L.) cultivars. *J. Stress Physiol. Biochem.* **2012**, *8*, 157–168.
6. Miljuš-Djukić, J.; Stanisavljević, N.; Radović, S.; Jovanović, Ž.; Mikić, A.; Maksimović, V. Differential response of three contrasting pea (*Pisum arvense*, *P. sativum* and *P. fulvum*) species to salt stress: Assessment of variation in antioxidative defence and miRNA expression. *Aust. J. Crop Sci.* **2013**, *7*, 2145–2153.
7. Singh, P.K.; Shahi, S.K.; Singh, A.P. Effects of salt stress on physic-chemical changes in maize (*Zea mays* L.) plants in response to salicylic acid. *Indian J. Plant Sci.* **2015**, *4*, 69–77.
8. Saleh, B. Salt stress alters physiological indicators in cotton (*Gossypium hirsutum* L.). *Soil Environ.* **2012**, *31*, 113–118.
9. Jamil, M.; Shik Rha, E. NaCl stress–induced reduction in growth, photosynthesis and protein in mustard. *J. Agric. Sci.* **2013**, *5*, 114–127. [CrossRef]
10. Chandramohanan, K.T.; Radhakrishnan, V.V.; Abhilash Joseph, E.; Mohanan, K.V. A study on the effect of salinity stress on the chlorophyll content of certain rice cultivars of Kerala state of India. *Agric. For. Fish.* **2014**, *3*, 67–70. [CrossRef]
11. Dhanapackiam, S.; Ilyas, M.H.M. Effect of salinity on chlorophyll and carbohydrate contents of *Sesbania grandiflora* seedlings. *Indian J. Sci. Technol.* **2010**, *3*, 64–66. [CrossRef]
12. Abdul Qados, A.M.S. Effect of salt stress on plant growth and metabolism of bean plant *Vicia faba* (L.). *J. Saudi Soc. Agric. Sci.* **2011**, *10*, 7–15. [CrossRef]
13. Heidari, M. Effects of salinity stress on growth, chlorophyll content and osmotic components of two basil (*Ocimum basilicum* L.) genotypes. *Afr. J. Biotechnol.* **2012**, *11*, 379–384. [CrossRef]
14. Santos, C.V. Regulation of chlorophyll biosynthesis and degradation by salt stress in sunflower leaves. *Sci. Hortic.* **2004**, *103*, 93–99. [CrossRef]
15. Masoumzadeh, B.M.; Imani, A.A.; Khayamaim, S. Salinity stress effect on proline and chlorophyll rate in four beet cultivars. *Ann. Biol. Res.* **2012**, *3*, 5453–5456. [CrossRef]
16. Joshi, B.; Pandey, S.N.; Rao, P.B. Allelopathic effects of weeds extracts on germination, growth and biochemical aspects in different varieties of wheat (*Triticum aestivum*). *Indian J. Agric. Res.* **2009**, *43*, 79–87.
17. Das, C.R.; Mondal, N.K.; Aditya, P.; Datta, J.K.; Banerjee, A.; Das, K. Allelopathic potentialities of leachates of leaf litter of some selected tree species on gram seeds under laboratory conditions. *Asian J. Exp. Biol. Sci.* **2012**, *3*, 59–65.
18. Abdel-Farid, I.B.; El-Sayed, M.A.; Mohamed, E.A. Allelopathic potential of *Calotropis procera* and *Morettia philaeana*. *Int. J. Agric. Biol.* **2013**, *15*, 120–134.

19. Hussain, M.; Park, H.W.; Farooq, M.; Jabran, K.; Lee, D.J. Morphological and physiological basis of salt resistance in different rice genotypes. *Int.J. Agric. Biol.* **2013**, *15*, 113–118.

20. Abdel-Farid, I.B.; Jahangir, M.; van den Hondel, C.A.M.J.J.; Kim, H.K.; Choi, Y.H.; Verpoorte, R. Fungal infection-induced metabolites in *Brassica rapa*. *Plant Sci.* **2009**, *176*, 608–615. [CrossRef]

21. Haghighi, Z.; Karimi, N.; Modarresi, M.; Mollayi, S. Enhancement of compatible solute and secondary metabolites production in *Plantago ovata* Forsk. by salinity stress. *J. Med. Plants Res.* **2012**, *6*, 3495–3500. [CrossRef]

22. Ashraf, M.A.; Ashraf, M.; Ali, Q. Response of two genetically diverse wheat cultivars to salt stress at different growth stages: Leaf lipid peroxidation and phenolic contents. *Pak. J. Bot.* **2010**, *42*, 559–565.

23. Odjegba, V.J.; Alokolaro, A.A. Simulated drought and salinity modulates the production of phytochemicals in *Acalypha wilkesiana*. *J. Plant Stud.* **2013**, *2*, 105–112. [CrossRef]

24. Zapata, P.J.; Serrano, M.; Pretel, M.T.; Amorós, A.; Botella, M.Á. Polyamines and ethylene changes during germination of different plant species under salinity. *Plant Sci.* **2004**, *167*, 781–788. [CrossRef]

25. Jamil, M.; Ur Rehman, S.; Lee, K.J.; Kim, J.M.; Kim, H.-S.; Shik Rha, E. Salinity reduced growth PS2 photochemistry and chlorophyll content in radish. *Sci. Agric.* **2007**, *64*, 111–118. [CrossRef]

26. Bybordi, A. The influence of salt stress on seed germination, growth and yield of canola cultivars. *Not. Bot. Horti Agrobot. Cluj-Napoca* **2010**, *38*, 128–133. [CrossRef]

27. Turhan, A.; Kuşçu, H.; Şeniz, V. Effects of different salt concentrations (NaCl) on germination of some spinach cultivars. *J. Agric. Fac. Uludag Univ.* **2011**, *25*, 65–77.

28. Foolad, M.R.; Lin, G.Y. Relationships between cold-and salt-tolerance during seed germination in tomato: Germplasm evaluation. *Plant Breed.* **1999**, *118*, 45–48. [CrossRef]

29. Jones, R.A. High salt tolerance potential in Lycopersicon species during germination. *Euphytica* **1986**, *35*, 575–582. [CrossRef]

30. Doğan, M.; Avu, A.; Can, E.N.; Aktan, A. Farklı domates tohumlarının çimlenmesiüzerine tuz stresinin etkisi. *SDÜ Fen Edebiyat Fakültesi Fen Dergisi (E-DERGİ)* **2008**, *3*, 174–182. (In Turkish)

31. Passam, H.C.; Kakouriotis, D. The effects of osmoconditioning on the germination emergence and early plant growth of cucumber under saline conditions. *Sci. Hortic.* **1994**, *57*, 233–240. [CrossRef]

32. Roussos, P.A.; Pontikis, C.A. Long term effects of sodium chloride salinity on growing *in vitro*, proline and phenolic compound content of jojoba explants. *Eur. J. Hortic. Sci.* **2003**, *68*, 38–44.

33. Gharineh, M.H.; Khoddami, H.R.; Rafieian-kopaei, M. The influence of different levels of salt stress on germination of marigold (*Calendula officinalis* L.). *Int. J. Agric. Crop Sci.* **2013**, *5*, 1851–1854.

34. Andriolo, J.L.; Luz, G.L.; Witter, M.H.; Godoi, R.S.; Barros, G.T.; Bortolotto, O.C. Growth and yield of lettuce plants under salinity. *Hortic. Bras.* **2005**, *23*, 931–934. [CrossRef]

35. Dantus, B.F.; Ribeiro, L.; Aragao, C.A. Physiological response of cowpea seeds to salinity stress. *Rev. Bras.Sem.* **2005**, *27*, 144–148. [CrossRef]

36. Niaz, B.H.; Athar, M.; Salim, M.; Rozema, J. Growth and ionic relations of fodder beet and sea beet under saline. *Int. J. Environ. Sci. Technol.* **2005**, *2*, 113–120. [CrossRef]

37. Nedjimi, B.; Daoud, Y.; Touati, M. Growth, water relations, proline and ion content of in vitro cultured *Atriplex halimus* sub sp. *Schweinfurthii as affected by CaCl2. Commun. Biom. Crop Sci.* **2006**, *1*, 79–89.

38. Memon, S.A.; Hou, X.; Wang, L.J. Morphological analysis of salt stress response of pak Choi. *Electron. J. Environ. Agric. Food Chem.* **2010**, *9*, 248–254.

39. Sevengor, S.; Yaşar, F.; Kusvuran, S.; Ellialtıoğlu, S. The effect of salt stress on growth, chlorophyll content, lipid peroxidation and antioxidative enzymes of pumpkin seedling. *Afr. J. Agric. Res.* **2011**, *6*, 4920–4924. [CrossRef]

40. Al Hassan, M.; Fuertes, M.M.; Sánchez, F.J.R.; Vicente, O.; Boscaiu, M. Effects of salt and water stress on plant growth and on accumulation of osmolytes and antioxidant compounds in cherry tomato. *Not. Bot. Horti Agrobot. Cluj-Napoca* **2015**, *43*, 1–11. [CrossRef]

41. Gómez-Caravaca, A.M.; Segura-Carretero, A.; Fernández-Gutiérrez, A.; Caboni, M.F. Simultaneous determination of phenolic compounds and saponins in quinoa (*Chenopodium quinoa* Willd) by a liquid chromatography-diode array detection-electrospray ionization-time-of-flight mass spectrometry methodology. *J. Agric. Food Chem.* **2011**, *59*, 10815–10825. [CrossRef] [PubMed]

42. Lim, J.-H.; Park, K.-J.; Kim, B.-K.; Jeong, J.W.; Kim, H.-J. Effect of salinity stress on phenolic compounds and carotenoids in buckwheat (*Fagopyrum esculentum* M.) sprout. *Food Chem.* **2012**, *135*, 1065–1070. [CrossRef] [PubMed]

43. Radi, A.A.; Farghaly, F.A.; Hamada, A.M. Physiological and biochemical responses of salt–tolerant and salt–sensitive wheat and bean cultivars to salinity. *J. Biol. Earth Sci.* **2013**, *3*, B72–B88.

44. Reginato, M.A.; Castagna, A.; Furlán, A.; Castro, S.; Ranieri, A.; Luna, V. Physiological responses of a halophytic shrub to salt stress by Na_2SO_4 and NaCl: Oxidative damage and the role of polyphenols in antioxidant protection. *AOB Plants* **2014**, *6*, 1–13. [CrossRef]

45. Attaran, E. Canadian Society of Plant Physiologists/La Société Canadiene de Physiologie Végétale, Annual Meeting. In Proceedings of the Evaluation of the Effect of Salinity Stress on Saponin Contents in Bellis Perrenis L., Guelph, ON, Canada, 19–22 June 2004.

46. Docimo, T.; De Stefano, R.; Cappetta, E.; Piccinelli, A.L.; Celano, R.; De Palma, M.; Tucci, M. Physiological, biochemical, and metabolic responses to short and prolonged saline stress in two cultivated cardoon genotypes. *Plants* **2020**, *9*, 554. [CrossRef]

47. Abdel-Farid, I.B.; Sheded, M.G.; Mohamed, E.A. Metabolomic profiling and antioxidant activity of some Acacia species. *Saudi J. Biol. Sci.* **2014**, *21*, 400–408. [CrossRef]

48. Abdel-Farid, I.B.; Mahalel, U.A.; Jahangir, M.; Elgebaly, H.A.; El-Naggar, S.A. Metabolomic profiling and antioxidant activity of *Opophytum forsskalii*. *Aljouf Univ. Sci. Eng. J.* **2016**, *3*, 19–24. [CrossRef]

49. El-Naggar, S.; Abdel-Farid, I.B.; Elgebaly, H.A.; Germoush, M.O. Metabolomic profiling, antioxidant capacity and in vitro anticancer activity of some compositae plants growing in Saudi Arabia. *Afr. J. Pharm. Pharmacol.* **2015**, *9*, 764–774. [CrossRef]

50. Mittler, R. Abiotic stress, the field environment and stress combination. *Trends Plant Sci.* **2006**, *11*, 15–19. [CrossRef]

51. Kasim, W.A.; Dowidar, S. Amino acids and soluble protein profile of radish seedlings under salt stress as affected by GA_3 priming. *Indian J. Plant Physiol.* **2006**, *11*, 75–82. [CrossRef]

52. Misra, A.; Sahu, A.N.; Misra, M.; Singh, P.; Meera, I.; Das, N.; Kar, M.; Sahu, P. Sodium chloride induced changes in leaf growth, and pigment and protein contents in two rice cultivars. *Biol. Plant.* **1997**, *39*, 257–262. [CrossRef]

53. Bai, Y.; Kissoudis, C.; Yan, Z.; Visser, R.G.; van der Linden, G. Plant behaviour under combined stress: Tomato responses to combined salinity and pathogen stress. *Plant J.* **2018**, *93*, 781–793. [CrossRef]

54. El-Khatib, A.A.; Barakat, N.A.; Nazeir, H. Growth and physiological response of some cultivated species under allelopathic stress of *Calotropis procera* (Aiton) W.T. *Appl. Sci. Rep.* **2016**, *14*, 237–246. [CrossRef]

55. El-Khatib, A.A.; Abd-Elaah, G.A. Allelopathic potential of *Zilla spinosa* on growth of associate flowering plants and some rhizosphere fungi. *Biol. Plant.* **1998**, *41*, 461–467. [CrossRef]

56. Grange, S.L.; Leskovar, D.I.; Pike, L.M.; Cobb, B.G. Excess moisture and seed coat nicking influence germination of triploid watermelon. *HortScience* **2000**, *35*, 1355–1356. [CrossRef]

57. Singleton, V.L.; Orthifer, R.; Lamuela-Raventos, R.M. Analysis of total phenols and other oxidation substrates and antioxidants by means of Folin-ciocalteau reagent. *Methods Enzymol.* **1999**, *299*, 152–178. [CrossRef]

58. Zhishen, J.; Mengcheng, T.; Jianming, W. The determination of flavonoid contents in mulberry and their scavenging effects on superoxide radicals. *Food Chem.* **1999**, *64*, 555–559. [CrossRef]

59. Ebrahimzadeh, H.; Niknam, V. A revised spectrophotometric method for determination of triterpenoid saponins. *Indian Drugs* **1998**, *35*, 379–381.

60. Prieto, P.; Pineda, M.; Aguilar, M. Spectrophotometric quantitation of antioxidant capacity through the formation of a phosphomolybdenum complex: Specific application to the determination of vitamin E. *Anal. Biochem.* **1999**, *269*, 337–341. [CrossRef] [PubMed]

61. Bates, L.S.; Waldren, R.P.; Teare, T.D. Rapid determination of free proline for water-stress studies. *Plant Soil* **1973**, *39*, 205–207. [CrossRef]

62. Lichtenthaler, H.K.; Wellburn, A.R. Determination of total carotenoids and chlorophylls a and b of leaf in different solvents. *Biochem. Soc. Trans.* **1985**, *11*, 591–592. [CrossRef]

63. Dere, S.; Güneş, T.; Sivaci, R. Spectrophotometric determination of chlorophyll A, B and total carotenoid contents of some algae species using different solvents. *Tur. J. Bot.* **1998**, *22*, 13–17.

2

Exogenous Salicylic Acid Modulates the Response to Combined Salinity-Temperature Stress in Pepper Plants (*Capsicum annuum* L. var. Tamarin)

Ginés Otálora *, María Carmen Piñero, Jacinta Collado-González, Josefa López-Marín ⓘ
and Francisco M. del Amor * ⓘ

Department of Crop Production and Agri-Technology,
Murcia Institute of Agri-Food Research and Development (IMIDA), C/Mayor s/n, 30150 Murcia, Spain;
mariac.pinero2@carm.es (M.C.P.); jacinta.collado@carm.es (J.C.-G.); josefa.lopez38@carm.es (J.L.-M.)
* Correspondence: gines.oralora@carm.es (G.O.); franciscom.delamor@carm.es (F.M.d.A.)

Abstract: Growers in the cultivated areas where the climate change threatens the agricultural productivity and livelihoods are aware that the current constraints for good quality water are being worsened by heatwaves. We studied the combination of salinity (60 mM NaCl) and heat shock stress (43 °C) in pepper plants (*Capsicum annuum* L. var. Tamarin) since this can affect physiological and biochemical processes distinctly when compared to separate effects. Moreover, the exogenous application of 0.5 mM salicylic acid (SA) was studied to determine its impacts and the SA-mediated processes that confer tolerance of the combined or stand-alone stresses. Plant growth, leaf Cl^- and NO_3^- concentrations, carbohydrates, and polyamines were analyzed. Our results show that both salinity stress (SS) and heat stress (HS) reduced plant fresh weight, and SA only increased it for HS, with no effect for the combined stress (CS). While SA increased the concentration of Cl^- for SS or CS, it had no effect on NO_3^-. The carbohydrates concentrations were, in general, increased by HS, and were decreased by CS, and for glucose and fructose, by SA. Additionally, when CS was imposed, SA significantly increased the spermine and spermidine concentrations. Thus, SA did not always alleviate the CS and the plant response to CS cannot be directly attributed to the full or partial sum of the individual responses to each stress.

Keywords: heat shock stress; salinity; combined stress; NaCl; temperature; extreme weather

1. Introduction

Global warming is increasing the number, intensity, and duration of abiotic stress combinations worldwide, which impair crop growth, yield, and product quality [1]. The plant response to such combinations of stresses is complex, involving a multitude of molecular signaling pathways that control several responses, which, in turn, may interact with one another [2]. Therefore, it has been suggested that each abiotic stress combination requires new research as it should be studied as an entirely new stress [3]. Increases in both air temperature and the salinity of irrigation waters will be two of the major constraints to human food production in the coming years. Therefore, improved agronomic management is of paramount importance to develop key adaptation strategies intended to increase stress tolerance in crops [4]. Salinity is a serious concern, being an increasing problem in agriculture because of the competition of good-quality water from industry and the progressive salinization of aquifers and other water resources, especially in arid and semiarid regions. This scarcity of good-quality water is dramatically accentuated by the rainfall alterations provoked by climate change. In addition, extreme temperatures associated with prolonged heatwaves impact more than 10% of land surfaces [5].

A cautionary example is that of the summer of 2003, when a heatwave that impacted Europe resulted in a 30% reduction in ecosystem gross primary production [6], while a more devastating heatwave during 2010 in Russia resulted in an estimated 50% reduction in gross primary production [7].

Plant growth regulators (phytohormones) play an essential role in plant responses to biotic and abiotic stress [8,9]. Salicylic acid (SA) is an endogenous plant hormone and different studies have pointed out the positive effects of using this hormone as a treatment to stimulate plant growth under abiotic stress conditions, including combinations of drought, heat, and salinity [10,11], since SA acts as a key signaling molecule under such conditions. Its activity is essential for basal defense and systemic acquired resistance [12] and leads to the reprogramming of the expression of genes and the synthesis of proteins, affecting a number of metabolic processes [13]. However, the exact mode of action of SA is still poorly understood, as SA signaling is not a simple linear route and SA may interact with several other stress-related compounds [13].

Pepper (*Capsicum annuum* L.) is a salt-sensitive crop [14], and the use of poor-quality waters causes a significant reduction in yield, especially in the marketable yield [15]. Thus, pepper yield and quality are affected by temperature and salinity stresses [16], with the optimal temperature range for growth being 20–30 °C, while above 32 °C pepper shows pollination and fertilization problems and a significant reduction in fruit quality [17]. However, vegetative and generative growth can be differentially affected [18], with specific heat-stress-related genes involved in determining heat tolerance [19]. Pepper is an important horticultural crop due to its economic and nutraceutical values and is cultivated worldwide, especially in greenhouses of the Mediterranean-climate areas. To our knowledge, this is the first paper addressing combined drought and salinity stress in pepper, with the addition of SA. The purpose of this research was to gain insights into the combined effect of salinity and drought in pepper, and its response to exogenous (leaf-sprayed) SA. Thus, the specific objectives were: (i) to examine the extent to which the studied stresses had univocal effects and the degree of interaction when plants were submitted to both stresses, (ii) to evaluate growth and ion-specific responses, to characterize the sum of the stresses, (iii) to determine whether the combined stresses had a synergic or antagonist effect regarding the leaf carbohydrate concentration, (iv) to reveal whether polyamines modulated the general response, and (v) to determine whether the spraying of SA onto plants subjected to the combination of stresses, or the individual stresses, altered the response of the studied parameters, with mitigation of the effects of the stresses.

2. Results

2.1. Plant Growth

Plant fresh weight (aerial part) decreased with the exposure to salinity and heat (Figure 1a). Thus, the fresh weight values recorded for control plants were reduced by 20.4% and 32.4% when salinity or heat stress, respectively, was imposed, but salinity did not reduce the growth of plants already exposed to heat stress (combined stress). The spraying of SA boosted the fresh weight of non-stressed plants (20.3%) and of heat-stressed plants (39.6%); however, this effect was clearly diminished with salinity alone or with the combined stress.

The leaf water content (Figure 1b) was reduced by salinity, especially when no HS was imposed; thus, HS increased the water content of those leaves submitted to SS. Additionally, SA did not affect the response to HS or SS alone or to the combination of both stresses. Thus, in our controlled conditions, with no restrictions in the water supply and a nutrient balance in the rhizosphere, heat increased the leaf water content in salinized plants.

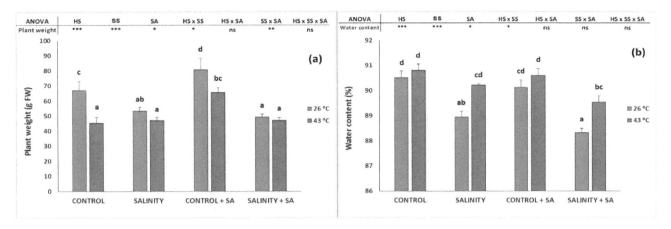

Figure 1. Effect of heat-shock and salinity, with or without salicylic acid, on the fresh weight of the aerial part (leaves and stems) of pepper plants (**a**), and on the leaf water content (**b**) in the leaves of pepper plants. The * refers to significant differences at the level of $p \leq 0.05$; ** $p \leq 0.005$; *** $p \leq 0.001$; n.s., not significant. Different lowercase letters denote significant differences between columns, $p < 0.05$ HS is heat stress; SS is salinity stress; SA is salicylic acid.

2.2. Ion Concentrations

As expected, the leaf Cl^- increased when Cl^- in the nutrient solution was increased (Figure 2a), and HS dramatically increased the Cl^- concentration in the leaves of SS plants. Surprisingly, we found a differential effect with respect to the application of SA; thus, in leaves sprayed with SA, the concentration of this ion significantly increased at both temperatures by 16.3% (26 °C) and 38.1% (43 °C). The concentration of nitrate in the leaves was increased by 37.6% by HS with respect to control plants, and HS increased nitrate by 22.1% when combined with SS. However, the application of SA had no effect on the nitrate concentration under HS or SS (Figure 2b).

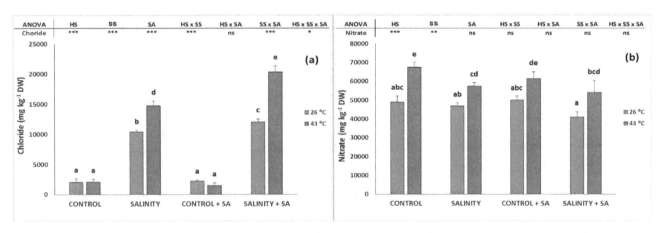

Figure 2. Effect of heat-shock and salinity, with or without salicylic acid, on the Cl^- (**a**) and NO_3^- (**b**) concentrations in the leaves of pepper plants. The * refers to significant differences at the level of $p \leq 0.05$; ** $p \leq 0.005$; *** $p \leq 0.001$; n.s., not significant. Different lowercase letters denote significant differences between columns, $p < 0.05$ HS is heat stress; SS is salinity stress; SA is salicylic acid.

2.3. Total Soluble Sugars

The carbohydrates concentrations in the leaves are presented in Figure 3. When HS was imposed on control plants, the glucose concentration (Figure 3a) was increased by 40.1%, but this effect of temperature was augmented by salinity, and with the CS, the glucose concentration increased by 55.75%. The SA application had a lower effect on non-salinized plants, but caused an increase at ambient temperature (SS plants), but not for HS. The patterns observed for the other carbohydrates measured (fructose (Figure 3b) and sucrose (Figure 3c)) were similar to those for glucose.

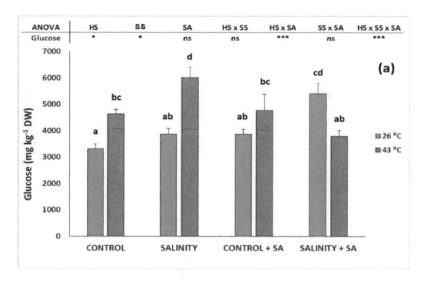

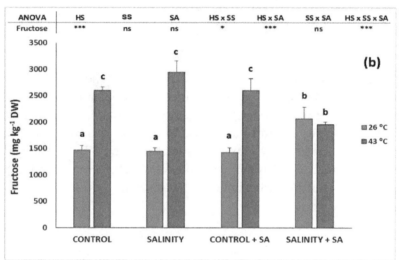

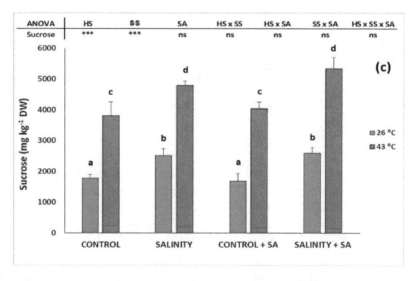

Figure 3. Effect of heat-shock and salinity, with or without salicylic acid, on the carbohydrates (glucose (**a**), fructose (**b**), and sucrose (**c**)) in the leaves of pepper plants. The * refers to significant differences at the level of $p \leq 0.05$; *** $p \leq 0.001$; n.s., not significant. Different lowercase letters denote significant differences between columns, $p < 0.05$ HS is heat stress; SS is salinity stress; SA is salicylic acid.

2.4. Polyamines Analysis

The concentration of putrescine was not affected by HS (Figure 4a); however, when only SS was applied, we found a dramatic increase in this polyamine, from 70.2 nmol g^{-1} FW to 144.5 nmol g^{-1} FW. Interestingly, with the CS, its concentration in the leaves was lower at 86.6 nmol g^{-1} FW. The application of SA did not alter this response in the leaves, and only in SS plants did it produce a significant change. Spermine was not significantly altered by heat or salinity or by the combined stress when no SA was sprayed (Figure 4b), but SA applied to plants that were submitted to salinity or to both stresses jointly originated a significant increase that was not observed with no HS and without application of SA. In this way, spermine was increased by 61.5% and 40.4% when SA was applied under HS and the combined stress, respectively. Spermidine (Figure 4c) had a similar behavior with the exception that significant differences were observed for HS plants, and SA gave increases of 36.4% and 27.7% when applied to the HS and CS plants, respectively. Note that the combined stress did not significantly increase the spermidine concentration with respect to non-salinized plants. Cadaverine had an interesting and distinct pattern with respect to spermine or spermidine (Figure 4d): SA increased the concentration of cadaverine in the absence of stresses, and moreover, it produced a reduction in the response when plants were submitted to CS, as compared to SA application under SS alone.

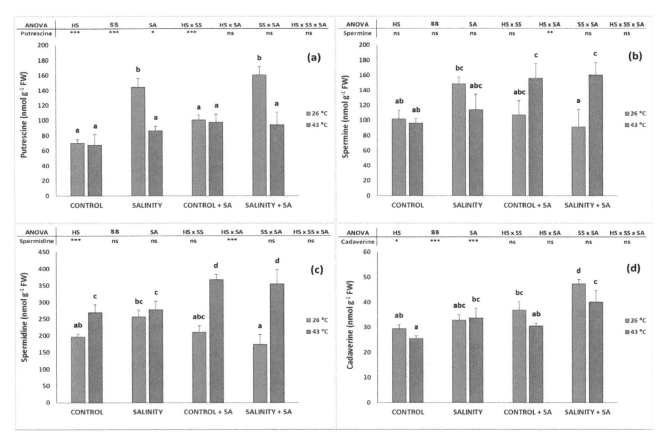

Figure 4. Effect of heat-shock and salinity, with or without salicylic acid, on the polyamines (putrescine (**a**), spermine (**b**), spermidine (**c**), and cadaverine (**d**)) in the leaves of pepper plants. The * refers to significant differences at the level of $p \leq 0.05$; ** $p \leq 0.005$; *** $p \leq 0.001$; n.s., not significant. Different lowercase letters denote significant differences between columns, $p < 0.05$. HS is heat stress; SS is salinity stress; SA is salicylic acid.

2.5. Principal Component Analysis (PCA)

Principal component analysis (PCA) was applied to our results to see the relationship of the variables studied with the temperature and salinity conditions, together with the effect of SA.

The first two main components explain 56% of the total variability of the 11 variables analyzed. The points represent the measurements and the arrows, whose length indicates the amount of variance explained by each variable, represent the variables. The abscissa axis "Dim1" is mainly a salinity gradient, where the variables located to the right of the axis are more affected by salinity than those located to the left (Figure 5b). In turn, the ordinate axis (Dim2) is a temperature gradient (Figure 5a). The highest temperature (heat stress) is at the top of the axis, while the lowest temperature is at the bottom of the axis.

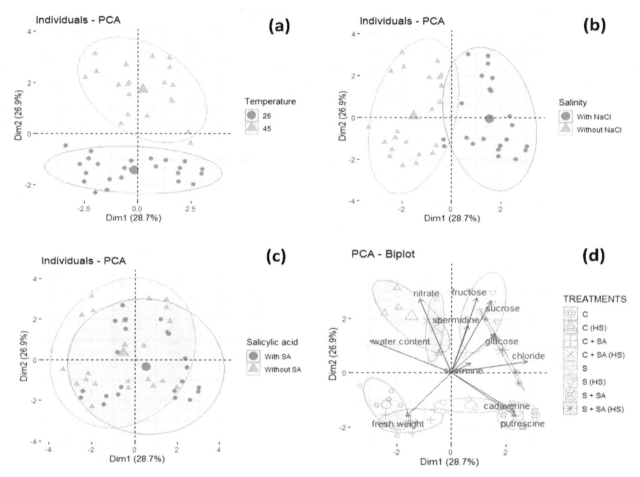

Figure 5. Principal component analysis (PCA) of the parameters analyzed in pepper plants cultivated with or without heat-shock and/or salinity, and with or without salicylic acid. (**a**) Cases are grouped by the temperature conditions applied; (**b**) cases are grouped by the salinity conditions applied; (**c**) cases are grouped by the salicylic acid conditions applied; and (**d**) PCA biplot where the cases are grouped by the treatments applied: C, control; S, salinity; SA, salicylic acid; HS, heat stress.

Figure 5d shows how the glucose, fructose, sucrose, spermine, and spermidine variables are grouped by the heat and salinity stress conditions. Moreover, the cadaverine and putrescine variables are affected by salinity but not heat stress. The control treatment and heat stress mainly affect the water content and nitrate. Conversely, the fresh weight variable is grouped in the control treatment.

3. Discussion

When plants are submitted to an unfavorable combination of stresses, the exogenous application of phytohormones has been pointed out as a method to assist their adaptation to these conditions [11]. SA is a plant hormone that regulates growth and several biochemical and physiological processes in plants [20]. Our data show that SA had an effect on shoot FW under control and HS conditions, but not for the combined stress (HS and SS). Khan et al. [21] reported that application of 0.5 mM SA to non-stressed wheat plants improved the photosynthetic characteristics significantly, resulting in greater growth. Additionally, SA directly influences the activity of enzymes that participate in mechanisms of heat tolerance or induces the genes involved [22]. This agrees with our data showing that SA ameliorated HS, giving an increase in fresh weight under HS. However, for this pepper cultivar and intensity of SS (60 mM NaCl), no beneficial effect of SA was observed for fresh weight. Pancheva et al. [23] reported a decrease in the growth of leaves and roots and delayed leaf emergence of barley plants when SA (0.1 mM–1 mM) was applied exogenously. However, Gunes et al. [24] reported interesting results for maize about the exogenous application of SA (0.1 to 1 mM) and its effect on nutrient concentrations; SA increased plant growth significantly, in both saline and non-saline conditions. In contrast, Hayat et al. [25] underlined that SA could exert deleterious effects on plants under normal growth conditions, as a decline in net photosynthetic rate, and the transpiration rate was observed in maize plants. Therefore, in our work, for the chosen cultivar, SA acted beneficially when salinity was applied, rather than under the control or HS conditions. Barba-Espin et al. [26] reported that SA (100 µM) negatively affected the growth parameters of pea plants subjected to salt stress (70 mM NaCl). This was attributed to an SA-induced increase in oxidative damage during NaCl stress [27], together with a decrease in the NO_3^- concentration; this latter effect of SA was not found in our study.

SA exacerbated the uptake of Cl^- by salinized plants at both ambient and extreme temperature. Intriguingly, some studies found that SA increased the root length by 45% compared to the control plants [28], an effect that could favor an increase in the uptake of this anion. Additionally, the observed effect was even bigger at higher temperatures, which would have induced a higher potential transpiration flux. Interestingly, Barba-Espín et al. [26] indicated that the percentage of leaves showing chlorosis symptoms was increased after application of SA under SS. However, in our experiment, the leaves did not have necrotic margins despite the fact that the leaf Cl^- concentration was boosted by the application of SA under SS. Moreover, the increase in Cl^- and these effects of SA on growth and ion concentrations may depend on the plant species [24]. In our study, N assimilation was not enhanced by SA, in contrast to the findings of Nazar et al. [29] and Khan et al. [21] for salinity (for mungbean and wheat, respectively), but HS caused an increase in the accumulation of this nutrient for our pepper cultivar. Hayat et al. [25] indicated that the treatment of maize plants with lower concentrations of SA enhanced N uptake and the activity of the enzyme NR, whereas higher concentrations were inhibitory (they did not lower Cl^- accumulation or raise the N concentration). Thus, in that crop, SA likely acts differentially depending on the concentration sprayed. It should be underlined that the accumulation of this toxic anion (Cl^-) in the leaves could have impaired the photosynthetic machinery, leading to the reduction in growth observed for CS, but clearly, the response of growth to that stress combination is qualitatively different to the sum of the individual stress responses. Silva et al. [30] found that the combined stress (HS and SS) was more harmful to *Jatropha curcas* than the stresses applied individually, with a main role for the detrimental effect of Cl^- (enhanced uptake of this toxic ion), whereas in tomato such interaction produced a higher degree of tolerance [31]. However, the work of Silva et al. [30] and ours were performed at significantly higher temperatures.

Sugars are a major source of carbon and energy in plants, but also play an important signaling role in many physiological processes [32]. Recent works have shown that, under abiotic stress, sucrose import is blocked, and this can have a critical effect on generative plant growth, causing the abortion of fruits [33]. Our data for the leaves indicate that the sugar contents were, in general, enhanced by the HS imposed, but SA had an effect in the plants subjected to the CS, decreasing glucose and fructose but not sucrose. Elwan et al. [34] observed that spraying pepper plants with a low concentration of SA (1 µM) decreased significantly the sugars content of leaves but increased it in fruits, and attributed this to the role of SA in energy status balancing, translocation, and storage of assimilates. The research of Zhou et al. [35] agrees with our data in that heat stress significantly increased the glucose, fructose, and sucrose in the leaves of tomato. In general, it is stated that high temperatures can cause a rapid consumption of carbohydrates for the maintenance of respiration [36], but we can envisage that, in our conditions, the photosynthetic metabolism was increased and able to counteract the demand for carbohydrates of this cultivar. Thus, Zhang et al. [32] indicated that plants supplemented with SA under HS showed a significant enhancement in the content of photosynthetic pigments. Carbohydrates have a fundamental role in cell osmotic adjustment and in the maintenance of membrane integrity [37], and an increase in the total soluble sugars was observed in heat-tolerant genotypes [19], whilst for sensitive cultivars, HS has been associated with disturbed carbohydrate metabolism due to an increase in oxidative damage [38].

Polyamines have an important role in the regulation of the plant response to abiotic stresses [39–41]. Recently, Tajti et al. [42] identified, in Arabidopsis SA-deficient mutants, a possible crosstalk between the SA and polyamine signaling pathways, whilst Collado-González et al. [43] reported in cauliflower that extreme heat increased the content of all polyamines. Gupta et al. [44] indicated that the principal enzymes involved in the polyamine biosynthesis pathway are under complex metabolic and developmental control. Tanou et al. [45] reported that polyamines ameliorated NaCl stress in five-month-old sour orange, but in our recent work in melon heat stress provoked a significant effect on the polyamines response [46], whilst Upadhyay et al. [47] indicated in tomato leaves that heat stress (42 °C) led to a decrease in the levels of free putrescine. As indicated previously, we did not find a general summative response to the combined stress or a general amelioration effect of the SA against the stresses applied jointly or individually. Such a combination of stresses should be studied as a new state of abiotic stress that demands new, complex acclimation responses [48] in which SA may also trigger new signaling pathways.

4. Materials and Methods

4.1. Growth Conditions and Treatments Applied

Pepper plants (*Capsicum annuum* L.) var. Tamarin (Enza Zaden España S.L., Almeria, Spain), which are bloky pepper type, were pregerminated in commercial seed trays (El Jimenado S.A., Torre-Pacheco, Murcia). Seedlings were transplanted into 40 5-L pots filled with coconut fiber and watered with a modified Hoagland solution. They were grown in a climate chamber with the characteristics described by del Amor et al. (2010), with fully controlled environmental conditions. Photosynthetically active radiation (PAR) of 250 µmol m^{-2} s^{-1} was provided by a combination of fluorescent lamps (TL-D Master reflex 830 and 840, Koninklijke Philips Electronics N.V., Amsterdam, The Netherlands) and high pressure lamps (Son-T Agro, Koninklijke Philips Electronics N.V., The Netherlands). The initial day/night conditions (14/10 h) in the chamber were 26/22/18 °C (14/4/6 h), with a relative humidity of 60%. The plants were kept for 15 days under these conditions. After the acclimatization period, half of the plants were watered with a modified Hoagland solution (control) and 60 mM NaCl was added to this solution for the other half. In turn, half of the plants in each irrigation treatment were treated with 0.5 mM salicylic acid (plus 0.01% Tween-20 as a surfactant) every 3 days for 15 days. Furthermore, the effect of a thermal shock was tested by subjecting the plants to a temperature of 43 °C daily for 6 h. The irrigation was increased to maintain 35% drainage, thus avoiding nutrient imbalance in the roots. Thus, there were 8 treatments with 5 repetitions each: plants watered with the control solution, plants watered with the control solution + 60 mM NaCl, plants watered with the control solution + 0.5 mM salicylic acid, and plants watered with the control solution + 60 mM NaCl + 0.5 mM salicylic acid; all these treatments were submitted to ambient temperature (26 °C) or to heat shock stress (43 °C).

4.2. Plant Growth

Forty plants were analyzed, and the aerial parts were weighed and separated into leaves and stems (including petioles). The water content was calculated from the fresh weights and the dry weights that were determined after a minimum of 72 h at 65 °C.

4.3. Ion Concentrations

Leaf NO_3^- and Cl^- were extracted from ground material (0.4 g) with 20 mL of deionized water. They were analyzed in an ion chromatograph (Metrohm 861 Advanced Compact IC; Metrohm 838 Advanced Sampler). The column used was a Metrohm Metrosep A Supp7 250/4.0 mm.

4.4. Total Soluble Sugars

The contents of sugars (fructose, sucrose, and glucose) in pepper leaves were measured using an 817 Bioscan ion chromatography system (Metrohm® Ltd., Herisau, Switzerland) equipped with a pulsed amperometric detector (PAD) and a gold electrode. The column used was a METROHM Metrosep Carb 1–150 IC column (4.6 mm × 250 mm), heated to 32 °C.

4.5. Polyamines Analysis

The polyamine (PA) contents were studied according to [46], with minor modifications. Leaf samples (2 g) were mixed with 5 mL of perchloric acid (5%), homogenized for 2 min, kept for 1 h under refrigeration with periodic stirring, and then centrifuged at 5000× g for 8 min. Each supernatant, containing free PAs, was placed in a plastic jar and kept in a freezer at −20 °C until used. The free PAs were benzoylated by taking 1 mL of each sample and mixing it with 1mL of 2 mol L^{-1} NaOH and 20 µL of benzoyl chloride. The mixture was stirred in a vortex mixer for 15 s and was allowed to rest for 20 min at room temperature. Subsequent to this, 4 mL of a saturated solution of sodium chloride were added, and the system was stirred while 2 mL of diethyl ether were added. The system was left at rest for 30 min, at −20 °C. Then, 1 mL of the diethyl ether phase was taken and evaporated. The residue was resuspended in 0.5 mL of acetonitrile/water (56:44 v/v). The PAs present were determined with an ACQUITY UPLC system (Waters, Milford, MA, USA) equipped with a UV detector (230 nm) and a reversed-phase column (ACQUITY UPLC HSS T3 1.8 µm, 2.1 mm × 100 mm) maintained at 40 °C. An acetonitrile/water mixture (42:58 v/v) was used as the elution solvent, with a flow rate of 0.55 mL min^{-1}.

4.6. Statistical Analysis

The data from the studied parameters were tested for homogeneity of variance and normality of distribution. The SPSS statistical package (IBM SPSS Statistics 25.0, Armonk, NY, USA) was used for the analysis of variance (ANOVA), to determine the effects of temperature, salinity, and salicylic acid, and their interactions. Additionally, Duncan's multiple range test was used to determine the significance ($p < 0.05$) of the differences among means. Principal component analysis (PCA) was carried out in R, version 4.0.2 [49].

5. Conclusions

Exogenous application of SA seems to be a promising management tool to confer heat and salinity stress tolerance. However, when combined, these two stresses elicited a markedly different response in the pepper plants. Logically, the species and variety, the dose and frequency of application, phenological development, the target organ/tissue, and the type of combined stresses considered all the influences of the ability of SA to provide tolerance and thus maintain the global food supply. This study provides new insights into SA-mediated processes; however, future research is of paramount importance to gain a better knowledge of the efficiency and effectiveness as well as the mechanism of action of this hormone in pepper crops under combined stresses.

Author Contributions: Conceptualization, G.O. and F.M.d.A.; methodology, G.O.; software, G.O.; validation, M.C.P., J.C.-G. and J.L.-M.; formal analysis, G.O. and F.M.d.A.; investigation, G.O.; resources, G.O.; data curation, F.M.d.A.; writing—original draft preparation, G.O. and F.M.d.A.; writing—review and editing, G.O.; visualization, M.C.P. and J.C.-G.; supervision, F.M.d.A. and J.L.-M.; project administration, F.M.d.A.; funding acquisition, F.M.d.A. All authors have read and agreed to the published version of the manuscript.

Acknowledgments: We thank José Sáez Sironi, José Manuel Gambín, Miguel Marín, and Raquel Roca, for technical assistance, and David J. Walker, for his assistance with the correction of the English.

References

1.   Pandey, P.; Irulappan, V.; Bagavathiannan, M.V.; Senthil-Kumar, M. Impact of Combined Abiotic and Biotic Stresses on Plant Growth and Avenues for Crop Improvement by Exploiting Physio-morphological Traits. *Front. Plant Sci.* **2017**, *8*, 537. [CrossRef] [PubMed]

2.   Atkinson, N.J.; Lilley, C.J.; Urwin, P.E. Identification of Genes Involved in the Response of Arabidopsis to Simultaneous Biotic and Abiotic Stresses. *Plant Physiol.* **2013**, *162*, 2028–2041. [CrossRef] [PubMed]

3.   Mittler, R.; Blumwald, E. Genetic Engineering for Modern Agriculture: Challenges and Perspectives. *Annu. Rev. Plant Biol.* **2010**, *61*, 443–462. [CrossRef] [PubMed]

4.   Francini, A.; Sebastiani, L. Abiotic Stress Effects on Performance of Horticultural Crops. *Horticulturae* **2019**, *5*, 67. [CrossRef]

5.   E Hansen, J.; Sato, M.; Ruedy, R. Perception of climate change. *Proc. Natl. Acad. Sci. USA* **2012**, *109*, E2415–E2423. [CrossRef]

6.   Ciais, P.; Reichstein, M.; Viovy, N.; Granier, A.; Ogée, J.; Allard, V.; Aubinet, M.; Buchmann, N.; Bernhofer, C.; Carrara, A.; et al. Europe-wide reduction in primary productivity caused by the heat and drought in 2003. *Nature* **2005**, *437*, 529–533. [CrossRef]

7.   Bastos, A.; Gouveia, C.; Trigo, R.M.; Running, S.W. Comparing the impacts of 2003 and 2010 heatwaves in NPP over Europe. *Biogeosci. Discuss.* **2013**, *10*, 15879–15911. [CrossRef]

8.   Ding, P.; Ding, Y. Stories of Salicylic Acid: A Plant Defense Hormone. *Trends Plant Sci.* **2020**, *25*, 549–565. [CrossRef]

9.   Rekhter, D.; Lüdke, D.; Ding, Y.; Feussner, K.; Zienkiewicz, K.; Lipka, V.; Wiermer, M.; Zhang, Y.; Petit-Houdenot, Y. Isochorismate-derived biosynthesis of the plant stress hormone salicylic acid. *Science* **2019**, *365*, 498–502. [CrossRef]

10.   Shaar-Moshe, L.; Blumwald, E.; Peleg, Z. Unique Physiological and Transcriptional Shifts under Combinations of Salinity, Drought, and Heat. *Plant Physiol.* **2017**, *174*, 421–434. [CrossRef]

11.   Torun, H. Time-course analysis of salicylic acid effects on ROS regulation and antioxidant defense in roots of hulled and hulless barley under combined stress of drought, heat and salinity. *Physiol. Plant.* **2018**, *165*, 169–182. [CrossRef] [PubMed]

12.   Huang, W.; Wang, Y.; Li, X.; Zhang, Y. Biosynthesis and Regulation of Salicylic Acid and N-Hydroxypipecolic Acid in Plant Immunity. *Mol. Plant* **2020**, *13*, 31–41. [CrossRef] [PubMed]

13.   Janda, T.; Szalai, G.; Pál, M. Salicylic Acid Signalling in Plants. *Int. J. Mol. Sci.* **2020**, *21*, 2655. [CrossRef] [PubMed]

14.   Del Amor, F.M.; Cuadra-Crespo, P. Plant growth-promoting bacteria as a tool to improve salinity tolerance in sweet pepper. *Funct. Plant Biol.* **2012**, *39*, 82–90. [CrossRef] [PubMed]

15.   Rameshwaran, P.; Tepe, A.; Yazar, A.; Ragab, R. Effects of drip-irrigation regimes with saline water on pepper productivity and soil salinity under greenhouse conditions. *Sci. Hortic.* **2016**, *199*, 114–123. [CrossRef]

16.   Erickson, A.N.; Markhart, A.H. Flower developmental stage and organ sensitivity of bell pepper (*Capsicum annuum* L.) to elevated temperature. *Plant Cell Environ.* **2002**, *25*, 123–130. [CrossRef]

17.   Guo, M.; Yin, Y.-X.; Ji, J.-J.; Ma, B.-P.; Lu, M.-H.; Gong, Z.-H. Cloning and expression analysis of heat-shock transcription factor gene CaHsfA2 from pepper (*Capsicum annuum* L.). *Genet. Mol. Res.* **2014**, *13*, 1865–1875. [CrossRef]

18.   Kafizadeh, N.; Carapetian, J.; Manouchehri Kalantari, K. Effects of Heat Stress on Pollen Viability and Pollen Tube Growth in Pepper. *Res. J. Biol. Sci.* **2008**, *3*, 1159–1162.

19.   Li, T.; Xu, X.; Li, Y.; Wang, H.; Li, Z.; Li, Z. Comparative transcriptome analysis reveals differential transcription in heat-susceptible and heat-tolerant pepper (*Capsicum annum* L.) cultivars under heat stress. *J. Plant Biol.* **2015**, *58*, 411–424. [CrossRef]

20.   Vicente, M.R.-S.; Plasencia, J. Salicylic acid beyond defence: Its role in plant growth and development. *J. Exp. Bot.* **2011**, *62*, 3321–3338. [CrossRef]

21.   Khan, M.I.R.; Iqbal, N.; Masood, A.; Per, T.S.; A Khan, N. Salicylic acid alleviates adverse effects of heat stress on photosynthesis through changes in proline production and ethylene formation. *Plant Signal. Behav.* **2013**, *8*, e26374. [CrossRef] [PubMed]

22.   Horváth, E.; Szalai, G.; Janda, T. Induction of Abiotic Stress Tolerance by Salicylic Acid Signaling. *J. Plant Growth Regul.* **2007**, *26*, 290–300. [CrossRef]

23. Pancheva, T.; Popova, L.; Uzunova, A. Effects of salicylic acid on growth and photosynthesis in barley plants. *J. Plant Physiol.* **1996**, *149*, 57–63. [CrossRef]

24. Gunes, A.; Inal, A.; Alpaslan, M.; Eraslan, F.; Bagci, E.G.; Cicek, N. Salicylic acid induced changes on some physiological parameters symptomatic for oxidative stress and mineral nutrition in maize (*Zea mays* L.) grown under salinity. *J. Plant Physiol.* **2007**, *164*, 728–736. [CrossRef] [PubMed]

25. Hayat, Q.; Hayat, S.; Irfan, M.; Ahmad, A. Effect of exogenous salicylic acid under changing environment: A review. *Environ. Exp. Bot.* **2010**, *68*, 14–25. [CrossRef]

26. Barba-Espín, G.; Clemente-Moreno, M.J.; Álvarez, S.; García-Legaz, M.F.; Hernández, J.A.; Diaz-Vivancos, P. Salicylic acid negatively affects the response to salt stress in pea plants. *Plant Biol.* **2011**, *13*, 909–917. [CrossRef] [PubMed]

27. Borsani, O.; Valpuesta, V.; Botella, J.R. Evidence for a Role of Salicylic Acid in the Oxidative Damage Generated by NaCl and Osmotic Stress in Arabidopsis Seedlings. *Plant Physiol.* **2001**, *126*, 1024–1030. [CrossRef]

28. Gutiérrez-Coronado, M.A.; Trejo-López, C.; Larqué-Saavedra, A. Effects of salicylic acid on the growth of roots and shoots in soybean. *Plant Physiol. Biochem.* **1998**, *36*, 563–565. [CrossRef]

29. Nazar, R.; Iqbal, N.; Syeed, S.; Khan, N.A. Salicylic acid alleviates decreases in photosynthesis under salt stress by enhancing nitrogen and sulfur assimilation and antioxidant metabolism differentially in two mungbean cultivars. *J. Plant Physiol.* **2011**, *168*, 807–815. [CrossRef]

30. Silva, E.N.; Vieira, S.A.; Ribeiro, R.V.; Ponte, L.F.A.; Ferreira-Silva, S.L.; Silveira, J.A.G. Contrasting Physiological Responses of Jatropha curcas Plants to Single and Combined Stresses of Salinity and Heat. *J. Plant Growth Regul.* **2013**, *32*, 159–169. [CrossRef]

31. Rivero, R.M.; Mestre, T.C.; Mittler, R.; Rubio, F.; Garcia-Sanchez, F.; Martinez, V. The combined effect of salinity and heat reveals a specific physiological, biochemical and molecular response in tomato plants. *Plant Cell Environ.* **2013**, *37*, 1059–1073. [CrossRef] [PubMed]

32. Yi, Z.; Li, S.; Liang, Y.; Zhao, H.; Hou, L.; Shi, Y.; Ahammed, G.J. Effects of Exogenous Spermidine and Elevated CO2 on Physiological and Biochemical Changes in Tomato Plants Under Iso-osmotic Salt Stress. *J. Plant Growth Regul.* **2018**, *37*, 1222–1234. [CrossRef]

33. Ruan, Y.-L. Sucrose Metabolism: Gateway to Diverse Carbon Use and Sugar Signaling. *Annu. Rev. Plant Biol.* **2014**, *65*, 33–67. [CrossRef]

34. Elwan, M.; El-Hamahmy, M. Improved productivity and quality associated with salicylic acid application in greenhouse pepper. *Sci. Hortic.* **2009**, *122*, 521–526. [CrossRef]

35. Zhou, R.; Yu, X.; Kjær, K.H.; Rosenqvist, E.; Ottosen, C.-O.; Wu, Z. Screening and validation of tomato genotypes under heat stress using Fv/Fm to reveal the physiological mechanism of heat tolerance. *Environ. Exp. Bot.* **2015**, *118*, 1–11. [CrossRef]

36. Teskey, R.; Wertin, T.; Bauweraerts, I.; Ameye, M.; McGuire, M.A.; Steppe, K. Responses of tree species to heat waves and extreme heat events. *Plant Cell Environ.* **2015**, *38*, 1699–1712. [CrossRef]

37. Chen, H.; Jiang, J.G. Osmotic adjustment and plant adaptation to environmental changes related to drought and salinity. *Environ. Rev.* **2010**, *18*, 309–319. [CrossRef]

38. Mittler, R. Abiotic stress, the field environment and stress combination. *Trends Plant Sci.* **2006**, *11*, 15–19. [CrossRef]

39. Mostafaei, E.; Zehtab-Salmasi, S.; Salehi-Lisar, Y.; Ghassemi-Golezani, K. Changes in photosynthetic pigments, osmolytes and antioxidants of Indian mustard by drought and exogenous polyamines. *Acta Biol. Hung.* **2018**, *69*, 313–324. [CrossRef]

40. Piñero, M.C.; Otálora, G.; Porras, M.E.; Sánchez-Guerrero, M.C.; Lorenzo, P.; Medrano, E.; Del Amor, F.M. The Form in Which Nitrogen Is Supplied Affects the Polyamines, Amino Acids, and Mineral Composition of Sweet Pepper Fruit under an Elevated CO2Concentration. *J. Agric. Food Chem.* **2017**, *65*, 711–717. [CrossRef]

41. Piñero, M.C.; Porras, M.E.; López-Marín, J.; Sánchez-Guerrero, M.C.; Medrano, E.; Lorenzo, P.; Del Amor, F.M. Differential Nitrogen Nutrition Modifies Polyamines and the Amino-Acid Profile of Sweet Pepper Under Salinity Stress. *Front. Plant Sci.* **2019**, *10*. [CrossRef]

42. Tajti, J.; Hamow, K.Á.; Majláth, I.; Gierczik, K.; Németh, E.; Janda, T.; Pál, M. Polyamine-Induced Hormonal Changes in eds5 and sid2 Mutant Arabidopsis Plants. *Int. J. Mol. Sci.* **2019**, *20*, 5746. [CrossRef] [PubMed]

43. Collado-González, J.; Piñero, M.C.; Otálora, G.; López-Marín, J.; Del Amor, F.M. Exogenous spermidine modifies nutritional and bioactive constituents of cauliflower (*Brassica oleracea* var. botrytis L.) florets under heat stress. *Sci. Hortic.* **2021**, *277*, 109818. [CrossRef]

44. Gupta, K.; Dey, A.; Gupta, B. Plant polyamines in abiotic stress responses. *Acta Physiol. Plant.* **2013**, *35*, 2015–2036. [CrossRef]

45. Tanou, G.; Ziogas, V.; Belghazi, M.; Christou, A.; Filippou, P.; Job, D.; Fotopoulos, V.; Molassiotis, A. Polyamines reprogram oxidative and nitrosative status and the proteome of citrus plants exposed to salinity stress. *Plant Cell Environ.* **2013**, *37*, 864–885. [CrossRef]

46. Piñero, M.C.; Otálora, G.; Collado, J.; López-Marín, J.; Del Amor, F.M. Foliar application of putrescine before a short-term heat stress improves the quality of melon fruits (*Cucumis melo* L.). *J. Sci. Food Agric.* **2020**. [CrossRef]

47. Upadhyay, R.; Fatima, T.; Handa, A.K.; Mattoo, A.K. Polyamines and Their Biosynthesis/Catabolism Genes are Differentially Modulated in Response to Heat Versus Cold Stress in Tomato Leaves (*Solanum lycopersicum* L.). *Cells* **2020**, *9*, 1749. [CrossRef]

48. Colmenero-Flores, J.M.; Rosales, M.A. Interaction between salt and heat stress: When two wrongs make a right. *Plant Cell Environ.* **2014**, *37*, 1042–1045. [CrossRef]

49. R Core Team. *R: A Language and Environment for Statistical Computing*; R Foundation for Statistical Computing: Vienna, Austria, 2020. Available online: https://www.R-project.org/ (accessed on 12 November 2020).

3

CAX1a TILLING Mutations Modify the Hormonal Balance Controlling Growth and Ion Homeostasis in *Brassica rapa* Plants Subjected to Salinity

Eloy Navarro-León [1,*], Francisco Javier López-Moreno [2], Santiago Atero-Calvo [1],
Alfonso Albacete [3,4] ⓘ, Juan Manuel Ruiz [1] and Begoña Blasco [1]

[1] Department of Plant Physiology, Faculty of Sciences, University of Granada, 18071 Granada, Spain;
atero98@correo.ugr.es (S.A.-C.); jmrs@ugr.es (J.M.R.); bblasco@ugr.es (B.B.)

[2] IFAPA, Institute of Research and Training in Agriculture and Fisheries, 18004 Granada, Spain;
franciscoj.lopez.moreno@juntadeandalucia.es

[3] Department of Plant Nutrition, CEBAS-CSIC, Campus Universitario de Espinardo, 30100 Murcia, Spain;
alfonsoa.albacete@carm.es

[4] Department of Plant Production and Agrotechnology, Institute for Agri-Food Research and Development
of Murcia (IMIDA), C/Mayor s/n, 30150 La Alberca, Murcia, Spain

* Correspondence: enleon@ugr.es

Abstract: Salinity is a serious issue for crops, as it causes remarkable yield losses. The accumulation of Na^+ affects plant physiology and produces nutrient imbalances. Plants trigger signaling cascades in response to stresses in which phytohormones and Ca^{2+} are key components. Cation/H^+ exchangers (CAXs) transporters are involved in Ca^{2+} fluxes in cells. Thus, enhanced CAX activity could improve tolerance to salinity stress. Using the TILLING (targeting induced local lesions in genomes) technique, three *Brassica rapa* mutants were generated through a single amino acidic modification in the CAX1a transporter. We hypothesized that *BraA.cax1a* mutations could modify the hormonal balance, leading to improved salinity tolerance. To test this hypothesis, the mutants and the parental line R-o-18 were grown under saline conditions (150 mM NaCl), and leaf and root biomass, ion concentrations, and phytohormone profile were analyzed. Under saline conditions, *BraA.cax1a-4* mutant plants increased growth compared to the parental line, which was associated with reduced Na^+ accumulation. Further, it increased K^+ concentration and changed the hormonal balance. Specifically, our results show that higher indole-3-acetic acid (IAA) and gibberellin (GA) concentrations in mutant plants could promote growth under saline conditions, while abscisic acid (ABA), ethylene, and jasmonic acid (JA) led to better signaling stress responses and water use efficiency. Therefore, CAX1 mutations directly influence the hormonal balance of the plant controlling growth and ion homeostasis under salinity. Thus, Ca^{2+} signaling manipulation can be used as a strategy to improve salinity tolerance in breeding programs.

Keywords: *Brassica rapa*; calcium; phytohormones; potassium; salinity; sodium

---

1. Introduction

Saline soils represent 3.1% of the total land area of the Earth. Thereby, salinity is a serious issue for crops because it causes remarkable yield losses [1]. This problem has become more important over the last years and it is expected to be even more important in the future because of climate change [2]. Most crop species are affected by salinity including species from the Brassicaceae family, such as cabbage, broccoli, and rapeseed [3]. The most common and soluble salt compound is NaCl. The high concentration of Na^+ and Cl^- ions in saline soils cause osmotic potential imbalances hampering water and nutrients uptake. The accumulation of Na^+ in plants alters the osmotic potential and causes direct toxicity and nutrient imbalances affecting plant physiology [1]. The similarity of Na^+ with K^+ hinders K^+ uptake and activity in the plant. Thus, Na^+ accumulates in the cytosol displacing K^+ and also Ca^{2+} ions from their active sites and inhibits enzyme activities. In addition, the altered K^+/Na^+ impairs photosynthesis processes [4] and causes oxidative stress, as indicated by a high reactive oxygen species (ROS) accumulation triggering the activation of antioxidant responses [5,6].

As in other stresses, plants trigger phytohormone-mediated responses to cope with saline stress [2]. Plant hormones are compounds from different chemical groups involved in numerous processes in plants. They are crucial for plant adaptation to stress because they mediate adaptive responses that modulate growth, development, and plant nutrition. The resilience of plants to stress is highly dependent on the regulation of hormone signaling pathways [7,8]. Abscisic acid (ABA) and ethylene have been classically considered stress-related hormones. ABA is an important hormone in salinity response because it regulates the water status via stomata closure and the expression of ABA-responsive genes for long-term responses [9]. In addition, ABA regulates the synthesis and accumulation of osmoprotectants, such as proline and some proteins [3,10]. Alternatively, ethylene synthesis is usually activated as a response to stress and is considered as the main senescence-related hormone. Under salinity, senescence is favored by ABA and the ethylene precursor 1-aminocyclopropane-1-carboxylic acid (ACC) accumulation, as well as by a decrease of indole-3-acetic acid (IAA) and cytokinins (CKs) levels [2]. Indeed, other hormones have been demonstrated to regulate stress responses. Auxins, such as IAA, are involved in the physiological response that modulates oxidative stress and prevents oxidative damage, while CKs protect plants against salinity that maintains growth and delays senescence [9]. Other important phytohormones affected by salinity are gibberellins (GAs) associated with sugar signaling and antioxidant system modulation [11]. Furthermore, jasmonic acid (JA) and salicylic acid (SA) are usually related to biotic stresses, although both hormones are involved in stress signaling in response to salinity and other abiotic stresses [9,12].

Besides phytohormones in cooperation with them, Ca^{2+} is a second messenger that fulfills a crucial role in signaling cascades in response to stress [13]. During stress signaling, the Ca^{2+} signal is very fast and occurs much earlier than ABA accumulation and, thereby, Ca^{2+} acts in ABA signaling processes [4]. In response to salinity, Ca^{2+} is involved in salt sensing, Na^+ extrusion/sequestration, pH regulation, and cellular barriers synthesis [14]. Indeed, the supplementation of plants with $CaCl_2$ has been demonstrated to improve salinity tolerance [4]. This positive effect has been observed in many crops including the Brassicaceae species [15,16]. Alternatively, seed priming with $CaCl_2$ has beneficial effects on the hormonal balance of the plant alleviating salinity stress symptoms. $CaCl_2$ application reduced ABA and SA levels and increased IAA, promoting plant growth [17]. Besides, Ca^{2+} application maintains membrane integrity, reducing K^+ leakage and preventing Na^+ accumulation, and, thereby, sustaining K^+/Na^+ selectivity [1].

Cation/H^+ exchangers (CAXs) are a family of Ca^{2+}/H^+ antiporters situated on plasma and organelle membranes including vacuoles. CAXs transporters remove Ca^{2+} from the cytosol to generate different Ca^{2+} profiles in the cell. Thus, CAX transporters fulfill a key role in the generation of Ca^{2+} gradients involved in stress signaling [18]. Adequate Ca^{2+} homeostasis driven by CAX and other transporters could be crucial to improving Ca^{2+} fluxes and stress tolerance [19], as it was observed in the halophyte species *Suaeda salsa* [20]. Furthermore, CAX1 is the CAX transporter with the highest Ca^{2+}/H^+ activity [18], thus the modification of CAX1 activity could be useful to improve the tolerance to salinity stress [19]. TILLING (targeting induced local lesions in genomes) is a promising technique that generates new variants in target genes [21]. Using this technique, three new variants were produced in *Brassica rapa* ssp. *trilocularis* 'R-o-18' CAX1a transporter: *BraA.cax1a-4*, *BraA.cax1a-7*, and *BraA.cax1a-12* [22]. These mutations change amino acids that could affect protein conformation and thereby improve CAX1 function [23]. As observed in a previous experiment, *BraA.cax1a* mutations induce changes in phytohormone profile [24]. Given the role of phytohormones in salinity stress, this study aims to test whether changes in Ca^{2+} signaling through *BraA.cax1a* mutations could modify the hormonal balance of the plant leading to improved growth under salinity.

2. Materials and Methods

2.1. Plant Material and Growth Conditions

As plant material, we used three *Brassica rapa* ssp. *trilocularis* 'R-o-18' mutants (*BraA.cax1a-4*, *BraA.cax1a-7*, and *BraA.cax1a-12*) and the parent line R-o-18 (without changes in BraA.CAX1a). The amino acidic changes produced in BraA.CAX1a transporter were: *BraA.cax1a-4* (A-to-T change at amino acid 77), *BraA.cax1a-7* (R-to-K change at amino acid 44), and *BraA.cax1a-12* (P-to-S change at amino acid 56). Mutant plants were generated as described by Lochlainn et al. [22] and Graham et al. [23]. Seeds were germinated on filter paper in Petri dishes and then transplanted to pots (13 cm × 13 cm × 12.5 cm) filled with vermiculite. Pots were placed in trays (55 cm × 40 cm × 8.5 cm). Plants were grown in a chamber with controlled conditions: relative humidity (60–80%), temperature (23/18 °C; day/night), photoperiod (14/10 h; day/night), and a photosynthetic photon flux density of 350 μmol m^{-2} s^{-1} registered at the top of plants using a 190 SB quantum sensor (LI-COR Inc., Lincoln, NE, USA). Plants were supplied with a nutritive solution composed of 6 mM KH_2PO_4, 4 mM KNO_3, 4 mM $Ca(NO_3)_2 \cdot 4H_2O$, 2 mM $MgSO_4 \cdot 7H_2O$, 1 mM $NaH_2PO_4 \cdot 2H_2O$, 10 μM H_3BO_3, 5 μM Fe-chelate (Sequestrene; 138FeG100), 2 μM $MnCl_2 \cdot 4H_2O$, 1 μM $ZnSO_4$, 0.25 μM $CuSO_4 \cdot 5H_2O$, and 0.1 μM $Na_2MoO_4 \cdot 2H_2O$. The pH of the nutritive solution was kept between 5.5 and 6.0.

2.2. Experimental Design, Treatments, and Plant Sampling

Treatments were applied 30 days after sowing and were maintained for 21 days. Plants received two different treatments: control (without NaCl added to the nutrient solution) and salinity (150 mM NaCl supplemented to the nutrient solution). The factors considered in the experiment were the salinity (S) and the mutation (M). The experimental design comprised a randomized complete block with 8 treatments, 3 trays per treatment, and 8 plants per tray, thus a total of 24 plants were grown for each treatment. At the end of the experiment, plant leaves and roots were rinsed, dried, and weighed to obtain the fresh weight (FW). Then, leaves and roots were lyophilized to determine the dry weight (DW) and a part of the lyophilized leaf material was used to determine the phytohormone concentrations. Nine independent replicates from each treatment ($n = 9$) were finally used for the analytical determinations.

2.3. Analysis of Na^+, Ca^{2+}, and K^+ Concentrations

Ca^{2+}, Na^+, and K^+ were determined subjecting the leaf samples to a mineralization process by wet digestion [25]. Next, 150 mg of dry leaves were milled and mineralized with a combination of nitric acid and hydrogen peroxide at 30%. Then, 20 mL of Milli-Q water were added and element concentrations were measured using ICP-MS (X-Series II; Thermo Fisher Scientific Inc., Waltham, MA, USA).

2.4. Hormone Extraction and Analysis

Phytohormone concentrations were determined in leaves according to Albacete et al. [26] with some modifications. Lyophilized samples (30 mg) were mixed with 1 mL of cold (−20 °C) extraction mixture of methanol/water (80/20, *v/v*). Samples were centrifuged (20,000× *g*, 15 min, 4 °C) and re-extracted for 30 min at 4 °C in 1 mL of the same extraction solution. Supernatants were passed through Sep-Pak Plus †C18 cartridges (SepPak Plus, Waters, Milford, MA 01757 USA) and evaporated at 40 °C under vacuum. The residue was dissolved in 1 mL methanol/water (20/80, *v/v*) using an ultrasonic bath. The dissolved samples were filtered through Millex nylon membrane filters 13 mm diameter of 0.22 μm pore size (Millipore, Bedford, MA, USA). Filtered extracts (10 μL) were injected in a U-HPLC-MS system consisting of an Accela Series U-HPLC coupled to an Exactive mass spectrometer (Thermo Fisher Scientific Inc., Waltham, MA, USA) with a heated electrospray ionization (HESI) interface. The mass spectra were measured using the Xcalibur software version 2.2. For phytohormones quantification, calibration curves were constructed (1, 10, 50, and 100 $\mu g\ L^{-1}$) and corrected for 10 $\mu g\ L^{-1}$ deuterated internal standards. Total CKs were calculated as the sum of trans-zeatin (tZ) and isopentenyl adenine (iP) concentrations. Total GAs were calculated as the sum of gibberellin A1 (GA1), gibberellin A3 (GA3), and gibberellin A4 (GA4) concentrations.

2.5. Statistical Analysis

The mean and standard error of each treatment was calculated from the 9 individual data of each parameter analyzed. To assess the differences between treatments we performed a one-way analysis of variance (ANOVA) with 95% confidence. A two-tailed ANOVA was used to determine whether the NaCl treatment (S), the *BraA.cax1a* mutations (M), or their interaction (S × M) significantly influenced the results. Means were compared using Fisher's least significant differences (LSD). The significance levels were stated as * $p < 0.05$, ** $p < 0.01$, *** $p < 0.001$, or NS (not significant). A principal components analysis (PCA) was performed to assess relationships between treatments and all parameters analyzed. All statistical analyses were carried out using the Statgraphics Centurion 16.1.03 software.

3. Results

3.1. Plant Biomass and Cation Concentration

Plants grown under salinity conditions presented a remarkable decrease in leaf and root DW (Figure 1). However, *BraA.cax1a-4* plants grown under salinity showed significantly higher leaf biomass in comparison to the other mutants and the parent line. Indeed, this mutant presented 41% higher leaf DW than R-o-18 plants (Figure 1a). Nonetheless, the four lines analyzed did not show significant differences in root DW under saline conditions (Figure 1b). Regarding cation concentrations, salinity reduced leaf Ca^{2+} concentration in comparison to control conditions in all lines, but no differences were observed between lines under salinity conditions. As expected, NaCl application strongly increased Na^+ concentration in leaves. However, this increment was lower in *BraA.cax1a-4* plants, which showed the lowest Na^+ concentration. Additionally, NaCl application reduced K^+ concentration, although this reduction was lower in *BraA.cax1a-4* plants, which presented the highest K^+ concentration in comparison to the other genotypes. Consequently, *BraA.cax1a-7* plants presented the highest Na^+/K^+ ratio, followed by *BraA.cax1a-12*, R-o-18, and *BraA.cax1a-4* plants. Specifically, *BraA.cax1a-4* mutant plants showed 42% lower Na^+/K^+ ratio than R-o-18 plants (Table 1).

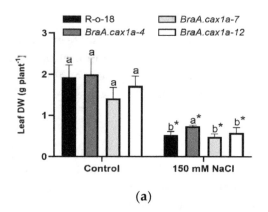
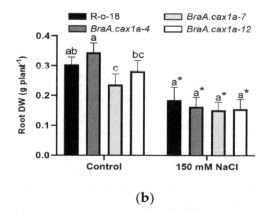

(a)   (b)

Figure 1. Effect of *BraA.cax1a* mutations and salinity on leaf dry weight (DW) (**a**) and root DW (**b**). Values are expressed as means ± standard error ($n = 9$). Bars marked with different letters indicate significant differences among genotypes based on the LSD test ($p < 0.05$). Asterisk (*) indicates significant differences between control and 150 mM NaCl treatments.

Table 1. Effect of *BraA.cax1a* mutations and salinity on analyzed cation concentration and Na^+/K^+ ratio in leaves.

		Ca^{2+}	Na^+	K^+	Na^+/K^+
Control	R-o-18	15.09 b	3.29 a	39.24 ab	0.08 a
	BraA.cax1a-4	17.61 a	3.64 a	38.81 ab	0.09 a
	BraA.cax1a-7	17.85 a	3.23 a	36.69 b	0.09 a
	BraA.cax1a-12	18.51 a	3.76 a	41.14 a	0.09 a
	p-value	*	NS	*	NS
	$LSD_{0.05}$	2.22	1.05	4.11	0.02
150 mM NaCl	R-o-18	9.50 a	36.74 a	22.17 b	1.65 ab
	BraA.cax1a-4	8.54 a	26.65 b	28.15 a	0.96 c
	BraA.cax1a-7	9.57 a	41.07 a	22.15 b	1.85 a
	BraA.cax1a-12	10.19 a	40.68 a	25.85 a	1.57 b
	p-value	NS	*	**	***
	$LSD_{0.05}$	2.22	9.81	2.55	0.26
Analysis of variance					
Salinity (S)		***	***	***	***
Mutation (M)		*	*	**	***
S × M		NS	*	*	***
$LSD_{0.05}$		1.44	4.54	2.22	0.12

Values are expressed as $mg\,g^{-1}$ DW and differences between means ($n = 9$) were compared by Fisher's least-significance test (LSD; $p = 0.05$). Values with different letters indicate significant differences among genotypes. The levels of significance were represented by $p > 0.05$: NS (not significant), $p < 0.05$ (*), $p < 0.01$ (**), and $p < 0.001$ (***).

3.2. Phytohormone Concentrations

Salinity increased total CKs, total GAs, and provoked differential changes in the concentrations of other hormones in R-o-18 and mutant plants (Figure 2; Table 2). Under saline conditions, *BraA.cax1a-4* was the only mutant that showed significantly higher IAA levels (55%) in comparison to R-o-18 plants (Table 2). Regarding GAs under saline conditions, *BraA.cax1a-4* plants presented the highest GA concentrations. The other two mutants presented similar values than R-o-18 plants (Figure 2a). Particularly, *BraA.cax1a-4* showed significant increments in all GAs (4-fold higher than R-o-18), while in *BraA.cax1a-7* only GA1 increased in comparison to the parent line (Table 2). Concerning CKs, iP decreased in the *BraA.cax1a-4* mutant in comparison to R-o-18, whereas incremented in *BraA.cax1a-7* plants. Importantly, tZ increased in both *BraA.cax1a-4* and *BraA.cax1a-7* mutants, and its absolute concentrations were much higher than those of iP, leading to increased total CK content (Figure 2b). ABA concentration was significantly higher in *BraA.cax1a-4* plants (57%) in comparison to R-o-18 plants. All mutants showed higher ACC levels in comparison to the parent line. SA increased in *BraA.cax1a-4* and *BraA.cax1a-7* mutants, whereas JA concentration increased in *BraA.cax1a-4* and *BraA.cax1a-12* in comparison to the parental R-o-18 (Table 2).

3.3. Principal Component Analysis

A principal component analysis (PCA) was performed to detect general trends in the data and to evaluate the relationships among parameters. The first principal component (PC1) of the score plot clearly separated *BraA.cax1a-4* from the rest of the lines and accounted for 50.55% of the variance within the data. The second principal component (PC2) separated *BraA.cax1a-7* from the other lines and accounted for 23.73% of the variance (Figure 3a). The PCA loading plot revealed three clusters (Figure 3b). The first cluster associated leaf DW with K^+, GA3, GA4, ACC, GAs, and ABA levels. The second cluster related Na^+, Na^+/K^+ ratio, Ca^{2+}, and iP levels. Finally, the third cluster, grouped tZ, total CK, GA1, and SA concentrations (Figure 3b).

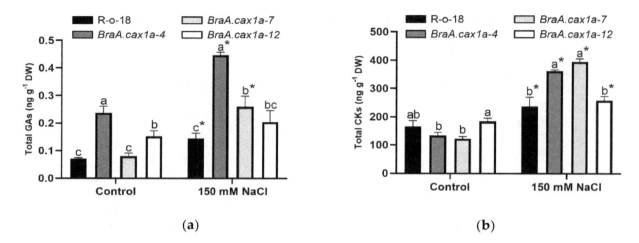

Figure 2. Effect of *BraA.cax1a* mutations and salinity on total GA (**a**) and CK (**b**) concentrations in the leaves. Values are expressed as means ± standard error ($n = 9$). Bars marked with different letters indicate significant differences among genotypes based on the LSD test ($p < 0.05$). Asterisk (*) indicates significant differences between control and 150 mM NaCl treatments.

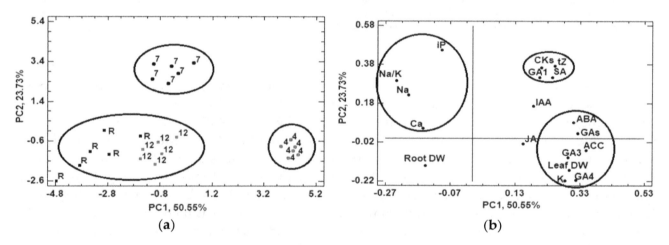

Figure 3. Scores (**a**) and corresponding loadings plot (**b**) of principal component analysis (PCA) using all parameters analyzed in R-o-18 (R), *BraA.cax1a-4* (4), *BraA.cax1a-7* (7), and *BraA.cax1a-12* (12) plants grown under salinity conditions.

Table 2. Effect of *BraA.cax1a* mutations and salinity on leaf phytohormone concentrations.

		IAA	GA1	GA3	GA4	iP	tZ	ABA	ACC	SA	JA
Control	R-o-18	2.40 bc	0.05 b	nd	0.02 b	6.83 a	180.47 a	6.43 a	489.79 c	282.53 a	131.42 b
	BraA.cax1a-4	3.43 b	0.10 a	0.08	0.05 a	4.55 b	129.43 ab	4.38 ab	848.95 a	283.45 a	68.98 c
	BraA.cax1a-7	1.62 c	0.07 b	nd	0.01 b	2.94 b	119.26 b	4.13 b	343.59 d	317.96 a	102.67 bc
	BraA.cax1a-12	5.42 a	0.06 b	0.05	0.04 a	8.54 a	176.00 a	6.49 a	623.31 b	306.47 a	190.93 a
	p-value	**	**		*	***	*	*	***	NS	**
	LSD$_{0.05}$	1.55	0.02		0.03	1.29	56.09	2.12	59.33	165.18	54.96
150 mM NaCl	R-o-18	2.03 b	0.08 c	0.03 c	0.04 b	5.10 b	231.93 b	13.93 b	515.55 c	611.82 c	125.07 c
	BraA.cax1a-4	3.14 a	0.16 ab	0.18 a	0.11 a	2.91 c	355.02 a	21.86 a	1234.98 a	942.16 ab	201.57 ab
	BraA.cax1a-7	2.95 ab	0.18 a	0.05 bc	0.03 b	7.82 a	382.54 a	16.86 ab	695.43 b	1035.42 a	168.46 bc
	BraA.cax1a-12	2.79 ab	0.09 bc	0.07 b	0.04 b	2.94 c	255.38 b	15.01 b	722.12 b	795.16 bc	224.16 a
	p-value	*	*	***	**	**	**	*	***	**	**
	LSD$_{0.05}$	1.04	0.07	0.04	0.04	2.05	63.88	5.62	164.17	206.85	50.51
Analysis of variance											
Salinity (S)		NS	***		**	*	***	***	***	***	***
Mutation (M)		***	***		***	**	NS	NS	***	**	***
S × M		**	NS		NS	***	***	**	***	*	**
LSD$_{0.05}$		0.86	0.03		0.02	1.11	39.07	2.76	80.24	121.67	34.31

Values are expressed as ng g^{-1} DW and differences between means ($n = 9$) were compared by Fisher's least-significance test (LSD; $p = 0.05$). Values with different letters indicate significant differences among genotypes. The levels of significance were represented by $p > 0.05$: NS (not significant), $p < 0.05$ (*), $p < 0.01$ (**), and $p < 0.001$ (***).

4. Discussion

Improved growth responses under salinity are associated with salinity tolerance. According to leaf DW results (Figure 1a), *BraA.cax1a-4* presented higher performance under salinity in comparison to other lines. Several studies observed that a better Ca^{2+} and K^+ nutrition and homeostasis provide salt tolerance due to Ca^{2+}-associated signaling processes in response to stress and because it is a Na^+ antagonist [4,27]. In addition, Ca^{2+} improves the accumulation of other nutrients and reduces Na^+/K^+ ratio [11,13]. Although *BraA.cax1a-4* plants did not present higher Ca^{2+} concentration, they registered lower Na^+ and higher K^+ concentrations, leading to a better Na^+/K^+ ratio (Table 1). A previous study showed that this fact could be due to *BraA.cax1a-4* plants favors the transport of K^+ over Na^+ to the shoot from roots [28]. Hence, the modification of CAX1 activity could result in different ion accumulation, because of changes in Ca^{2+} fluxes (but not absolute concentrations) in *BraA.cax1a-4*, improving Na^+ and K^+ homeostasis and thereby growth [29]. This is supported by the close relationship between leaf DW and K^+ concentration observed in loading plot analysis (Figure 3b). Alternatively, Ca^{2+} is involved in modulating ROS levels that are generated by oxidative stress caused by salinity [5]. Thus, as observed in a previous study, the higher tolerance of *BraA.cax1a-4* plants could also be related to this ROS modulation [28].

The possible alteration of Ca^{2+} fluxes by *BraA.cax1a* mutations could affect the function of Ca^{2+} sensors, such as calmodulins and protein kinases, that are crucial for hormone synthesis and signaling. Thus, the improved growth response under salinity of *BraA.cax1a-4* plants could be also related to changes in the hormonal balance, as it has been proposed in the present study. Indeed, auxins, and particularly the active compound IAA, have crucial roles in stress signaling responses in *B. rapa* seedlings [3] and also participate in redox and antioxidative metabolism [9]. However, IAA concentration usually decreases and, thereby, senescence is promoted in plants grown under salt stress [30]. In the present study, we did not observe a significant reduction in IAA concentration. Thus, *BraA.cax1a-4* was the line with the highest IAA concentration, which could contribute to the higher growth observed in this mutant under salinity conditions (Table 2). In fact, the greater IAA levels might enhance ROS detoxification under saline stress which could increase the tolerance to NaCl [31], as previously demonstrated in *BraA.cax1a-4* mutants [28].

CKs regulate several physiological processes, promote plant growth, and play important roles in salinity tolerance [32–34]. A decrease in CKs usually is an early response to salt stress [10], although the contrary was also reported by Ghanem et al. [35]. These authors observed that CK concentrations increased in plants grown under salt stress as a response to increased salinity tolerance in tomato. This study agrees with our findings as we observed an increment in the CK concentrations in the plants subjected to salinity. Total CK concentrations significantly increased in the mutant, which presented the highest biomass under salinity, *BraA.cax1a-4*, compared to the parent line (Figure 2b). However, their role in the control of growth of *BraA.cax1a* mutants seems to be limited since the PCA analysis revealed that CKs do not associate with any of the growth-related parameters recorded in this assay (Figure 3b).

Some studies have stated that GA accumulation in plants grown under abiotic stress provide salinity tolerance [36], whereas other studies have shown GA reduction because of repressor protein accumulation, leading to plant growth reduction [4]. Our results show that total GAs—especially GA3—markedly increased in *BraA.cax1a-4* plants under salinity (Table 2, Figure 2a). GA3 is the main GA that regulates important steps in plant growth and development and alleviates salt stress [10]. Furthermore, Khan et al. [11] proved that the exogenous application of GA3 to linseed, alone or in combination with Ca^{2+}, reduced the damage to membranes and improved water status. Therefore, this particular increment in GA3 and GA4 could also explain the higher biomass of the *BraA.cax1a-4* mutants under salinity stress, as suggested by the close association between GAs and leaf DW in the PCA (Figure 3b).

Despite auxins, CKs, and GAs play a role in the response of the plants to salinity and other abiotic stresses, the phytohormone more broadly studied in relation to water and salinity stress is ABA, since this hormone is over-produced as a consequence of both stresses. Thus, genes related to ABA synthesis and accumulation are up-regulated by NaCl application, and in turn, ABA is the main hormone that activates salt responsive genes [37]. Our data reflects an increase in ABA concentration under salinity, which was especially strong in the *BraA.cax1a-4* mutant (Table 2). ABA triggers stress responses, such as water balance and osmotic stress regulation, and leads to stomatal closure to avoid excessive water loss [38]. This fact could be important in *BraA.cax1a-4* plants where the high ABA levels provoke the closure of the stomata, thus enhancing water use efficiency (WUE) (data not shown). This is in agreement with previous studies in maize and tomato in which salinity-tolerant genotypes presented greater WUE and stomatal regulation, and lower Na^+ accumulation than the sensitive ones, associated with higher ABA concentrations [39,40]. Furthermore, Iqbal et al. [10] proved that Ca^{2+} induces ABA accumulation, thus reducing Na^+ and Cl^- content and increasing K^+ content. Therefore, ABA could contribute to better ion homeostasis via Na^+/K^+ reduction, as observed in *BraA.cax1a-4* plants (Table 1), maintain growth in plants grown under salinity conditions. This is further supported by the linkage among ABA, K^+, and leaf FW shown in the PCA plot (Figure 3b).

Another hormone traditionally associated with stress responses is ethylene, especially in relation to leaf senescence and abscission. Salinity stress triggers the accumulation of the ethylene precursor ACC, leading to *de-novo* synthesis of ethylene, which induces cell death and leaf senescence, and reduces plant growth [9]. In our study, ACC concentration was higher in plants subjected to salinity which could contribute to biomass reduction (Table 2, Figure 1). Likewise, de la Torre-González et al. [31] observed that a salinity-tolerant tomato genotype presented much lower ACC concentration than the sensitive one under saline conditions. However, some studies proved that plants supplied with ACC or that overexpress ethylene-responsive factors presented tolerance to salinity stress [36,41]. Thus, all mutants evaluated in the present study showed a significant increase in ACC concentrations, and importantly, the *BraA.cax1a-4* mutant presented the highest concentrations (Table 2) which may be associated with its improved growth-response due to better regulation of Ca^{2+} homeostasis under salinity. This conclusion is additionally endorsed by the link of ACC with leaf FW and K^+ (Figure 3b). Indeed, the increased growth of *BraA.cax1a-4* mutant under salinity stress could be partially associated with improved ethylene regulation of the high-affinity K^+ transporters, which increase K^+ tissue accumulation [41], as observed in this mutant (Table 1).

SA and JA regulate biotic stress responses but also are involved in regulating abiotic stresses, such as salinity, in cooperation with other hormones [12,42]. JA appears to provide tolerance to salt stress via plant signal transduction [7], whereas SA induces antioxidant defenses, protects photosynthesis, and reduces membrane damage [10,43]. Iqbal et al. [10] observed that JA reduces membrane depolarization and, thereby, K^+ loss. Hence, the highest JA concentrations of *BraA.cax1a-4* and *BraA.cax1a-12* (Table 2) could prevent K^+ leakage and contribute to the higher K^+ concentrations detected in these two mutants (Table 1). Additionally, this also may be explained by the lower lipid peroxidation levels previously observed in *BraA.cax1a-4* and *BraA.cax1a-12* plants [28]. Regarding SA, we observed a significant increment in SA concentration in all mutant plants grown under salinity in comparison to the parental R-o-18 plants (Table 2). Ku et al. [7] proved that increments in cytosolic Ca^{2+} are a necessary step in SA accumulation in response to salinity. Thus, *BraA.cax1a* mutations could induce a higher SA accumulation through a greater cytosolic Ca^{2+} response.

5. Conclusions

The present study demonstrates that the control of Ca^{2+} homeostasis through *BraA.cax1a* mutants modifies the phytohormone balance of *B. rapa* plants controlling growth responses under salinity stress conditions. Specifically, the growth improvement of *BraA.cax1a-4* mutant plants under salinity can be primarily associated with higher GA, IAA, and CK concentrations. Besides, ABA accumulation activates signaling stress responses under saline conditions, leading to better control of Na^+/K^+ homeostasis. This study confirms the key role of the CAX1 transporter in phytohormone regulation and as a potential target for improving growth under salinity stress. However, further research is still necessary to mechanistically demonstrate the relationship between Ca^{2+} fluxes and phytohormone accumulation in the control of the growth and productivity under salinity conditions.

Author Contributions: Conceptualization, J.M.R. and B.B.; Methodology, F.J.L.-M., S.A.-C., A.A. and E.N.-L.; Validation, A.A., J.M.R. and B.B.; Formal analysis, B.B. and E.N.-L.; Data curation, E.N.-L.; Writing—original draft preparation, E.N.-L.; Writing—review and editing, J.M.R., B.B. and A.A. All authors have read and agreed to the published version of the manuscript.

Acknowledgments: We thank Martin R. Broadley and Neil Graham from Nottingham University for providing us the seeds utilized in this experiment. We also thank María del Puerto Sánchez-Iglesias for technical support with hormone analyses.

References

1. Wu, H. Plant salt tolerance and Na$^+$ sensing and transport. *Crop J.* **2018**, *6*, 215–225. [CrossRef]
2. Acosta-Motos, J.; Ortuño, M.; Bernal-Vicente, A.; Diaz-Vivancos, P.; Sanchez-Blanco, M.; Hernandez, J. Plant responses to salt stress: Adaptive mechanisms. *Agronomy* **2017**, *7*, 18. [CrossRef]
3. Pavlović, I.; Pěnčík, A.; Novák, O.; Vujčić, V.; Radić Brkanac, S.; Lepeduš, H.; Strnad, M.; Salopek-Sondi, B. Short-term salt stress in *Brassica rapa* seedlings causes alterations in auxin metabolism. *Plant Physiol. Biochem.* **2018**, *125*, 74–84. [CrossRef]
4. Köster, P.; Wallrad, L.; Edel, K.H.; Faisal, M.; Alatar, A.A.; Kudla, J. The battle of two ions: Ca^{2+} signalling against Na$^+$ stress. *Plant Biol.* **2018**, *21*, 39–48. [CrossRef] [PubMed]
5. Shoresh, M.; Spivak, M.; Bernstein, N. Involvement of calcium-mediated effects on ROS metabolism in the regulation of growth improvement under salinity. *Free Radic. Biol. Med.* **2011**, *51*, 1221–1234. [CrossRef] [PubMed]
6. Kravchik, M.; Bernstein, N. Effects of salinity on the transcriptome of growing maize leaf cells point at cell-age specificity in the involvement of the antioxidative response in cell growth restriction. *BMC Genom.* **2013**, *14*, 24. [CrossRef]
7. Ku, Y.-S.; Sintaha, M.; Cheung, M.-Y.; Lam, H.-M. Plant hormone signaling crosstalks between biotic and abiotic stress responses. *Int. J. Mol. Sci.* **2018**, *19*, 3206. [CrossRef]
8. Sharma, A.; Shahzad, B.; Kumar, V.; Kohli, S.K.; Sidhu, G.P.S.; Bali, A.S.; Handa, N.; Kapoor, D.; Bhardwaj, R.; Zheng, B. Phytohormones regulate accumulation of osmolytes under abiotic stress. *Biomolecules* **2019**, *9*, 285. [CrossRef]
9. Fahad, S.; Hussain, S.; Matloob, A.; Khan, F.A.; Khaliq, A.; Saud, S.; Hassan, S.; Shan, D.; Khan, F.; Ullah, N.; et al. Phytohormones and plant responses to salinity stress: A review. *Plant Growth Regul.* **2015**, *75*, 391–404. [CrossRef]
10. Iqbal, N.; Umar, S.; Khan, N.A.; Khan, M.I.R. A new perspective of phytohormones in salinity tolerance: Regulation of proline metabolism. *Environ. Exp. Bot.* **2014**, *100*, 34–42. [CrossRef]
11. Khan, M.N.; Siddiqui, M.H.; Mohammad, F.; Naeem, M.; Khan, M.M.A. Calcium chloride and gibberellic acid protect linseed (*Linum usitatissimum* L.) from NaCl stress by inducing antioxidative defence system and osmoprotectant accumulation. *Acta Physiol. Plant.* **2010**, *32*, 121–132. [CrossRef]
12. Dar, T.A.; Uddin, M.; Khan, M.M.A.; Hakeem, K.R.; Jaleel, H. Jasmonates counter plant stress: A Review. *Environ. Exp. Bot.* **2015**, *115*, 49–57. [CrossRef]
13. White, P.J.; Broadley, M.R. Calcium in plants. *Ann. Bot.* **2003**, *92*, 487–511. [CrossRef] [PubMed]
14. Manishankar, P.; Wang, N.; Köster, P.; Alatar, A.A.; Kudla, J. Calcium signaling during salt stress and in the regulation of ion homeostasis. *J. Exp. Bot.* **2018**, *69*, 4215–4226. [CrossRef]
15. Yousuf, P.Y.; Ahmad, A.; Hemant, M.; Ganie, A.H.; Aref, I.M.; Iqbal, M. Potassium and calcium application ameliorates growth and oxidative homeostasis in salt-stressed indian mustard (*Brassica Juncea*) plants. *Pak. J. Bot.* **2015**, *47*, 1629–1639.
16. Latef, A.; Hamed, A.A. Ameliorative effect of calcium chloride on growth, antioxidant enzymes, protein patterns and some metabolic activities of canola (*Brassica napus* L.) under seawater stress. *J. Plant Nutr.* **2011**, *34*, 1303–1320. [CrossRef]
17. Iqbal, M.; Ashraf, M.; Jamil, A.; Ur-Rehman, S. Does seed priming induce changes in the levels of some endogenous plant hormones in hexaploid wheat plants under salt stress? *J. Integr. Plant Biol.* **2006**, *48*, 181–189. [CrossRef]

18. Pittman, J.K.; Hirschi, K.D. CAX-ing a wide net: Cation/H^+ transporters in metal remediation and abiotic stress signaling. *Plant Biol.* **2016**, *18*, 741–749. [CrossRef]

19. Pokotylo, I.V.; Kretinin, S.V.; Kravets, V.S. Role of phospholipase D in metabolic reactions of transgenic tobacco cax1 cells under the influence of salt stress. *Cytol. Genet.* **2012**, *46*, 131–135. [CrossRef]

20. Han, N.; Shao, Q.; Bao, H.; Wang, B. Cloning and characterization of a Ca^{2+}/H^+ antiporter from halophyte *Suaeda salsa* L. *Plant Mol. Biol. Rep.* **2011**, *29*, 449–457. [CrossRef]

21. Till, B.J.; Reynolds, S.H.; Greene, E.A.; Codomo, C.A.; Enns, L.C.; Johnson, J.E.; Burtner, C.; Odden, A.R.; Young, K.; Taylor, N.E.; et al. Large-scale discovery of induced point mutations with high-throughput TILLING. *Genome Res.* **2003**, *13*, 524–530. [CrossRef]

22. Lochlainn, S.Ó.; Amoah, S.; Graham, N.S.; Alamer, K.; Rios, J.J.; Kurup, S.; Stoute, A.; Hammond, J.P.; Østergaard, L.; King, G.J.; et al. High Resolution Melt (HRM) analysis is an efficient tool to genotype EMS mutants in complex crop genomes. *Plant Methods* **2011**, *7*, 43. [CrossRef] [PubMed]

23. Graham, N.S.; Hammond, J.P.; Lysenko, A.; Mayes, S.; Lochlainn, S.O.; Blasco, B.; Bowen, H.C.; Rawlings, C.J.; Rios, J.J.; Welham, S.; et al. Genetical and comparative genomics of *Brassica* under altered Ca supply identifies *Arabidopsis* Ca-transporter orthologs. *Plant Cell* **2014**, *26*, 2818–2830. [CrossRef]

24. Navarro-León, E.; Ruiz, J.M.; Albacete, A.; Blasco, B. Effect of CAX1a TILLING mutations and calcium concentration on some primary metabolism processes in *Brassica rapa* plants. *J. Plant Physiol.* **2019**, *237*, 51–60. [CrossRef] [PubMed]

25. Wolf, B. A comprehensive system of leaf analyses and its use for diagnosing crop nutrient status. *Commun. Soil Sci. Plant Anal.* **1982**, *13*, 1035–1059. [CrossRef]

26. Albacete, A.; Ghanem, M.E.; Martinez-Andujar, C.; Acosta, M.; Sanchez-Bravo, J.; Martinez, V.; Lutts, S.; Dodd, I.C.; Perez-Alfocea, F. Hormonal changes in relation to biomass partitioning and shoot growth impairment in salinized tomato (*Solanum lycopersicum* L.) plants. *J. Exp. Bot.* **2008**, *59*, 4119–4131. [CrossRef]

27. Wan, H.; Chen, L.; Guo, J.; Li, Q.; Wen, J.; Yi, B.; Ma, C.; Tu, J.; Fu, T.; Shen, J. Genome-wide association study reveals the genetic architecture underlying salt tolerance-related traits in rapeseed (*Brassica napus* L.). *Front. Plant Sci.* **2017**, *8*, 593. [CrossRef]

28. Navarro-León, E.; López-Moreno, F.J.; de la Torre-González, A.; Ruiz, J.M.; Esposito, S.; Blasco, B. Study of salt-stress tolerance and defensive mechanisms in *Brassica rapa* CAX1a TILLING mutants. *Environ. Exp. Bot.* **2020**, *175*, 104061. [CrossRef]

29. Mei, H.; Zhao, J.; Pittman, J.K.; Lachmansingh, J.; Park, S.; Hirschi, K.D. In planta regulation of the Arabidopsis Ca^{2+}/H^+ antiporter CAX1. *J. Exp. Bot.* **2007**, *58*, 3419–3427. [CrossRef]

30. Ghanem, M.E.; Albacete, A.; Martínez-Andújar, C.; Acosta, M.; Romero-Aranda, R.; Dodd, I.C.; Lutts, S.; Pérez-Alfocea, F. Hormonal changes during salinity-induced leaf senescence in tomato (*Solanum lycopersicum* L.). *J. Exp. Bot.* **2008**, *59*, 3039–3050. [CrossRef]

31. de la Torre-González, A.; Navarro-León, E.; Albacete, A.; Blasco, B.; Ruiz, J.M. Study of phytohormone profile and oxidative metabolism as key process to identification of salinity response in tomato commercial genotypes. *J. Plant Physiol.* **2017**, *216*, 164–173. [CrossRef]

32. Ma, X.; Zhang, J.; Huang, B. Cytokinin-mitigation of salt-induced leaf senescence in perennial ryegrass involving the activation of antioxidant systems and ionic balance. *Environ. Exp. Bot.* **2016**, *125*, 1–11. [CrossRef]

33. Pavlů, J.; Novák, J.; Koukalová, V.; Luklová, M.; Brzobohatý, B.; Černý, M. Cytokinin at the crossroads of abiotic stress signalling pathways. *Int. J. Mol. Sci.* **2018**, *19*, 2450. [CrossRef]

34. Wu, X.; He, J.; Chen, J.; Yang, S.; Zha, D. Alleviation of exogenous 6-benzyladenine on two genotypes of eggplant (*Solanum melongena* Mill.) growth under salt stress. *Protoplasma* **2014**, *251*, 169–176. [CrossRef]

35. Ghanem, M.E.; Albacete, A.; Smigocki, A.C.; Frébort, I.; Pospíšilová, H.; Martínez-Andújar, C.; Acosta, M.; Sánchez-Bravo, J.; Lutts, S.; Dodd, I.C.; et al. Root-synthesized cytokinins improve shoot growth and fruit yield in salinized tomato (*Solanum lycopersicum* L.) plants. *J. Exp. Bot.* **2011**, *62*, 125–140. [CrossRef]

36. Park, H.J.; Kim, W.-Y.; Yun, D.-J. A new insight of salt stress signaling in plant. *Mol. Cells* **2016**, *39*, 447–459. [CrossRef]

37. Parida, A.K.; Das, A.B. Salt tolerance and salinity effects on plants: A review. *Ecotoxicol. Environ. Saf.* **2005**, *60*, 324–349. [CrossRef]

38. Zörb, C.; Geilfus, C.-M.; Dietz, K.-J. Salinity and crop yield. *Plant Biol.* **2019**, *21*, 31–38. [CrossRef]

39. Zörb, C.; Geilfus, C.-M.; Mühling, K.H.; Ludwig-Müller, J. The influence of salt stress on ABA and auxin concentrations in two maize cultivars differing in salt resistance. *J. Plant Physiol.* **2013**, *170*, 220–224. [CrossRef] [PubMed]

40. Amjad, M.; Akhtar, J.; Anwar-ul-Haq, M.; Yang, A.; Akhtar, S.S.; Jacobsen, S.-E. Integrating role of ethylene and ABA in tomato plants adaptation to salt stress. *Sci. Hortic. (Amsterdam)* **2014**, *172*, 109–116. [CrossRef]

41. Trivellini, A.; Lucchesini, M.; Ferrante, A.; Carmassi, G.; Scatena, G.; Vernieri, P.; Mensuali-Sodi, A. Survive or die? A molecular insight into salt-dependant signaling network. *Environ. Exp. Bot.* **2016**, *132*, 140–153. [CrossRef]

42. Arif, Y.; Sami, F.; Siddiqui, H.; Bajguz, A.; Hayat, S. Salicylic acid in relation to other phytohormones in plant: A study towards physiology and signal transduction under challenging environment. *Environ. Exp. Bot.* **2020**, *175*, 104040. [CrossRef]

43. Farhangi-Abriz, S.; Ghassemi-Golezani, K. How can salicylic acid and jasmonic acid mitigate salt toxicity in soybean plants? *Ecotoxicol. Environ. Saf.* **2018**, *147*, 1010–1016. [CrossRef]

4

Association Analysis of Salt Tolerance in Asiatic cotton (*Gossypium arboretum*) with SNP Markers

Tussipkan Dilnur [†], Zhen Peng [†], Zhaoe Pan, Koffi Kibalou Palanga, Yinhua Jia, Wenfang Gong and Xiongming Du *

State Key Laboratory of Cotton Biology, Institute of Cotton Research, Chinese Academy of Agricultural Sciences, Anyang 455000, China; tdilnur@mail.ru (T.D.); cripengzhen09@126.com (Z.P.); panzhaoe@163.com (Z.P.); palangaeddieh@yahoo.fr (K.K.P.); jiayinhua_0@sina.com (Y.J.); gwf018@126.com (W.G.)

* Correspondence: dujeffrey8848@hotmail.com

† These authors contributed equally to this work.

Abstract: Salinity is not only a major environmental factor which limits plant growth and productivity, but it has also become a worldwide problem. However, little is known about the genetic basis underlying salt tolerance in cotton. This study was carried out to identify marker-trait association signals of seven salt-tolerance-related traits and one salt tolerance index using association analysis for 215 accessions of Asiatic cotton. According to a comprehensive index of salt tolerance (CIST), 215 accessions were mainly categorized into four groups, and 11 accessions with high salinity tolerance were selected for breeding. Genome-wide association studies (GWAS) revealed nine SNP rich regions significantly associated with relative fresh weight (RFW), relative stem length (RSL), relative water content (RWC) and CIST. The nine SNP rich regions analysis revealed 143 polymorphisms that distributed 40 candidate genes and significantly associated with salt tolerance. Notably, two SNP rich regions on chromosome 7 were found to be significantly associated with two salinity related traits, RFW and RSL, by the threshold of $-\log_{10}P \geq 6.0$, and two candidate genes (Cotton_A_37775 and Cotton_A_35901) related to two key SNPs (Ca7_33607751 and Ca7_77004962) were possibly associated with salt tolerance in *G. arboreum*. These can provide fundamental information which will be useful for future molecular breeding of cotton, in order to release novel salt tolerant cultivars.

Keywords: *Gossypium arboretum*; salt tolerance; single nucleotide polymorphisms; association mapping

1. Introduction

Soil salinity accumulation has become a serious environmental problem [1] that could negatively affect plant growth, geographical distribution, and agricultural products [2,3]. Salinization consists of the accumulation of water-soluble salts in the soil, including ions of potassium (K^+), magnesium (Mg^{2+}), calcium (Ca^{2+}), chloride (Cl^-), sulfate (SO_4^{2-}), carbonate (CO_3^{2-}), bicarbonate (HCO_3^-) and sodium (Na^+). The causes of land salinization can be divided into two categories: 1) primary (natural) and 2) secondary (anthropogenic) [4]. The primary reason includes arid climates, high underground water levels, seawater infiltration, and so on [5]. The secondary reason is irrigation practices. Soil salinization is reducing the area that can be used for agriculture by 1%–2% every year, hitting hardest in the arid and semi-arid regions. Therefore, the development of salt-tolerant crops is a pressing scientific goal [6], but the ability of plants to deal with these adverse factors is different [7]. Cotton is one of the advantageous salt-tolerant crop with a threshold salinity level of 7.7 dS·m^{-1} [8]. However, high salt concentrations can still hinder growth during the germination and seedling stages, which are the two most susceptible stages of plants [2,9].

The mechanisms involved in the response to salinity in cotton have been well described by Peng et al. [3,10]. Under salinity stress conditions, soluble salts are accumulated in the root zone of plants, then causing osmotic and ionic stress and mineral perturbations [3,11], leading to dramatic reductions in crop quality and yield [5]. However, the genetic control of salt tolerance is only partially understood, because of the diversity of the regulation mechanisms, and the complexity of the genetic architecture of salt tolerance [9]. As the fundamental aim of genetics is to connect genotype to phenotype, the identification and characterization of genes associated with agronomical important traits is essential for both understanding the genetic basis of phenotypic variation and efficient crop improvement. Modern molecular biology techniques and new statistical methods have opened new horizons for the cotton breeders; thus, linkage mapping and association mapping are the two important methods employed for QTL analysis. Molecular marker-quantitative trait association is one of the powerful approaches for exploring the molecular basis of phenotypic variations in plant [12], and could be used to increase the efficiency of a breeding program, especially for salinity tolerance [13,14]. The present studies of genetic map construction are mainly reflected in three different DNA based molecular markers such as simple sequence repeats (SSR) [15], single-nucleotide polymorphism (SNP) markers [16,17] and Intron length polymorphisms (ILD) markers [18].

Single Nucleotide Polymorphism is often abbreviated to SNP; it describes a variation in a single nucleotidae that occurs at a specific position in the genome [19]. Genome-wide association studies based on linkage disequilibrium (LD) is an effective strategy tool to study phenotype-genotype association. Compared with traditional QTL mapping, GWAS can use SNPs obtained by genome re-sequencing as molecular markers to dissect complex traits [20]. One SNP occurs every 100–300 bp in any genome; therefore SNPs markers have higher polymorphism than SSRs and other molecular markers [19]. GWAS has been successfully applied in rice, Arabidopsis, maize, wheat, barley and other crops to identify characteristic-related SNP markers of their important trait [21–24]. In cotton, various genetic maps based on SSR and SNP markers have been constructed using bi-parental mapping populations and natural population of *Gossypium hirsutum*; however, there are fewer studies, and no causal genes responsible for the salt tolerance traits from *Gossypium arboretum* have been identified [25–28].

Asiatic cotton (*Gossypium arboretum*) was introduced into China from ancient India, Burma or Vietnam over 2000 years ago [29]. Du et al (2018) reported that the natural population of Gossypium. arboretum was classified into three main groups represented South China, Yangtze River region, and Yellow River region groups respectively that exhibited strong geographical distribution [30]. A draft genome of cotton diploid *Gossypium arboretum* (the size is 1.7 Gb, $2n = 2 \times = 26$) was recently reported by Li et al. (2015) [31]. The genetic basis of Asiatic cotton will provide a fundamental resource for genetics research of the important agronomic traits for cotton breeding. Therefore, the present study was performed in consideration of the following objectives: (i) to screen salinity tolerance at germination stage; (ii) to analyze the marker-trait associations by using SNP markers; (iii) to identify the causal genes which are responsible for the salt tolerance traits from *G. arboreum*.

2. Results

2.1. Phenotypic Diversity of G. arboretum Population

Seven salt tolerance related traits, including GR, FW, SL, WC, ChlC, EC, and MDA, were measured for all 215 *G. arboretum* accessions under 0 mM (C) and 150 mM (S) NaCl treatment (Figure 1 and Table S1). ANOVA analysis of seven salt-tolerance-related traits as measured for genetic diversity shows significant difference among the accessions ($P < 0.0001$) (Table 1). Correlation of GR with FW and SL was highly significant ($P < 0.001$), while GR with ChlC was also significant ($P < 0.01$). Correlation results of FW with SL, ChlC and MDA were also highly significant with $P < 0.001$. Correlation between SL and ChlC was highly significant ($P < 0.001$). A positive correlation was found between ChlC and EC ($P < 0.01$), while a negative correlation was found between ChlC and MDA ($P < 0.05$). The correlation between related EC and MDA was significant at $P < 0.05$. Interestingly, correlation analysis revealed that WC had significant correlation ($P < 0.05$) with MDA (Table 2). According to the CIST, 215 accessions were mainly categorized into four clusters. Cluster 1 contained 12 accessions that were highly sensitive to salt treatment (<0.6), Cluster 2 contained 26 accessions that were moderate tolerant to salt treatment (0.6–1.5), Cluster 3 included 153 accessions that were tolerant (1.5–2.5), and Cluster 4 had 24 accessions that were highly tolerant to salt treatment (>2.5) (Figure 1, Figure S2 and Table S1). Based on this result, the high tolerant accessions (24) were selected (Table S1). Basing on the comprehensive index of salt tolerance, we finally selected 11 high tolerant accessions (top 5%) for breeding using, including GuangXiZuoXianZhongMian, LiaoYang-1, ZhaoXianHongJieMian, PingLeXiaoHua, KaiYuanTuMian, YuXi33, ChangShuXiaoBaiZi, PingGuoJiuPingZhongMian, FuChuanJiangTangZhongMian, TangShanBaiZiZhongMian, and ShiJiaZhuangJianMian (Table S1).

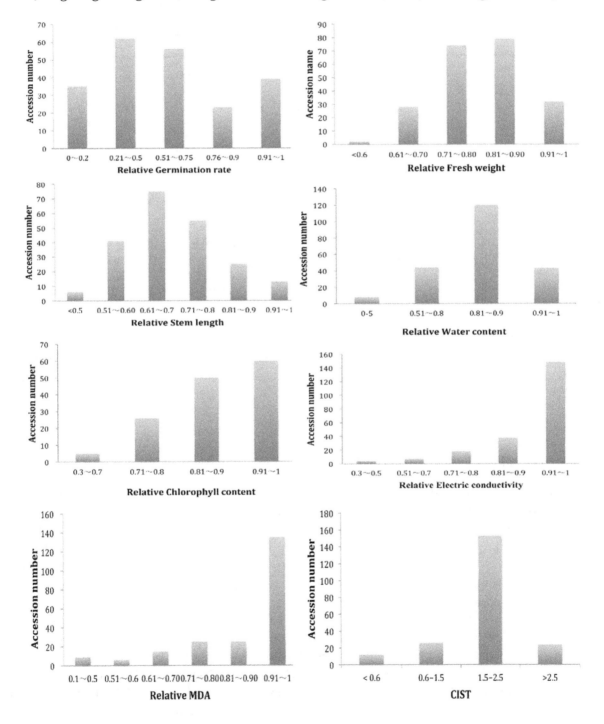

Figure 1. Relative value frequency distribution diagram of seven salt tolerance traits and one salt tolerance index of 215 *G. arboreum* accessions.

Table 1. Analysis of the traits related salt treatment in *G. arboretum* accessions.

Traits [1]	Mean	SD	Min	Max	CV	Mean Square	F	P > F
GR	19.936	10.77	0	40	54.02	338.55	8.16	<0.0001
FW	0.371	0.089	0.02	0.77	24.2	0.128	34.16	<0.0001
SL	4.969	1.481	0.40	10.2	29.84	40.77	49.76	<0.0001
WC	1.112	3.481	0.36	96.97	312.9	31.83	4.01	<0.0001
ChlC	39.47	9.96	3.50	50.2	25.53	6763.5	8.31	<0.0001
EC	31.47	42.54	0.116	844.4	135.19	541.2	6.49	<0.0001
MDA	0.007	0.0046	0	0.04	59.8	0.000095	14.36	<0.0001

[1] *GR* germination rate; *FW* fresh weight; *SL* stem length; *WC* water content; *ChlC* relative chlorophyll content; *EC* electric conduct; *MDA* methylene dioxyamphetamine.

Table 2. Analysis of traits related salt treatment of *G. arboretum* accessions (Pearson correlation coefficient).

Trait [1]	GR	FW	SL	WC	ChlC	EC	MDA
GR	1	0.22 ***	0.295 ***	−0.0178	0.112 **	0.0841	0.012
FW		1	0.575 ***	−0.043	0.070 ***	0.0613	0.135 ***
SL			1	−0.017	0.115 ***	−0.051	0.030
WC				1	0.048	0.002	0.059 *
ChlC					1	0.1 **	−0.073 *
EC						1	0.079 *
MDA							1

[1] For trait abb. Look at Table 1. * Significant at $P < 0.05$; ** Significant at $P < 0.01$; *** Significant at $P < 0.001$ for the correlation coefficient.

2.2. Association Mapping of Salt Tolerance Related Traits Using SNP Markers

The total of 1,568,133 high-quality SNPs (MAF > 0.05, missing rate < 40%) in 215 *G. arboreum* accessions were used for GWAS of the salt traits. The SNP markers associated with the seven salt-tolerance-related traits and one salt tolerance index were identified based on the threshold value, $\log_{10}(P) \geq 4.0$, using the MLMM model in the EMMAX software (Table S2 and Figure S3). The threshold of $-\log_{10}P \geq 4.0$ was also derived from the quantile–quantile (QQ) plots, For RGR, $-\log_{10}(P)$ values and QQ plots suggested relatively weak genetic association (Figure 2a). Most of the upward deviation from the linear line occurred at around $-\log_{10}(P) = 4.0$, which presumably indicates true positives (Figure 2b–h). For RGR, $-\log_{10}(P)$ values and QQ plots suggested relatively weak genetic association (Figure 2a). By applying the threshold of $-\log_{10}(P) \geq 4.0$, the 2062 SNP markers covered all 13 chromosomes and 100 SNP markers that were unknown location (Table 3). Chromosome 3 had the maximum number of SNPs (332 SNPs), and Chromosome 12 had the minimum (57) number of SNPs. Among the nucleotide polymorphisms, 1708 SNPs were interginic, 96 SNPs were intronic, 68 SNPs were exonic, 112 SNPs were upstream, 69 SNPs were downstream and 9 SNPs were upstream and downstream (Table S3).

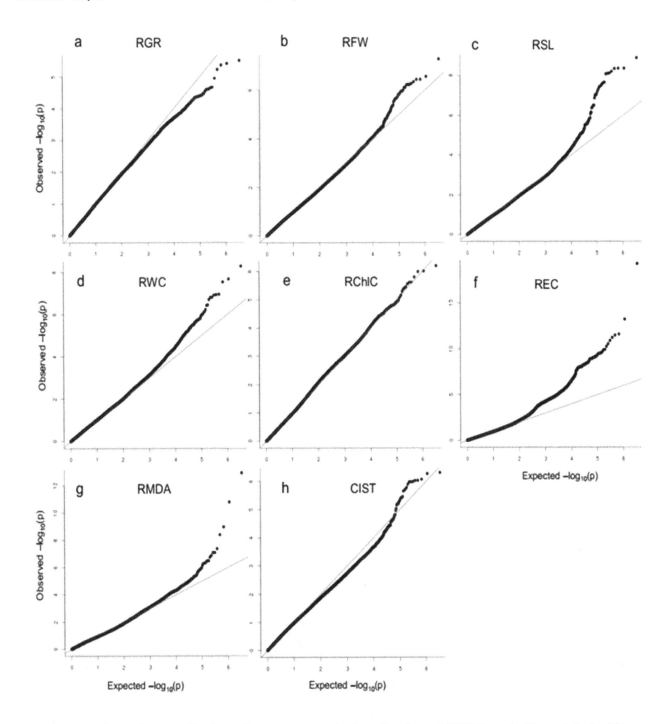

Figure 2. Quantile-quantile plots of versus expected −log$_{10}P$ values of GWAS result. The red dashed line in each plot represents an idealized case where theoretical test statistics quantile match simulated test statistic quantile. (**a**) Relative germination rate (RGR). (**b**) Relative fresh weight (RFW). (**c**) Relative stem length (RSL). (**d**) Relative water content (RWC). (**e**) Relative chlorophyll content (RChlC). (**f**) Relative electric conductivity (REC). (**g**) Relative MDA (RMDA). (**h**) Comprehensive index of salt tolerance (CIST).

Among these 2062 marker-trait associations, 61 markers were associated with RGR, 187 markers were associated with RFW, 255 markers were associated with RSL, 370 markers were associated with RWC, 190 markers were associated with RChlC, 583 markers were associated with REC, 335 markers were associated with RMDA and 81 markers were associated with CIST (Table 3, Table S2) The most SNPs related positive salt tolerance indicators (RFW, RSL and RWC) were on chromosome 7. The most SNPs related negative salt tolerance indicators (REC, RMDA) were on chromosome 3 (Table 3).

Table 3. Associated SNPs of different salinity traits distribution on chromosome [1.]

Chromosome	Total	RGR	RFW	RSL	RWC	RChlC	REC	RMDA	CIST
Chr-1	163	1	26	22	12	15	14	70	3
Chr-2	152	5	10	10	39	3	12	49	24
Chr-3	332	1	3	8	7	11	278	24	0
Chr-4	201	3	12	26	27	51	59	19	4
Chr-5	155	2	7	17	21	4	94	8	2
Chr-6	112	10	16	11	22	2	21	23	7
Chr-7	295	2	67	108	67	6	24	14	7
Chr-8	99	15	10	8	27	4	9	24	2
Chr-9	104	4	2	8	9	47	7	24	3
Chr-10	80	1	7	10	34	4	11	7	6
Chr-11	112	5	5	8	17	20	24	18	15
Chr-12	57	8	4	2	31	3	5	3	1
Chr-13	100	1	4	10	29	16	6	32	2
Chr-UN	100	3	14	7	28	4	19	20	5
Total	**2062**	**61**	**187**	**255**	**370**	**190**	**583**	**335**	**81**

[1] the SNP exceeding a significant threshold($-\log_{10}(P) \geq 4.0$); 2 For trait abb. Look at Table 1; CIST comprehensive index of salt tolerance.

2.3. SNP Rich Regions Associated with RFW and RSL

In the following, we focused on nine SNP rich regions associated RFW, RSL, RChlC, RWC and CIST for which MLMM analysis yield more significant associations considering $-\log_{10}P \geq 4.0$ values and the position of strong peaks in the Manhattan plots (Table S4).

Two SNP rich regions on chromosome 7 were found to associate with two biomass-related traits such as RFW and RSL (Figure 3a,b) when setting the threshold of $-\log_{10}P \geq 6.0$ on the Manhattan plots. The first candidate region (Group 1) starting at 33,513,007 bp and ending at 33,616,148 bp (103,141 bp), on chromosome 7, which contained 6 polymorphism SNPs (Figure 3a,b) and located within five genes, Cotton_A_37774, Cotton_A_37775, Cotton_A_37776, Cotton_A_37777 and Cotton_A_40811. All six SNPs were related to RSL, and three SNPs were related to related fresh weight (RFW) (Tables S4 and S6). Two intronic SNPs (Ca7_33606785 and Ca7_33607751) related with both of RSL and RFW), including key SNP Ca7_33607751 ($-\log_{10}(P) = 7.14$, possession 33693051 bp) in this region, were in gene of Cotton_A_37775 (Table S5 Group 1). Four intergenic SNPs were related to related fresh weight (RFW) (Table S5 Group 1). Cotton_A_37775 was annotated as heat shock protein, a homolog of Arabidopsis thaliana AT5G52640. Haplotype analysis showed a low level of linkage disequilibrium (LD) (lowest $r2 = 30$, highest $r2 = 100$) between the associated SNPs in Group 1 (Figure 3c,d). There were three genotypes for the key SNP Ca7_33607751. Genotypes CC, TT and CT contained 7, 135, and 6 accessions respectively, whereas the accessions carrying CC genotype showed the highest RFW and RSL; the accessions carrying CT genotype showed medium-higher RWF (0.909) and RSL (0.775); And the accessions carrying TT genotype showed the lowest RFW (0.81) and RSL (0.698) (Figure 3e,f).

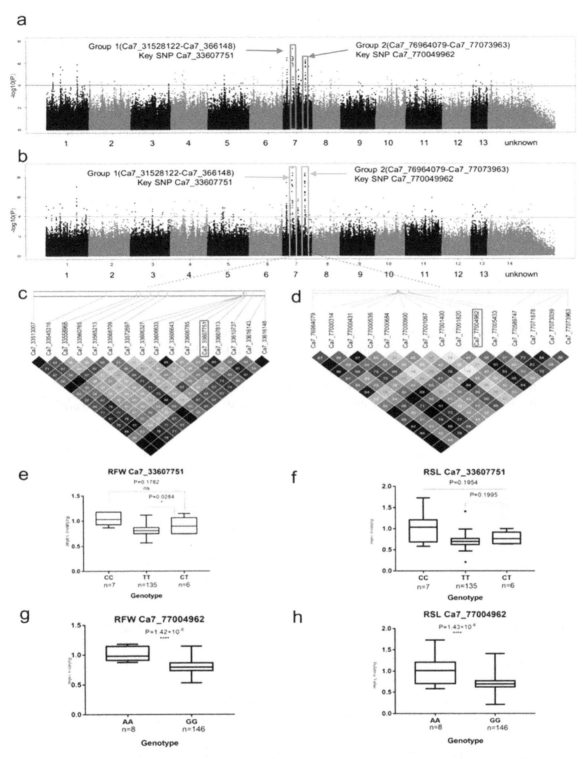

Figure 3. GWAS results for RFW and RSL, and analysis of the peaks on chromosome 7 (Group 1 and Group 2). (**a**) Manhattan plot for RFW. The horizontal line represents the significant threshold ($-\log_{10}P$ = 4). The pink color surrounds represent SNP rich regions (Group 1 and Group 2). (**b**) Manhattan plot for RSL. The horizontal line represents the significant threshold ($-\log_{10}P$ = 4). The pink color surrounds represent SNP rich regions (Group 1 and Group 2). (**c**) LD surrounding the peak of Group 1. (**d**) LD surrounding the peak of Group 2. (**e**) Phenotypic differences for RFW based on the key SNP (Ca7_33607751) of Group 1. (**f**) Phenotypic differences for RSL based on the key SNP (Ca7_33607751) of Group 1. (**g**) Phenotypic differences for RFW based on the key SNP (Ca7_77004962) of Group 2. (**h**) Phenotypic differences for RSL based on the key SNP (Ca7_77004962) of Group 2.

We then focused on the second highest peak (Group 2) on chromosome 7, which were common related to two biomass-related traits such as RFW and RSL. Group 2 was estimated to be 76,964,079 –77,073,963 bp (109,884 bp), and to contain 9 polymorphisms, which were located within 6 genes (Figure 3a,b and Tables S4 and S6). Among the 9 polymorphisms, 2 SNPs (Ca7_77000431 and Ca7_77004962) were related with both traits such as RFW and RSL, 7 SNPs were related only RSL. Most of these SNPs (6 of 9), including key SNP Ca7_77004962 ($-\log_{10}(P) = 8.36$) were located between two genes *Cotton_A_35901* and *Cotton_A_35900* (Table S6). Haplotype analysis showed a low level of LD (lowest $r^2 = 15$, highest $r^2 = 100$) between the associated SNPs in Group 2 (Figure 3d). There were two genotypes for the key SNP Ca7_77004962. Genotypes AA and GG contained 8 and 146 accessions, respectively, whereas the accessions carrying AA genotype showed the highest RFW (1) and RSL (1); the accessions carrying GG genotype showed lower RWF (0.808) and RSL (0.697) (Figure 3g,h).

2.4. SNP Rich Regions Associated with RChlC

Similarly, we analyzed four SNP rich regions on chromosome 4, 9 and 11, which closely associated with RChlC (Figure 4a). The candidate region on chromosome 4 (Group 3) was predicted to map from 32,124,549 bp to 32,259,755 bp (135,206 bp), and contained 14 polymorphisms, which were located within 6 genes (Figure 4a, Table S4 and Table S5 Group 3). Eight SNPs were distributed near the gene of *Cotton_A_26219* (Table S5). We found that two haplotypes in three SNPs (Ca4_32185154, Ca4_32191704 and Ca4_32197265) by using high pairwise LD correlation ($r^2 \geq 96$) (Figure 4b,f). These three SNPs were near *Cotton_A_26219*. Haplotype A and Haplotype B contained 27 and 157 accessions respectively (Figure 4f). The accessions carrying haplotype A showed lower RChlC (0.80) than haplotype B (0.97) (Figure 4i).

We estimated the candidate region on chromosome 9 (Group 4) to be 56,869,491–56,879,931 bp (10,440 bp) and assigned 14 polymorphisms, which were located within 3 genes (Figure 4a, Table S4 and Table S5 Group 4). Most of these polymorphisms (8 of 14), including the key SNP Ca9_56878752 (exonic) were in *Cotton_A_10864*. Five coding region SNPs in *Cotton_A_10865*, which is annotated as F-Box protein and leucine-rich repeat protein 14 (Table S5 Group 4). We found that two haplotypes in seven SNPs (Ca9_56875436, Ca9-56875607, Ca9-56876686, Ca9-56876694, Ca9-56877663 in Cotton_A_10865, and Ca9-56878677, Ca9-56878752 in *Cotton_A_10864*) by using high pairwise LD correlation ($r^2 > 80$) (Figure 4c,g). Haplotype A and Haplotype B contained 9 and 84 accessions, respectively (Figure 4g). The accessions carrying haplotype A showed higher RChlC (1) than haplotype B (0.93) (Figure 4j).

The candidate region on chromosome 9 (Group 5) was predicted to map from 92,702,808 bp to 92,735,184 bp (32,376 bp), and contained 19 polymorphisms (Figure 4a, Table S4 and Table S5 Group 5). All 19 significant SNP markers (key SNP Ca9_92711930, $-\log_{10}P = 5.80$) was located between two pathogenesis-related thaumatin superfamily protein genes: *Cotton_A_15275* (average distance 13,817 bp) and *Cotton_A_15276* (average distance 31,662 bp) (Table S5 Group 5). We found that two haplotypes in thirteen SNPs (from Ca9_92710379 to Ca9_92717732) by using high pairwise LD correlation ($r^2 \geq 91$) (Figure 4d,h). Haplotype A and Haplotype B contained 12 and 97 accessions, respectively (Figure 4h). The accessions carrying haplotype A showed higher RChlC (1) than haplotype B (0.92) (Figure 4k).

The candidate region on chromosome 11 (Group 6) was predicted to map from 47,006,642 bp to 47,011,718 bp (5076 bp), and contained 11 polymorphisms (Figure 5a, Table S4 and Table S5 Group 6). All 11 significant SNP markers (key SNP Ca11_47011718, $-\log_{10}P \geq 5.61$) was located between two genes: *Cotton_A_28249* (average distance from SNPs 7649 bp) and *Cotton_A_28248* (average distance 5083 bp from SNPs). There were three genotypes for the key Ca11_47011718 (Table S5 Group 6). Genotypes AA, GG and AG contained 190, 13, and 2 accessions respectively, whereas the accessions carrying AA (0.933) genotype showed lower RChlC; And the accessions carrying GG (1) and AG (1) genotype showed the highest RChlC (Figure 5i).

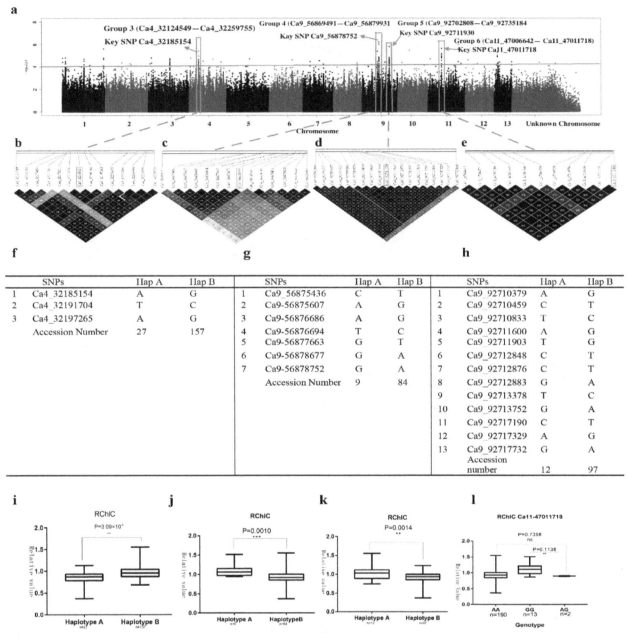

Figure 4. GWAS results for RChlC and analysis of the peaks on chromosome 4, 9 and 11. (**a**) Manhattan plot for RChlC. The horizontal line represents the significant threshold ($-\log_{10}P = 4$). The pink color surrounds represent SNP rich regions (Group 3, Group 4, Group 5 and Goup 6). (**b**) LD surrounding the peak of Group 3. (**c**) LD surrounding the peak of Group 4. (**d**) LD surrounding the peak of Group 5. (**e**) LD surrounding the peak of Group 6. (**f**) Haplotypes in Group 3. (**g**) Haplotypes in Group 4 (**h**) Haplotypes in Group 5. (**i**) Phenotypic differences of RChlC between two haplotypes in Group 3. (**j**) Phenotypic differences of RChlC between two haplotypesin Group 4. (**k**) Phenotypic differences of RChlC between two haplotypesin Group 5. (**l**) Phenotypic differences for RChlC based on the key SNP (Ca11_47011718) of Group 6.

2.5. SNP Rich Region Associated with RWC

The candidate region on chromosome 7 (Group 7) was predicted to map from 39,823,916 bp to 39,843,478 bp (19,562 bp), and contained 29 polymorphisms in four genes (*Cotton_A_05854*, *Cotton_A_05853*, *Cotton_A_05852* and *Cotton_A_05852*) (Figure 5a, Table S4 and Table S5 Group 7). Among these four genes, most important gene is *Cotton_A_05853*, because, 15 of these SNPs (11 intronic and 3 intergenic), including intronic key SNP Ca7_39832729 ($-\log_{10}(P) > 5.35$) were located in *Cotton_A_05853* (*AT3G12800*), annotated in The Arabidopsis Information Resource (TAIR) as short-chain dehydrogenase-reductase B and oxidation-reduction process (Table S5 Group 7). We found that two haplotypes of eight SNPs in *Cotton_A_05853* (from Ca7_39832407 to Ca7_39832920) by using high pairwise LD correlation ($r^2 \geq 90$) (Figure 5b). Haplotype A and Haplotype B contained 42 and 6 accessions, respectively (Figure 5c). The accessions carrying haplotype A showed higher RWC (0.86) than haplotype B (0.50) (Figure 5d).

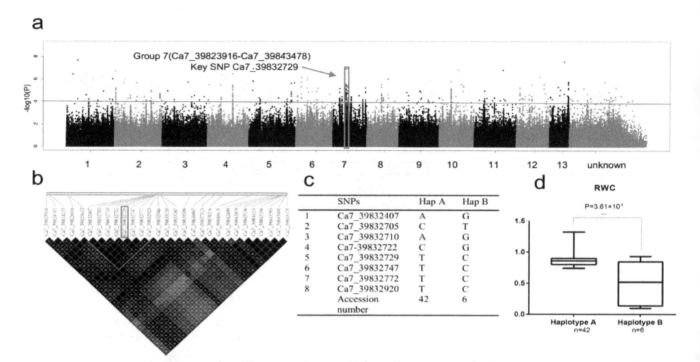

Figure 5. GWAS results for RWC and analysis of the peak on chromosome 7. (**a**) Manhattan plot for RWC. The horizontal line represents the significant threshold ($-\log_{10}P = 4$). The pink color surround represents SNP rich region (Group 7). (**b**) LD surrounding the peak of Group 7. (**c**) Haplotypes in Group 7. (**d**) Phenotypic differences of RWC between two haplotypes.

2.6. SNP Rich Regions Associated with CIST

The candidate region on chromosome 2 (Group 8) was predicted to map from 86,947,129 to 87,017,559 bp (11,697 bp) and to contain 16 polymorphisms in two genes (*Cotton_A_22673*, *Cotton_A_22672*) (Figure 6a, Table S4 and Table S5 Group 8). Fifteen significant intergenic SNPs, including key SNP Ca2_86954790 ($-\log_{10}(P) \geq 6.32$) were located near the gene of *Cotton_A_22673* (average distance 63,317 bp). Only one significant non-synonymous SNP is in *Cotton_A_22672*, which without annotation (Table S5 Group 8). We found that two haplotypes of six SNPs near to *Cotton_A_22673* (from Ca2_86947129 to Ca2_86956247) by using high pairwise LD correlation ($r^2 \geq 93$) (Figure 6b). Haplotype A and Haplotype B contained 27 and 157 accessions, respectively (Figure 6d). The accessions carrying haplotype A showed lower CIST (1.042) than haplotype B (1.935) (Figure 6f).

The candidate region on chromosome 11 (Group 9) was predicted to map from 39,493,066 to 39,504,763 bp (11,697 bp) and to contain 9 polymorphisms in two genes (*Cotton_A_21725* and *Cotton_A_21726*) (Figure 6a, Table S4 and Table S5 Group 9). *Cotton_A_21726* contained eight significant intronic SNPs, including key SNP Ca11_39504708 ($-\log_{10}P = 5.98$), annotated Glycosyl hydrolase family 10 proteins. Only one significant downstream SNP was in *Cotton_A_21725*, annotated as DNA/RNA polymerases superfamily protein (Table S5 Group 9). We found that three haplotypes in all nine SNPs in this region by using very high pairwise LD correlation ($r^2 = 100$) (Figure 6c,e). Haplotype A, Haplotype B and Haplotype C contained 35, 5 and 36 accessions, respectively (Figure 6e).

The accessions carrying haplotype A (1.973) and C (2.056) showed similar CIST, but the accessions carrying haplotype B showed the lowest CIST (−0.1398) (Figure 6g).

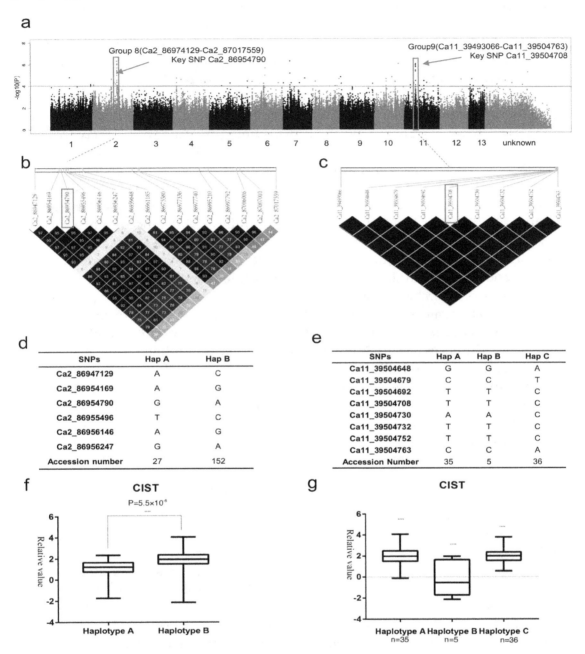

Figure 6. GWAS results for CIST and analysis of the peaks on chromosome 2 and 11. (**a**) Manhattan plot for CIST. The horizontal line represents the significant threshold ($-\log_{10}P = 4$). The pink color surrounds represent SNP rich regions (Group 8 and Group 9). (**b**) LD surrounding the peak of Group 8. (**c**) LD surrounding the peak of Group 9. (**d**) Haplotypes in Group 8. (**e**) Haplotypes in Group 9. (**f**) Phenotypic differences of CIST between two haplotypes in Group 8. (**g**) Phenotypic differences of CIST among three haplotypes in Group 9.

3. Discussion

3.1. Genetic Variation in Salt Tolerance Related Traits of G. arboretum Accessions

G. arboretum is considered superior to upland cotton varieties based on the following traits: precociousness, wide adaptability, drought tolerance and disease resistance from fusarium wilt, insects and bell mild disease [32]. However, it has not been well characterized at the molecular level. Thus, our current study mainly focused on identification and screening of salt tolerant germplasm during seedling stage, to find genetic relationships and to study marker-trait associations SNP markers. The *G. arboretum* accessions, considered as an invaluable gene pool for cotton improvement, were used in this study. Understanding of the genetic diversity of *G. arboretum* can facilitate the efficient use of these resources in the development of superior cotton cultivars with favorable agronomic traits.

Abiotic stress leads to a series of morphological, physiological, biochemical and molecular changes that have adverse effects on the plant growth, development and productivity. In fact, the salinity is a major abiotic stress that limits cotton growth and development at the germination and seedling stages [27]. In this study, the suitable optimum NaCl concentration was determined by monitoring germination rates of accessions. Shixiya No. 1, Sichuansuining and Chaoxianjinhuaxiaozi (included in the 215 accessions) under different NaCl concentrations (0, 50, 100, 150, 200, 250, and 300 mM). Under 200, 250, and 300 mM NaCl treatments, most seeds were unable to geminate. In addition, most of the seeds germinated under 100 and 150 mM NaCl treatments (Figure S1). Based on these results and the reports of Chen et al. (2008), 150 mM NaCl concentration was considered suitable salt treatment for *G. arboretum* [29]. Under 150 mM NaCl treatment, significant differences among different cotton accessions were observed during germination and seeding stages. The assessment of diversity in a species is important in plant breeding programs, and for effective conservation, management and utilization of genetic resources of the species [30]. The salt tolerance related traits (GR, FW, SL, WC, ChlC, EC, MDA) were comparable to that of the salt tolerance study and reported previous works [33]. Notably, plant germination rate, stem length, fresh weight are the major components of plant yield and were used as selection criteria in breeding. The ANOVA for the important salt tolerance parameters revealed significant differences ($P < 0.0001$) among the genotypes, implying that sufficient phenotypic polymorphism existed between individual *G. arboretum* accessions in this study. The correlation analysis is important to identify the mutual associations among the traits [34]. There was efficiency correlation between seven salt-tolerance-related traits; it is useful for multiple trait selection at one time for development of improved cotton varieties. The classification information derived from these studies may be used to facilitate the development of salt tolerant cotton accessions that could give economic yield in salinity prone areas.

3.2. Association Mapping of Salt Tolerance Traits Using SNP Markers

Genome-wide association study (GWAS) can effectively associate genotypes with phenotypes in natural populations and simultaneously detect many natural allelic variations and candidate genes in a single study, in contrast to QTL linkage mapping [35]. To cope with environmental stress, plants activate a large set of genes leading to the accumulation of specific stress-associated proteins. In this study, we first performed a genome-wide association analysis of salt tolerance related traits with 215 of natural accessions of *G. arboretum*. This study uncovered 2062 loci ($-\log_{10}P \geq 4.0$) associated salt tolerance traits and identified a set of candidate genes that could be exploited to alter salt tolerance development to improve *G. arboretum* accessions. The analysis of genomic distribution of SNPs in this study revealed that the most of SNPs related positive salt tolerance indicators (RFW, RSL and RWC) were on chromosome 7 and the most of SNPs related negative salt tolerance indicators (REC, RMDA) were on chromosome 3.

The nine SNP rich regions analysis revealed 143 polymorphisms that distributed 40 candidate genes and significantly associated RFW, RSL, RChlC, RWC and CIST. We found that twelve and eight SNPs in Group 1 and Group 2 were associated with two biomass traits such as RFW and RSL, because there is highly significant correlation ($P < 0.001$) between FW and SL (Table S5 Group 1 and Group 2).

In the first SNP rich region (Group 1), the most plausible candidate indented in the peak on chromosome 7 was *Cotton_A_37775* that was involved in Heat shock protein (Hsps), which plays a crucial role in salt stress response. *Cotton_A_37775* contains 1,951 amino acid and shares 79% identify at the amino acid level with a homolog of *Arabidopsis thaliana* AT5G52640, which annotated Heat shock protein (Hsp90.1). The 90 kDa heat shock protein (Hsp90) is a widespread family of molecular chaperones found in prokaryotes and all eukaryotes [36]. Hsp90 chaperone machinery is a key regulator of proteostasis under both physiological and stress conditions in eukaryotic cells. A large number of co-chaperones interact with HSP90 and regulate the ATPase-associated conformational changes of the HSP90 dimer that occur during the processing of clients [37]. The basic functions of HSP90s are of assisting protein folding, protein degradation and protein trafficking, and they also play an important role in signal transduction networks, cell cycle control and morphological evolution. Although HSP90s are constitutively expressed in most organisms, their expression is up-regulated in response to stresses such as cold, heat, salt stress, heavy metals, phytohormones, light and dark transitions [38]. The most prominent response of plants under high temperature stress is the rapid production of heat shock proteins (HSPs). Ding (2006) found that heat shock treatment in early growing period benefited the abundance of HSPs in cotton leaves in high temperature season, and then increased the ability thermo-tolerance [39]. In the plant *Arabidopsis thaliana (A. thaliana)*, *HSP90* homologs are encoded by seven different genetic loci. Of these, one is expressed in the endoplasmic reticulum (*HSP90.7*), one in the mitochondrion (*HSP90.6*), one in the chloroplast (*HSP90.5*), and four in the cytosol. The gene encoding one cytosolic protein (*HSP90.1/At5g52640*) is highly stress-inducible, whereas the other three (*HSP90.2/At5g56030*, *HSP90.3/At5g56010*, and *HSP90.4/At5g56000*) are constitutively expressed and are the products of very recent duplication events [40]. The homolog *At5g52640* is found up-regulated in response to viruses stresses by playing a role in cell migration [41].

In the second SNP rich region (Group 2), we identified a gene, *Cotton_A_35901* (301 amino acid), encoding a SNARE-like superfamily protein homologue, which has not been previously reported in cotton. Complexes of SNARE proteins mediate intracellular membrane fusion between vesicles and organelles to facilitate transport cargo proteins in plant cells [36]

In the third SNP rich region (Group 3), we found that two haplotypes in three SNPs (Ca4_32185154, Ca4_32191704 and Ca4_32197265) by using high pairwise LD correlation ($r^2 \geq 96$). These three intergenic haplotype SNPs located between two genes *Cotton_A_26219* and *Cotton_A_26218*. *Cotton_A_26219* have no annotation. Anther gene *Cotton_A_26218* is a homologue of *A.thaliana AT1G18800*, which encoding nucleosome assembly protein (nrp1-1 nrp2-1). Juan (2012) show that the nucleosome assembly protein (NAP1) family histone chaperones are required for somatic homologous recombination (HR) in *A. thaliana* [42]. HR is essential for maintaining genome integrity and variability. To orchestrate HR in the context of chromatin is a challenge, both in terms of DNA accessibility and restoration of chromatin organization after DNA repair. Histone chaperones function in nucleosome assembly/disassembly and could play a role in HR.

Depletion of either the NAP1 group or NAP1-RELATED PROTEIN (NRP) group proteins caused a reduction in HR in plants under normal growth conditions as well as under a wide range of genotoxic or abiotic stresses. In *Arabidopsis thaliana*, *AT1G18800* is required for maintaining cell proliferation and cellular organization in root tips [43].

The analysis in the fourth SNP rich region (Group 4) revealed two haplotypes in seven SNPs distributed in two candidate genes, Cotton_A_10865 and Cotton_A_10864 on Chr9, were associated with RChlC. Cotton_A_10865 is homologous to *AT4G08980* (F-BOX WITH WD-40), encodes an F-box gene that is a novel negative regulator of AGO1 protein levels and may play a role in abscisic aci (ABA) signaling and/or response. ABA signaling also plays a major role in mediating physiological responses to environmental stresses such as salt, osmotic, and cold stress. The accumulation of ABA in response to water or salt stress is a cell signaling process, encompassing initial stress signal perception, cellular signal transduction and regulation of expression of genes encoding key enzymes in ABA biosynthesis and catabolism [44]. In addition, there are a lot of studies to prove ABA response to salt stress in different crops such as *A. thaliana* [44–47], rice [48], wheat [49], corn [50]. The homologue of *Cotton_A_10864* is *AT5G65270* (RAB GTPASE HOMOLOG A4A) in *A. thaliana*. Several genes in the Rab GTPase family have been shown to be responsive to abiotic stress, including response of *SsRab2* to water stress in *Sporobolus stapfianus* [51], *MfARL1* to salt stress in *A. thaliana* [47], OsRab7 to cold stress in rice [52] and *AtRabG3e* to salt/osmotic stress in *A. thaliana* [53].

In the fifth SNP rich region (Group 5), the two genes (*Cotton_A_15275* and *Cotton_A_15276*) in Group 5 were pathogenesis-related thaumatin superfamily protein. Pathogenesis-related (PR) proteins play an important role in plants as a protein-based defensive system against abiotic and biotic stress, particularly pathogen infections. It is also named thaumatin-like proteins (TLPs), because it has sequence similarity with thaumatin [54,55].

In the six SNP rich region (Group 6), we found two genes *Cotton_A_28248* and *Cotton_A_28249*. *Cotton_A_28248* has no annotation. *Cotton_A_28249* is homologous to *AT4G32050* and encodes neurochondrin family protein, which is an atypical RIIα-specific A-kinase anchoring protein. In the seventh SNP rich region (Group 7), we found two haplotypes in seven SNPs distributed in one gene, *Cotton_A_05853*, has no annotation.

In the eighth SNP rich region (Group 8), we found two candidate genes *Cotton_A_22673* and *Cotton_A_22672*. *Cotton_A_22673* is annotated carbon-sulfur lyases, homologous to *A.thaliana AT5G09970*. *AT5G09970* locate in mitochondrion and encodes a DYW-class PPR protein required for RNA editing at four sites in mitochondria of *A. thaliana*.

In the ninth SNP rich region (Group 9), we found two candidate genes *Cotton_A_21725* and *Cotton_A_21726*. *Cotton_A_21725* encodes DNA/RNA polymerases superfamily protein. Several genes are induced under the influence of various abiotic stresses. Among these are DNA repair genes, which are induced in response to the DNA damage. Since the stresses affect the cellular gene expression machinery, it is possible that molecules involved in nucleic acid metabolism including helicases are likely to be affected. The light-driven shifts in redox-potential can also initiate the helicase gene expression. Helicases are ubiquitous enzymes that catalyze the unwinding of energetically stable duplex DNA (DNA helicases) or duplex RNA secondary structures (RNA helicases). Most helicases are members of DEAD-box protein superfamily and play essential roles in basic cellular processes such as DNA replication, repair, recombination, transcription, ribosome biogenesis and translation initiation. Therefore, helicases might be playing an important role in regulating plant growth and development under stress conditions by regulating some stress-induced pathways [56]. *Cotton_A_21726* encodes glycoside hydrolasefamily 10. GHs (glycosyl hydrolases) enzymes that catalyze the hydrolysis of glycosidic bonds between sugars and other moieties, can be classified into more than 100 families [55]. We could not find any relationship between glycoside hydrolase family 10-prtain and salt tolerance. But the glycoside hydrolase family 5 gene is expressed in rice leaves and seedling shoots, whereas its expression is induced by stress-related hormones, submergence and salt in whole seedlings [57].

The identified genetic variation and candidate genes deepen our understanding of the molecular mechanisms underlying salt tolerance traits. The genes discussed in this study may be considered as candidate genes for salt tolerance in cotton.

4. Materials and Methods

4.1. Plant Materials and Sample Preparation

The genetic materials used in the present study includes 215 accessions of *G. arboretum*, among them 209 accessions belong to China, 3 accessions from the United States, and 3 accessions from India, Pakistan and Japan. The germplasm was assembled from Germplasm bank of Institute of Cotton Research of Chinese Academy of Agricultural Sciences (CAAS), Anyang, China. The detailed list of accessions along with their origin is described in Table S1.

The phenotypic analysis and genetic experiment for these selected accessions were performed in the laboratory of Cotton Research Institute of CAAS, Anyang, China. Seedlings were grown in a phytotron incubating chamber under 14 h/10 h (light/dark) cycle, 28/14 °C (day/night), 450 μmol·m^{-2}·s light intensity and a relative humidity of 60–80% conditions [34]. For each genotype, 200 hand-selected seeds for each variety were sterilized with 15% hydrogen peroxide for 4 h, and subsequently rinsed with sterile distilled water at least 4 times, followed by seed submersion in sterile water for 12 h at room temperature. For identification of the seed germination rates (GR), 120 healthy seeds from each accession were selected and placed in germination boxes (200 × 150 mm diameter), containing two sheets of filter paper and soaked with 20 mL of the NaCl solutions (0 or 150 mM) respectively. For identification of other traits, such as fresh weight (FW), stem length (SL), water content (WC), chlorophyll content (ChlC), electric conductivity (EC) and methylene dioxyamphetamine (MDA), twenty-five healthy seeds each were planted in the 300 mL volumetric flask (Figure S1), containing 140 g sand (sterilized in autoclave at 160 °C for 2 h), 40 mL sterilized water (for control 0 mM) and 40 mL NaCl solutions (for 150 mM).

4.2. Trait Evaluation

After seven days of the germination, the 215 G. arboretum accession seedlings grown in the soil were evaluated for seven salt-tolerance-related traits, such as GR, FW, SL, WC, ChlC, EC and MDA. The seeds were considered to have germinated when the radicle and plumule length was equivalent to the seed length or half of the seed length. The germination rate was calculated GR (%) = (number of germinated seeds/total seed number used in the test) × 100. Fifteen plants from each accession were harvested and cleaned by sterilized water and immediately used for determination of FW, SL and ChlC.

For leave water content (WC) estimation, detached leaves were floated on deionized water for 8 h at 4 °C and turgid weights (TW) were determined. Later, dry weight (DW) of leaves was measured after oven-dried at 105 °C for 10 min, and then 80 °C for 24 h respectively. WC was calculated WC (%) = (FW − DW)/(TW − DW) × 100 [33].

For EC measurement, 0.5 g of leaves were rinsed with double distilled water (ddH$_2$O) and put in volumetric flask containing 40 mL of ddH$_2$O. Afterwards, the test flasks were incubated at room temperature for 4 h. The electrical conductivity of the solution (C1) was measured using a conductivity meter EM38 (ICT, Australian). The test flasks were boiled for 10min and then cooled at room temperature to measure the electrical conductivity (C2). The REC was calculated using the formula C1/C2 × 100%. [33]

For MDA measurement, 0.5 g of leaves was rinsed with double distilled water (ddH$_2$O) and then crushed in Thiobarbituric acid extract solution (5 mL, 0.5%). Absorbances were monitored at three different wavelengths i.e., 450 nm, 532 nm and 600 nm. MDA was calculated according to Le et al. (2000) method. MDA (X) = [6.452*(OD532 − OD600 − 0.559*OD450] *Vt/(Fw × 1000) [33]. Where X is MDA in μmol/g, Vt is total volume of extraction solution in mL and FW is fresh weighting.

All phenotypic data and physiological indicators were performed at least three biological replicates. The relative value of GR phenotypic data was used for further GWAS. The formula is RGR = GR$_{150}$/GR$_{control}$, i.e., the same as that for other traits, RFW, RSL, RWC, RChlC, REC and RMDA.

4.3. DNA Extraction

Young leaves were collected from five plants of each accession and stored at −80 °C. The DNA was isolated from the frozen leave using CTAB method [34] with some modifications. DNA concentration was quantified using a NanoDrop2000 instrument (Thermo Scientific, USA), and normalized to 50 ng/mL.

4.4. Phenotypic Diversity

Analysis of variance (ANOVA) and phenotypic correlations between different salt related physiological traits were performed using statistical software package SAS 9.21. Relative value for each trait and CIST were calculated according to these formulae:

Relative value = value under stress treatment (S)/value under control treatment (C);

CIST = positive index (RGR + RSL + RFW + RChlC + RWC)/negative index (REC + RMDA).

4.5. Genome-Wide Association Analysis

The SNP markers associated with the seven traits and one salt tolerance index were identified using the MLMM model in the EMMAX software [58]. Key SNP analysis was identified by IGV_2.3.83_4 (Integrative genomic viewer, version 2.3.83_4). Lastly, ~1.57 million high-quality SNPs (MAF > 0.05, missing rate < 40%) were obtained and used for the further GWAS. The genome-wide significance thresholds of all tested traits were evaluated using $-\log10(P) \geq 4.0$.

4.6. Statistical Analysis

For phenotypic experiment, each data represented the mean of three or more biological replicate treatments, with each treatment consisting of at least five plants. The statistical analyses were performed using Tukey's two-way analysis of variance (ANOVA) in IBM SPSS Statistics v19.0 (SPSS Inc., Chicago, IL, USA). P-values < 0.001 were considered statistically significant. Pearson correlation also was used SPSS V19.0. The different star between two traits represent significance at P-value < 0.05, 0.01, 0.001.

5. Conclusions

We firstly investigated the genetic architecture of natural variation in Asian cotton salt-tolerance-related traits at the seedlings stage by GWAS mapping in 215 accessions. The SNP markers and candidate genes in this study may be used as references for other association mapping studies of salt tolerance. The salt tolerance related novel candidate genes will provide an important resource for molecular breeding and functional analysis of the salt tolerance during the cotton germination.

Supplementary Materials
s1. Supplementary Table S1. The source, Pedigree information and salt tolerance classification according to the comprehensive index of salt tolerance (CIST) of 215 *G. arboretum* accessions after 7 days of seed growth. Supplementary Table S2. The SNPs showing $-\log_{10}P$ value ≥ 4.0 related with seven phenotypic traits and comprehensive index of salt tolerance (CIST) in 215 *G. arboreum* accessions. Supplementary Table S3. Basic information about genic and intergenic SNP markers. Supplementary Table S4. The basic information of nine SNP rich regions analyzed in this study. Supplementary Table S5. The SNPs in the nine SNP rich regions analyzed in this study. Supplementary Table S6. SNPs and associated genes of two SNP rich regions on chromosome considering $-\log_{10}(P) \geq 6.0$.

Author Contributions: X.D., T.D. and Z.P. (Zhen Peng) initiated the research; D.X. and T.D. designed the experiments. T.D. performed the experiments, T.D., W.G., and Z.P. (Zhaoe Pan) collected the data from the greenhouse, T.D., X.D. and Y.J. performed the analysis, T.D. drafted the manuscript, X.D., Z.P. (Zhen Peng) and K.K.P. contributed to the final editing of manuscript. All authors contributed in the interpretation of results and approved the final manuscript.

Acknowledgments: We are grateful to the National mid-term GenBank for cotton in Institute of Cotton Research of Chinese Academy of Agricultural Sciences (ICR, CAAS) for providing the 215 *G. arboretum* accessions used in this study.

Abbreviations

RGR	Relative germination rate
RFW	Relative fresh weight
RSL	Relative stem length
RWC	Relative water content
RChlC	Relative chlorophyll content;
REC	Relative electric conduct;
RMDA	Relative methylene dioxyamphetamine
CIST	Comprehensive index of salt tolerance
MMLM	Multi-Locus Mixed Model
QTL	Quantitative trait loci;
SSR	Simple sequence repeats

SNP   Single-nucleotide polymorphism
TW   Turgid weights;
ANOVA   Analysis of variance;
QQ   Quantile–quantile plots
LD   Linkage disequilibrium;
GWAS   Genome-wide association study;
HR   Homologous recombination
ABA   Abscisic acid;
GHs   Glycosyl hydrolases

References

1. Wang, Y.; Deng, C.; Liu, Y.; Niu, Z.; Li, Y. Identifying change in spatial accumulation of soil salinity in an inland river watershed, China. *Sci. Total Environ.* **2018**, *621*, 177–185. [CrossRef] [PubMed]

2. Nematzadeh, G.A. Salt-related Genes Expression in Salt-Tolerant and Salt-Sensitive Cultivars of Cotton (*Gossypium* sp. L.) under NaCl Stress. *J. Plant Mol. Breed.* **2018**. [CrossRef]

3. Peng, Z.; He, S.; Sun, J.; Pan, Z.; Gong, W.; Lu, Y.; Du, X. Na+ compartmentalization related to salinity stress tolerance in upland cotton (*Gossypium hirsutum*) seedlings. *Sci. Rep.* **2016**, *6*, 34548. [CrossRef] [PubMed]

4. Paul, D.; Lade, H. Plant-growth-promoting rhizobacteria to improve crop growth in saline soils: A review. *Agron. Sustain. Dev.* **2014**, *34*, 737–752. [CrossRef]

5. Gao, W.; Xu, F.C.; Guo, D.D.; Zhao, J.R.; Liu, J.; Guo, Y.W.; Singh, P.K.; Ma, X.N.; Long, L.; Botella, J.R. Calcium-dependent protein kinases in cotton: Insights into early plant responses to salt stress. *BMC Plant Biol.* **2018**, *18*, 15. [CrossRef] [PubMed]

6. Munns, R.; Tester, M. Mechanisms of Salinity Tolerance. *Annu. Rev. Plant Biol.* **2008**, *59*, 651–681. [CrossRef] [PubMed]

7. Gray, S.B.; Brady, S.M. Plant Developmental Responses to Climate Change. *Dev. Biol.* **2016**, *419*, 64–77. [CrossRef]

8. Ahmad, S.; Ashraf, M.; Khan, N. Genetic basis of salt-tolerance in cotton (*Gossypium hirsutum* L.). *Sci. Technol. Dev.* **2004**, *23*, 45–50.

9. Frouin, J.; Languillaume, A.; Mas, J.; Mieulet, D.; Boisnard, A.; Labeyrie, A.; Bettembourg, M.; Bureau, C.; Lorenzini, E.; Portefaix, M. Tolerance to mild salinity stress in japonica rice: A genome-wide association mapping study highlights calcium signaling and metabolism genes. *PLoS ONE* **2018**, *13*, e0190964. [CrossRef] [PubMed]

10. Peng, Z.; He, S.; Gong, W.; Sun, J.; Pan, Z.; Xu, F.; Lu, Y.; Du, X. Comprehensive analysis of differentially expressed genes and transcriptional regulation induced by salt stress in two contrasting cotton genotypes. *BMC Genom.* **2014**, *15*, 760. [CrossRef]

11. Parida, A.K.; Das, A.B. Salt tolerant and salinity effects on plants. *Ecotoxicol. Environ. Saf.* **2005**, *60*, 324–349. [CrossRef] [PubMed]

12. Yang, X.; Gao, S.; Xu, S.; Zhang, Z.; Prasanna, B.M.; Li, L.; Li, J.; Yan, J. Characterization of a global germplasm collection and its potential utilization for analysis of complex quantitative traits in maize. *Mol. Breed.* **2011**, *28*, 511–526. [CrossRef]

13. Abbasi, Z.; Majidi, M.M.; Arzani, A.; Rajabi, A.; Mashayekhi, P.; Bocianowski, J. Association of SSR markers and morpho-physiological traits associated with salinity tolerance in sugar beet (*Beta vulgaris* L.). *Euphytica* **2015**, *205*, 785–797. [CrossRef]

14. Kantartzi, S.; Stewart, J.M. Association analysis of fibre traits in *Gossypium arboreum* accessions. *Plant Breed.* **2008**, *127*, 173–179. [CrossRef]

15. Du, L.; Cai, C.; Wu, S.; Zhang, F.; Hou, S.; Guo, W. Evaluation and exploration of favorable QTL alleles for salt stress related traits in cotton cultivars (*G. hirsutum* L.). *PLoS ONE* **2016**, *11*, e0151076. [CrossRef]

16. Cai, C.; Zhu, G.; Zhang, T.; Guo, W. High-density 80 K SNP array is a powerful tool for genotyping *G. hirsutum* accessions and genome analysis. *BMC Genom.* **2017**, *18*, 654. [CrossRef]

17. Sun, Z.; Li, H.; Zhang, Y.; Li, Z.; Ke, H.; Wu, L.; Zhang, G.; Wang, X.; Ma, Z. Identification of SNPs and Candidate Genes Associated with Salt Tolerance at the Seedling Stage in Cotton (*Gossypium hirsutum* L.). *Front. Plant Sci.* **2018**, *9*, 1011. [CrossRef] [PubMed]

18. Cai, C.; Wu, S.; Niu, E.; Cheng, C.; Guo, W. Identification of genes related to salt stress tolerance using intron-length polymorphic markers, association mapping and virus-induced gene silencing in cotton. *Sci. Rep.* **2017**, *7*, 528. [CrossRef]

19. Wang, S.; Chen, J.; Zhang, W.; Hu, Y.; Chang, L.; Fang, L.; Wang, Q.; Lv, F.; Wu, H.; Si, Z. Sequence-based ultra-dense genetic and physical maps reveal structural variations of allopolyploid cotton genomes. *Genome Biol.* **2015**, *16*, 108. [CrossRef] [PubMed]

20. Tan, Z.; Zhang, Z.; Sun, X.; Li, Q.; Sun, Y.; Yang, P.; Wang, W.; Liu, X.; Chen, C.; Liu, D. Genetic Map Construction and Fiber Quality QTL Mapping Using the CottonSNP80K Array in Upland Cotton. *Front. Plant Sci.* **2018**, *9*, 225. [CrossRef]

21. Atwell, S.; Huang, Y.S.; Vilhjálmsson, B.J.; Willems, G.; Horton, M.; Li, Y.; Meng, D.; Platt, A.; Tarone, A.M.; Hu, T.T. Genome-wide association study of 107 phenotypes in Arabidopsis thaliana inbred lines. *Nature* **2010**, *465*, 627. [CrossRef] [PubMed]

22. Xue, Y.; Warburton, M.L.; Sawkins, M.; Zhang, X.; Setter, T.; Xu, Y.; Grudloyma, P.; Gethi, J.; Ribaut, J.-M.; Li, W. Genome-wide association analysis for nine agronomic traits in maize under well-watered and water-stressed conditions. *Theor. Appl. Genet.* **2013**, *126*, 2587–2596. [CrossRef]

23. Chen, G.; Zhang, H.; Deng, Z.; Wu, R.; Li, D.; Wang, M.; Tian, J. Genome-wide association study for kernel weight-related traits using SNPs in a Chinese winter wheat population. *Euphytica* **2016**, *212*, 173–185. [CrossRef]

24. McCouch, S.R.; Wright, M.H.; Tung, C.-W.; Maron, L.G.; McNally, K.L.; Fitzgerald, M.; Singh, N.; DeClerck, G.; Agosto-Perez, F.; Korniliev, P. Open access resources for genome-wide association mapping in rice. *Nat. Commun.* **2016**, *7*, 10532. [CrossRef]

25. Abdelraheem, A.; Fang, D.D.; Zhang, J. Quantitative trait locus mapping of drought and salt tolerance in an introgressed recombinant inbred line population of Upland cotton under the greenhouse and field conditions. *Euphytica* **2018**, *214*, 8. [CrossRef]

26. Zhao, Y.L.; Wang, H.M.; Shao, B.X. SSR-based association mapping of salt tolerance in cotton (Gossypium hirsutum L.). *Genet. Mol. Res.* **2016**, *15*, gmr-15027370. [CrossRef] [PubMed]

27. Saeed, M.; Guo, W.Z.; Zhang, T.Z. Association mapping for salinity tolerance in cotton (*Gossypium hirsutum* L.) germplasm from US and diverse regions of China. *Aust. J. Crop Sci.* **2014**, *8*, 338–346.

28. Jia, Y.H.; Sun, J.L.; Wang, X.W.; Zhou, Z.L.; Pan, Z.E.; He, S.P.; Pang, B.Y. Molecular Diversity and Association Analysis of Drought and Salt Tolerance in *Gossypium hirsutum* L. Germplasm. *J. Integr. Agric.* **2014**, *13*, 1845–1853. [CrossRef]

29. Liu, F.; Zhou, Z.L.; Wang, C.Y.; Wang, Y.H.; Cai, X.Y.; Wang, X.X.; Zhang, Z.S.; Wang, K.B. Genetic diversity and relationship analysis of *Gossypium arboreum* accessions. *Genet. Mol. Res.* **2015**, *14*, gmr-14522. [CrossRef] [PubMed]

30. Du, X.; Huang, G.; He, S.; Yang, Z.; Sun, G.; Ma, X.; Li, N.; Zhang, X.; Sun, J.; Liu, M. Resequencing of 243 diploid cotton accessions based on an updated A genome identifies the genetic basis of key agronomic traits. *Nat. Genet.* **2018**, *50*, 796. [CrossRef] [PubMed]

31. Li, F.; Fan, G.; Wang, K.; Sun, F.; Yuan, Y.; Song, G.; Li, Q.; Ma, Z.; Lu, C.; Zou, C. Genome sequence of the cultivated cotton Gossypium arboreum. *Nat. Genet.* **2014**, *46*, 567. [CrossRef]

32. Mehetre, S.S.; Aher, A.R.; Gawande, V.L.; Patil, V.R.; Mokate, A.S. Induced polyploidy in Gossypium: A tool to overcome interspecific incompatibility of cultivated tetraploid and diploid cottons. *Curr. Sci.* **2003**, *84*, 1510–1512.

33. Peng, Z.; He, S.; Sun, J.; Xu, F.; Jia, Y.; Pan, Z.; Wang, L. An Efficient Approach to Identify Salt Tolerance of Upland Cotton at Seedling Stage. *Acta Agron. Sin.* **2014**, *40*, 476–486. [CrossRef]

34. Zhao, Y.; Wang, H.; Wei, C.; Li, Y. Genetic Structure, Linkage Disequilibrium and Association Mapping of Verticillium Wilt Resistance in Elite Cotton (*Gossypium hirsutum* L.) Germplasm Population. *PLoS ONE* **2014**, *9*, e86308. [CrossRef]

35. Sun, Z.; Wang, X.; Liu, Z.; Gu, Q.; Zhang, Y.; Li, Z.; Ke, H.; Yang, J.; Wu, J.; Wu, L. Genome-wide association study discovered genetic variation and candidate genes of fibre quality traits in *Gossypium hirsutum* L. *Plant Biotechnol. J.* **2017**, *15*, 982–996. [CrossRef] [PubMed]

36. Song, H.M.; Zhao, R.M.; Fan, P.X.; Wang, X.C.; Chen, X.Y.; Li, Y.X. Overexpression of *AtHsp90.2*, *AtHsp90.5* and *AtHsp90.7* in Arabidopsis thaliana enhances plant sensitivity to salt and drought stresses. *Planta* **2009**, *229*, 955–964. [CrossRef] [PubMed]

37. Schopf, F.H.; Biebl, M.M.; Buchner, J. The HSP90 chaperone machinery. *Nat. Rev. Mol. Cell Biol.* **2017**, *18*, 345. [CrossRef] [PubMed]

38. Zhou, X.H.; Li, X.S.; Wang, P.; Yan, B.L.; Teng, Y.J.; Yi, L.F. Molecular cloning and expression analysis of *HSP90* gene from Porphyra yezoensis Ueda (*Bangiales, Rhodophyta*). *J. Fish. China* **2010**, *34*, 1844–1852.

39. Ding, Z.; Yang, G.; Wu, J. Effect of Heat Shock at Germinating Period on Growing Developmental of Cotton. *J. Wuhan Bot. Res.* **2006**, *24*, 579–582.

40. Sangster, T.A.; Bahrami, A.; Wilczek, A.; Watanabe, E.; Schellenberg, K.; Mclellan, C.; Kelley, A.; Kong, S.W.; Queitsch, C.; Lindquist, S. Phenotypic Diversity and Altered Environmental Plasticity in *Arabidopsis thaliana* with Reduced Hsp90 Levels. *PLoS ONE* **2007**, *2*, e648. [CrossRef] [PubMed]

41. Busch, W.; Wunderlich, H.F. Identification of novel heat shock factor-dependent genes and biochemical pathways in *Arabidopsis thaliana*. *Plant J. Cell Mol. Biol.* **2010**, *41*, 1–14. [CrossRef]

42. Juan, G.; Yan, Z.; Wangbin, Z.; Jean, M.; Aiwu, D.; Wen-Hui, S. NAP1 family histone chaperones are required for somatic homologous recombination in Arabidopsis. *Plant Cell* **2012**, *24*, 1437–1447.

43. Zhu, Y.; Dong, A.; Meyer, D.; Pichon, O.; Renou, J.P.; Cao, K.; Shen, W.H. Arabidopsis NRP1 and NRP2 encode histone chaperones and are required for maintaining postembryonic root growth. *Plant Cell* **2006**, *18*, 2879–2892. [CrossRef]

44. Waśkiewicz, A.; Beszterda, M.; Goliński, P. ABA: Role in Plant Signaling Under Salt Stress. *Salt Stress Plants* **2013**, 175–196. [CrossRef]

45. Asaoka, R.; Uemura, T.; Nishida, S.; Fujiwara, T.; Ueda, T.; Nakano, A. New insights into the role of Arabidopsis RABA1 GTPases in salinity stress tolerance. *Plant Signal. Behav.* **2013**, *8*, e25377. [CrossRef]

46. Park, M.Y.; Chung, M.S.; Koh, H.S.; Lee, D.J.; Ahn, S.J.; Kim, C.S. Isolation and functional characterization of the Arabidopsis salt-tolerance 32 (*AtSAT32*) gene associated with salt tolerance and ABA signaling. *Physiol. Plant.* **2010**, *135*, 426–435. [CrossRef]

47. Wang, T.Z.; Xia, X.Z.; Zhao, M.G.; Tian, Q.Y.; Zhang, W.H. Expression of a Medicago falcata small GTPase gene, *MfARL1* enhanced tolerance to salt stress in Arabidopsis thaliana. *Plant Physiol. Biochem.* **2013**, *63*, 227–235. [CrossRef]

48. Sripinyowanich, S.; Klomsakul, P.; Boonburapong, B.; Bangyeekhun, T.; Asami, T.; Gu, H.; Buaboocha, T.; Chadchawan, S. Exogenous ABA induces salt tolerance in indica rice (*Oryza sativa* L.): The role of OsP5CS1 and OsP5CR gene expression during salt stress. *Environ. Exp. Bot.* **2013**, *86*, 94–105. [CrossRef]

49. Noaman, M.M.; Dvorak, J.; Dong, J.M. Genes inducing salt tolerance in wheat, Lophopyrum elongatum and amphiploid and their responses to ABA under salt stress. In *Prospects for Saline Agriculture*; Springer: Dordrecht, The Netherlands, 2002; pp. 139–144.

50. Zhao, K.F.; Fan, H.; Harris, P. Effect of exogenous aba on the salt tolerance of corn seedlings under salt stress. *Acta Bot. Sin.* **1995**, *37*, 295–300.

51. O'Mahony, P.J.; Oliver, M.J. Characterization of a desiccation-responsive small GTP-binding protein (*Rab2*) from the desiccation-tolerant grass Sporobolus stapfianus. *Plant Mol. Biol.* **1999**, *39*, 809–821. [CrossRef]

52. Nahm, M.Y.; Kim, S.W.; Yun, D.; Lee, S.Y.; Cho, M.J.; Bahk, J.D. Molecular and biochemical analyses of OsRab7, a rice Rab7 homolog. *Plant Cell Physiol.* **2003**, *44*, 1341–1349. [CrossRef] [PubMed]

53. Mazel, A.; Leshem, Y.; Tiwari, B.S.; Levine, A. Induction of salt and osmotic stress tolerance by overexpression of an intracellular vesicle trafficking protein AtRab7 (*AtRabG3e*). *Plant Physiol.* **2004**, *134*, 118–128. [CrossRef] [PubMed]

54. Wang, X.J.; Tang, C.L.; Lin, D.; Cai, G.L.; Liu, X.Y.; Bo, L.; Han, Q.M.; Buchenauer, H.; Wei, G.R.; Han, D.J. Characterization of a pathogenesis-related thaumatin-like protein gene TaPR5 from wheat induced by stripe rust fungus. *Physiol. Plant.* **2010**, *139*, 27–38. [CrossRef]

55. Liu, J.J.; Sturrock, R.; Ekramoddoullah AK, M. The superfamily of thaumatin-like proteins: Its origin, evolution, and expression towards biological function. *Plant Cell Rep.* **2010**, *29*, 419–436. [CrossRef]

56. Vashisht, A.A.; Tuteja, N. Stress responsive DEAD-box helicases: A new pathway to engineer plant stress tolerance. *J. Photochem. Photobiol. B Biol.* **2006**, *84*, 150–160. [CrossRef]

57. Opassiri, R.; Pomthong, B.; Akiyama, T.; Nakphaichit, M.; Onkoksoong, T.; Cairns, M.K.; Cairns, J.R.K. A stress-induced rice (*Oryza sativa* L.) β-glucosidase represents a new subfamily of glycosyl hydrolase family 5 containing a fascin-like domain. *Biochem. J.* **2007**, *408*, 241–249. [CrossRef] [PubMed]

5

Vascular Plant One-Zinc-Finger (VOZ) Transcription Factors are Positive Regulators of Salt Tolerance in Arabidopsis

Kasavajhala V. S. K. Prasad [1], **Denghui Xing** [1,2] **and Anireddy S. N. Reddy** [1,*]

[1] Department of Biology and Cell and Molecular Biology Program, Colorado State University, Fort Collins, CO 80523, USA; kpsatya@mail.colostate.edu (K.V.S.K.P.); david.xing@mso.umt.edu (D.X.)

[2] Genomics Core Lab, Division of Biological Sciences, University of Montana, Missoula, MT 59812, USA

* Correspondence: reddy@colostate.edu

Abstract: Soil salinity, a significant problem in agriculture, severely limits the productivity of crop plants. Plants respond to and cope with salt stress by reprogramming gene expression via multiple signaling pathways that converge on transcription factors. To develop strategies to generate salt-tolerant crops, it is necessary to identify transcription factors that modulate salt stress responses in plants. In this study, we investigated the role of VOZ (VASCULAR PLANT ONE-ZINC FINGER PROTEIN) transcription factors (VOZs) in salt stress response. Transcriptome analysis in WT (wild-type), *voz1-1*, *voz2-1* double mutant and a *VOZ2* complemented line revealed that many stress-responsive genes are regulated by VOZs. Enrichment analysis for gene ontology terms in misregulated genes in *voz* double mutant confirmed previously identified roles of VOZs and suggested a new role for them in salt stress. To confirm VOZs role in salt stress, we analyzed seed germination and seedling growth of WT, *voz1*, *voz2-1*, *voz2-2* single mutants, *voz1-1 voz2-1* double mutant and a complemented line under different concentrations of NaCl. Only the double mutant exhibited hypersensitivity to salt stress as compared to WT, single mutants, and a complemented line. Expression analysis showed that hypersensitivity of the double mutant was accompanied by reduced expression of salt-inducible genes. These results suggest that VOZ transcription factors act as positive regulators of several salt-responsive genes and that the two VOZs are functionally redundant in salt stress.

Keywords: *Arabidopsis thaliana*; VOZ; transcription factor; salt stress; transcriptional activator

1. Introduction

Throughout their life span, plants are constantly subjected to diverse abiotic and biotic stresses, which severely inhibit plant growth and development, and cause huge losses in crop yields [1,2]. As sessile organisms, plants respond to these stresses rapidly by altering their gene expression patterns, which ultimately change biochemical and physiological processes that enable them to survive under stress conditions [2–4]. Plant transcription factors (TFs) play a key role in reprogramming gene expression in response to stresses [3–6]. Many of these transcription factors act by regulating the expression of down-stream genes that are important for stress tolerance [2–4,7]. VOZ (VASCULAR PLANT ONE-ZINC FINGER PROTEIN) is a plant-specific TF family with two members, *VOZ1* and *VOZ2*. Previous studies have shown that *VOZ1* is specifically expressed in the phloem tissue while *VOZ2* is highly expressed in the roots [8]. Yasui et al. [9] have shown their localization in vascular bundles and predominant subcellular presence in the cytoplasm, while they function in the nucleus. VOZ TFs (for brevity we refers to them as VOZs) were identified as proteins that bind to a *cis*-element GCGTNx7ACGC in the promoter of *AVP1* (V-PPAse) [8]. VOZs have two conserved domains (viz., A and B) and share about 53% similarity. VOZ2 regulates the expression of the target genes by binding to *cis*-elements via Domain B as a dimer. The B domain has a zinc finger motif and a basic region [8].

VOZs were also classified into NAC (for NAM (no apical meristem)) subgroup VIII-2 as they share homology with NAC proteins in the C-terminal basic region [10]. VOZs regulate flowering through their interaction with *PHYB* and promote the expression of *FLC* and *FT* [9,11]. More recent genetic, biochemical and cell biological studies have shown that VOZs interact with and modulate the function of CONSTANS (CO) in promoting flowering [12]. VOZs also play a key role in plant responses to abiotic (cold, drought and heat) and biotic (pathogens) stresses. VOZs function as a positive regulator of plant responses against bacterial and fungal pathogens, and as a negative regulator of two abiotic stresses-old and drought [13,14]. The expression levels of both *VOZ1* and *VOZ2* were also altered in response to biotic and abiotic stresses in an opposite manner [13]. Overexpression of the VOZ2 conferred biotic stress tolerance, however it showed sensitivity to freezing and drought stress in Arabidopsis [13]. Recent reports indicate that VOZ1 and VOZ2 act as transcriptional repressors for *DREB2C* and *DREB2A* respectively, which mediate heat stress response in plants [15,16]. Despite a few reports describing the role of VOZs in some abiotic stresses [17–20], the full scope of VOZs' function in other stresses and their potential target genes are not well understood.

In the current study, analysis transcriptomes from WT, *voz1-1 voz2-1* double mutant and a complemented line suggested a new role for VOZs in salt stress. Analysis of seed germination and seedling growth of WT, *voz1*, *voz2-1*, *voz2-2* single mutants, *voz1-1 voz2-1* double mutant and a complemented line under different concentrations of NaCl revealed that only the double mutant is hypersensitive to salt stress. Through analysis of the upstream regions of genes regulated by VOZs for canonical and non-canonical binding sites, we have identified potential new targets of VOZ transcription factors. Furthermore, expression of salt-induced genes is impaired in the VOZ double mutant. Collectively, these results suggest that VOZs act as positive regulators of salt stress response and that the two VOZs are functionally redundant in salt stress.

2. Results

Arabidopsis VOZ TF family contains two members—*VOZ1* and *VOZ2*. Both these TFs were reported to be involved in flowering and response to abiotic (cold, drought and heat) and biotic stresses [9,11–16]. Here we performed RNA-Seq analysis of gene expression with RNA from wild type (WT), a *voz* double knockout (DKO—*voz1-1 voz2-1*) mutant and a complemented line (COMP2-4). Double mutant (DKO) lines exhibited suppressed growth and leaf vein clearing in older leaves in comparison to WT and COMP2-4 line (Figure 1a). Expression of *VOZ2* in *voz1-1 voz2-1* (COMP2-4) rescued the DKO phenotype (Figure 1a). Hence, we have chosen 30-day-old plants to identify new potential targets of the VOZs by comparing the transcriptomes of WT, DKO and complemented line (COMP2-4). Prior to RNA-Seq and phenotypic analysis, the genotypes of all the lines were verified by genomic PCR (Figure 1b and Figure S1a–c) with gene-specific and T-DNA or transposon-specific primers. RT-PCR analysis with *VOZ1* and *VOZ2* specific primers confirmed the absence of transcripts in *DKO* line (Figure 1c).

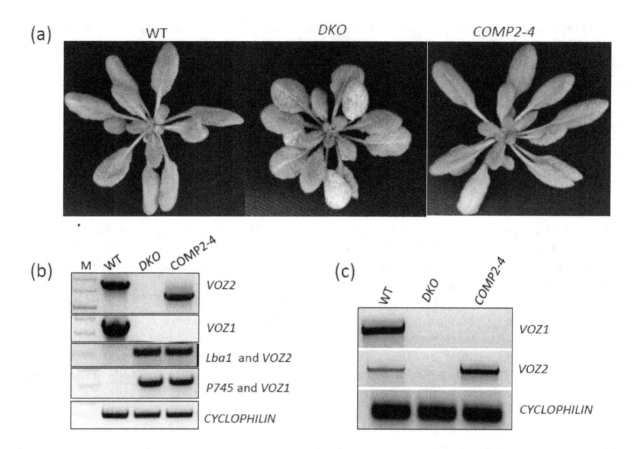

Figure 1. Validation of genotypes used for RNA-Seq. (**a**) Top panel: Phenotype of 30-day-old plants of wild-type (WT), double knockout (*DKO)* mutant (*voz1-1 voz2-1*) and *DKO* complemented line (COMP2-4) grown at 21 °C under day neutral conditions at 60% humidity. (**b**) Genomic PCR of three genotypes used for RNA-Seq. Top panel (PCR with *VOZ2*-specific primers); second panel (PCR with *VOZ1*-specific primers); third panel (PCR with T-DNA specific Lba1 and *VOZ2*-specific reverse primer); fourth panel (PCR with Tn insertion specific primer P745 and VOZ1-specific forward primer); bottom panel (PCR with *CYCLOPHILIN*-specific primers). In all cases expected size PCR product was obtained. (**c**) Analysis of expression of *VOZ1* (top panel), *VOZ2* (middle panel) and *CYCLOPHILIN* (bottom panel) using sqRT-PCR in 30-day-old seedlings of WT, *DKO* mutant (*voz1-1 voz2-1*) and *DKO* complemented line (COMP2-4).

2.1. Loss of Function of VOZs Resulted in Misregulation of Genes

For RNA-Seq, two biological replicates of WT, *DKO* and COMP2-4 were used. About 37 to 75 million short reads per sample were obtained (Table S1). The reads were mapped to the Arabidopsis genome (TAIR 10) and ~90% of these were uniquely mapped (Table S1). The expression levels of individual transcripts were determined by the number of reads per kilobase per million (RPKM). The expression patterns of the genes were well correlated among the replicates. However, the expression patterns were poorly correlated between WT and *DKO*, as indicated by an R^2 value of 0.77, suggesting a significant effect of VOZs on gene expression (Figure S2). The Cufflinks package was used to identify differentially expressed (DE) genes by comparing the transcriptome of WT with *DKO*. In the *DKO*, 112 genes were misregulated (significance adj. $p \leq 0.05$ and fold change >log2) as compared to WT (Additional File 1 sheet1). Further, expression levels of the majority of these were either partially or fully restored in the *VOZ2* complemented line (COMP2-4) (Figure 2a–c; Additional File 1 sheet2), suggesting that DE genes are either direct or indirect targets of VOZs and loss of these TFs caused significant effects on expression of many genes. The majority of the DE genes (101) were up-regulated, while only 11 were down-regulated in the *DKO* mutant. About 83% of up-regulated genes are partially complemented while ~27% of them are fully complemented by overexpression of *VOZ2*. In the case of down-regulated DE genes, ~27% of genes were fully complemented and 72% exhibited partial complementation (Figure 2). The misregulation of a number of DE genes was verified using RT-qPCR (Figure 3a,b), corroborating the RNA-Seq data.

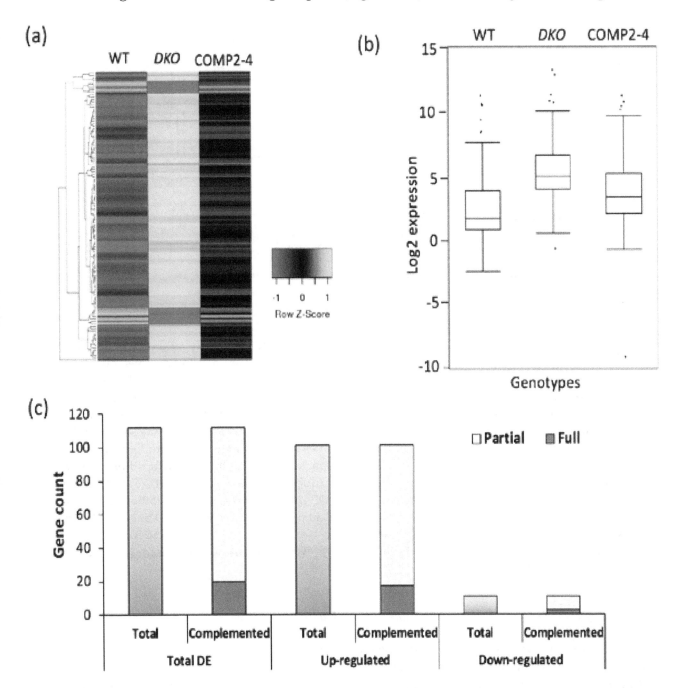

Figure 2. Analysis of differentially expressed genes. (**a**) Heatmap representation of differentially expressed genes in *WT, DKO* and *COMP2-4* (COMP) plants. Expression values were used to generate the heatmap using the Heatmapper. Columns represent samples and rows represent genes. Color scale indicates the gene expression level. Green indicates high expression and red Indicates low expression. (**b**) Box-and-whisker plots showing expression of differentially expressed (DE) genes in different genotypes. (**c**) Gene counts of total, up and down-regulated DE genes that are either fully or partially complemented in *COMP2-4* line.

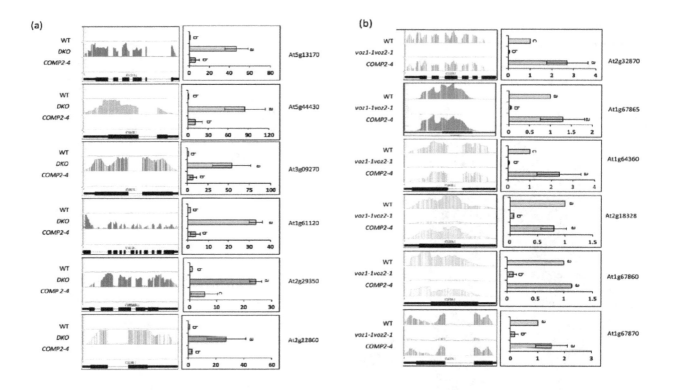

Figure 3. Validation of up- and down-regulated genes in *DKO*. (**a**) RT-qPCR validation of randomly selected up-regulated genes. (**b**) RT-qPCR of randomly selected down-regulated genes. Left panels in (**a,b**) show relative sequence read abundance (Integrated Genome Browser view) as histograms in WT, *DKO (voz1-1 voz2-1)* and COMP2-4 lines. The Y-axis indicates read depth with the same scale for all three lines. The gene structure is shown below the read depth profile. The lines represent introns and the boxes represent exons. The thinner boxes represent 5′ and 3′ UTRs. Right panels in (**a,b**) show fold change in expression level relative to WT. WT values were considered as 1. Student's *t*-test was performed and significant differences ($p < 0.05$) among samples are labeled with different letters. The error bars represent SD. The genes that were randomly picked include At1g61120 (terpene synthase 4), At1g64360 (enescence-associated and QQS-related), At1g67860 (hypothetical protein), At1g67865 (hypothetical protein), At1g67870 (hypothetical protein), At2g18328 (RAD-like4), At2g22860 (phytosulfokine 2 Precursor), At2g29350 (senescence-associated gene 13), At3g09270 (glutathione S-transferase TAU8), At2g32870 (TRAF-like protein), At5g13170 (senescence-associated gene29), and At5g44430 (plant defensing 1.2C).

2.2. VOZs Regulate Expression of Several Transcription Factors

It is possible that the effect of VOZs on the expression of its DE genes is mediated via its regulation of other TFs, hence we analyzed the DE genes for enrichment of TFs. Arabidopsis has over 1700 genes encoding TFs that are grouped into 58 families. Among DE genes, we observed eight TFs belonging to four families (Figure 4a, Additional File 2 and Figure S3). Of these families, bHLH (Basic-Helix-Loop-Helix) ($p \leq 0.03$), MYB (v-myb avian myeloblastosis viral oncogene homolog)-related ($p \leq 0.00009$) and NAC ($p \leq 0.00002$) are highly enriched (Figure 4a). The number of TFs in each family and the direction of their expression in the *DKO* are presented in Figure S3 and Additional File 2. Interestingly, members of bHLH, NAC and C_2H_2 TF families are up-regulated whereas the members of MYB-related families are down-regulated in *DKO*. Significantly, five out of eight members of TF families showed expression levels similar to that of WT in the complemented line, indicating that VOZs regulates the expression of these transcription factors.

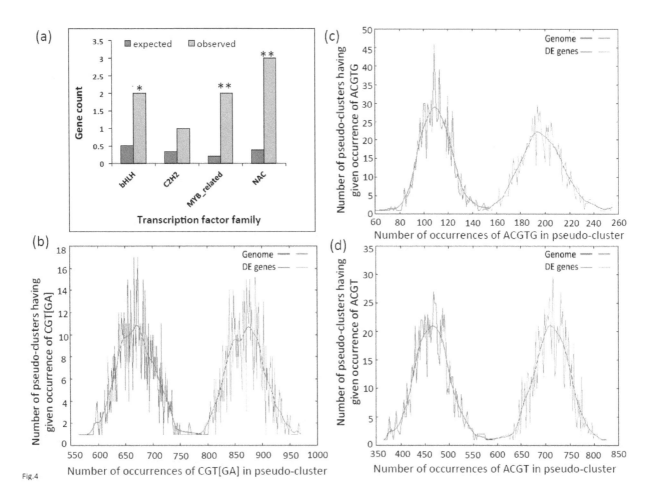

Fig.4

Figure 4. Enrichment of transcription factor families and VOZ binding sites in the promoters of DE genes. (**a**) DE genes are enriched for specific TF families. Observed: Number of genes associated with particular TF family in DE genes. Expected: Number of genes expected in each individual TF family in the genome. Asterisks on the bar represent significant overrepresentation of TFs with a (* $p \leq 0.05$) and (** $p \leq 0.0001$), respectively. (**b–d**) POBO analysis of NAC consensus sequence (*CGT[GA]*), G-box core sequence (*ACGTG*) and LS-7 *cis*-element (*ACGT*), respectively, in the −1000 bp upstream of TSS. One thousand pseudoclusters were generated from top 112 DE genes and genome background. The jagged lines show the motif frequencies from which the best-fit curve is derived. *CGT[GA]*, *ACGTG* and *ACGT* elements are significantly overrepresented (two-tailed $p < 0.0001$) in the upstream sequences of DE genes.

2.3. Promoters of Differentially Expressed Genes are Enriched for G-Box, NAC and LS-7 Elements

As VOZs share significant sequence similarity with NAC subgroup VIII-2 TFs (particularly in the C-terminal region) and also reported to bind to the palindromic NAC binding sequence (palNAC-BS), we analyzed the promoters of DE genes (−1000 bp upstream of TSS) for the enrichment for G-box core sequence (*ACGTG*), NAC-consensus sequence (*CGT[GA]*) and TGA TFs recognition LS-7 element (*ACGT*), which are thought to bind VOZs, using POBO analysis. This analysis revealed a significant enrichment of *cis*-elements ($p < 0.0001$) of *ACGTG*, *CGT[GA]* and *ACGT* in the promoter regions of DE genes (Figure 4b–d). Ninety percent of the DE genes contain *CGT[GA]* (1 to 12), 58% have *ACGTG* (1 to 8) and 84% have *ACGT* (1 to 12) binding sites in their upstream (−500 bp) region (Table S2). Furthermore, POBO analysis of up- and down-regulated DE genes separately also exhibited significant ($p \leq 0.0001$) enrichment of these binding sites (Figure 5). These results indicate VOZs might regulate the expression of some of these DE genes directly through these elements.

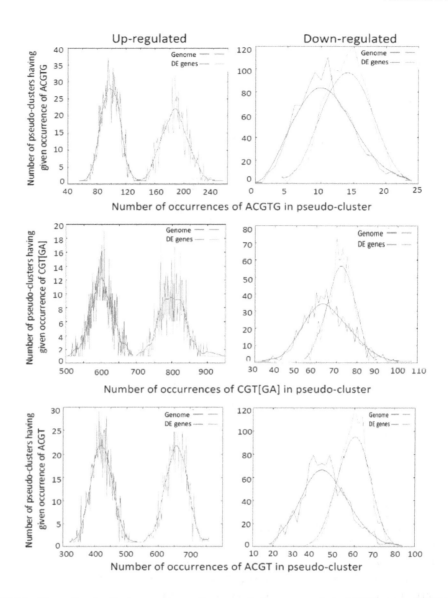

Figure 5. VOZ-binding sites in the promoters of up- and down-regulated DE genes. POBO analysis of VOZs binding motif, G-box core sequence (ACGTG) (top panels), NAC consensus sequence (CGT[GA]) (middle panels), and LS-7 *cis*-element (ACGT) (bottom panels) in the −1000 bp upstream of TSS. A total of 1000 pseudoclusters were generated from 101 up-regulated (left panels) and 11 down-regulated genes (right panels) and genome background. The jagged lines show the motif frequencies from which best-fitted curve is derived. VOZs binding sites are significantly (two-tailed $p < 0.0001$) over-represented in the upstream sequences of both up- and down-regulated genes.

2.4. GO Term Enrichment Analysis for Biological Processes in Differentially Expressed Genes

Previously, VOZ proteins have been shown to play an important role in flowering, plant immunity, cold, heat, and drought stresses [9,11–17]. We performed gene ontology (GO) analysis not only to verify if DE genes function in previously reported processes but also to gain insight into other functional roles of VOZs. A singular GO term enrichment analysis for biological processes was performed with GeneCodis separately for up-regulated and down-regulated DE genes. No significant enrichment of any GO term for down-regulated DE genes was found. However, a total of 38 GO terms for biological processes were enriched in up-regulated DE genes (Figure 6a and Additional File 3). Consistent with previously reported functions of VOZs, GO terms associated with the processes involving plant response to pathogen/pests and water deprivation, osmotic stress and oxidative stress were enriched. GO terms that are of special interest are "response to salt stress" and "hyperosmotic salinity response" for the following reasons: (i) both are among the top 10 most enriched GO terms; (ii) these two GO terms together have 10 genes (second most of all GO categories); (iii) role of VOZ proteins in salt stress is not known; and (iv) expression of the majority of these genes is altered in opposite direction in the complemented line (Table S3).

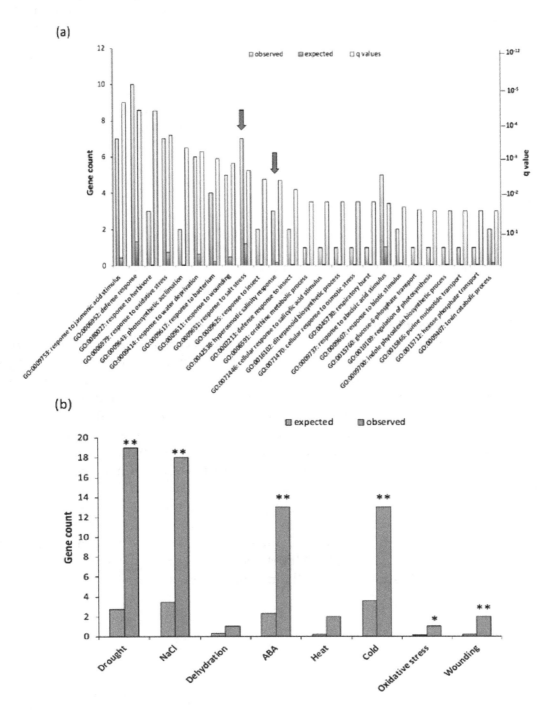

Figure 6. Gene ontology (GO) enrichment analysis of DE genes. (**a**) GO term enrichment analysis for biological processes of up-regulated genes. For each GO term, the expected and observed gene numbers along with the statistical significance (*p*-value) for the enrichment is presented. Observed: Number of DE genes associated with a GO term for biological processes. Expected: Number of genes expected for each GO term in the genome. "Response to salt stress" and "Hyperosmotic salinity response" GO terms are indicated with an arrow. (**b**) A significant number of DE genes are associated with abiotic stress response in comparison with genome background with a ** $p \leq 0.0001$ and * $p \leq 0.05$.

2.5. VOZs Regulate Expression of Many Abiotic Stress-Responsive Genes

Since G-box, NAC and TGA *cis*-elements occurred significantly in the promoter region of DE genes (Figures 4 and 5), we performed enrichment analysis to determine the number of DE genes associated with different stresses. For this analysis, we compared the DE gene list with all the listed abiotic stress genes at http://caps.ncbs.res.in/stifdb/browse.html#se. This analysis also indicated a substantial enrichment ($p \leq 0.001$) of different abiotic stress-responsive genes with a significant number of genes associated with drought and cold stress, which is consistent with the reported functions of VOZs (Figure 6b and Additional File 4) of Nakai et al. [14]. However, VOZ's function in salt stress is not known. Enrichment analysis indicated that about 16% of DE genes (18 genes) in *DKO* are associated with salt stress (Additional File 4). Furthermore, in the COMP2-4 line, the expression of salt stress-responsive genes found in both these analyses was partially or fully restored to WT levels (Figure S5 and Table S3a,b).

2.6. VOZs Regulate Salt Stress Tolerance

2.6.1. VOZ Double Mutant Exhibits Hypersensitivity to Salt Stress

Previous studies have shown that VOZs play an important role in drought, cold and heat [13,15,16], but their regulatory role in salt stress is not known. Since salt stress-responsive genes are enriched in DE genes, we investigated the role of VOZs in salt stress tolerance. Wild type, double mutant (DKO), COMP 2-4 and single mutants of VOZs (*voz1-1, voz2-1, voz2-2*) were tested for salt tolerance. Seed germination rate, seedling growth and root length were scored by growing them on different concentrations (0, 100 or 150 mM) of NaCl. In general, seed germination rate was significantly affected under salt stress in a NaCl concentration-dependent manner in all genotypes. However, irrespective of NaCl concentration, seeds of DKO genotype exhibited delayed germination in comparison to WT, COMP2-4 and single *voz* mutants (Figure 7a). Further, seedling growth (fresh weight) was also significantly affected in a NaCl concentration-dependent manner in all the genotypes. Similar to the rate of seed germination, the growth of DKO seedlings was severely suppressed when compared to that of WT, COMP2-4 and single mutants (Figure 7b,c). A suppression of the primary root length in a NaCl concentration-dependent manner was also observed (Figure 7b). Particularly at 100 and 150 mM NaCl, a difference in root growth was evident among the genotypes (Figure 7b—left bottom panel and 7c). A significant suppression in the primary root growth was observed in DKO lines as compared to WT, COMP2-4 and single mutant lines, indicating the increased sensitivity of DKO mutant to salt stress (Figure 7b,c). Even at 150 mM NaCl, WT, COMP2-4 and single mutants were found to be relatively more tolerant to salt stress, indicating DKO seedlings exhibit hypersensitivity to salt stress.

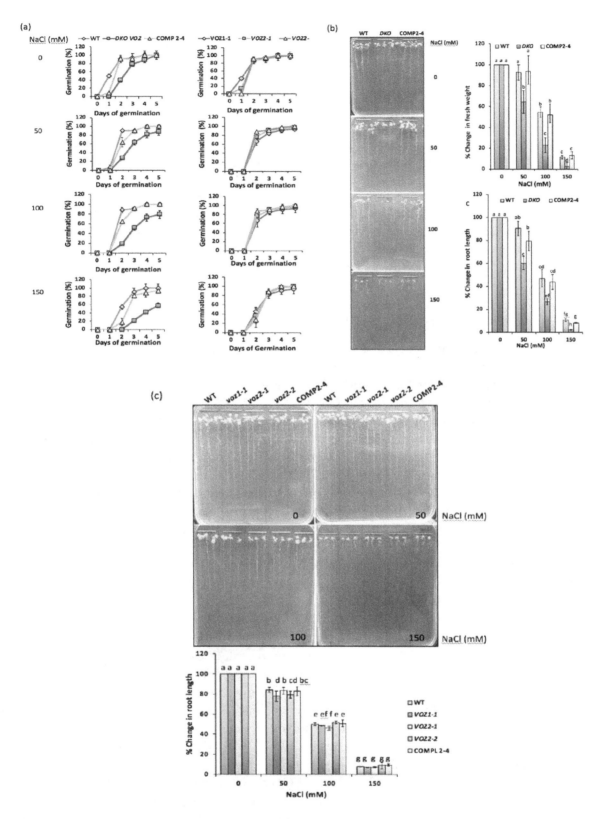

Figure 7. Germination and seedling growth of WT, mutants and complemented line in the presence of NaCl. (**a**) *VOZ* Double mutant (DKO) exhibits delayed germination under salt stress. The time course of seed germination of WT, DKO, COMP2-4 (left panels), *voz1-1*, *voz2-1* and *voz2-2* (right panels) in the presence of 0, 50, 100 and 150 mM NaCl. Each value shown here is mean of three biological replicates with *n* = 10. The error bars represent SD. (**b**) *VOZ* Double mutant (DKO) is hypersensitive to salt stress. Left panel: Growth of seedlings of WT, DKO and COMP2-4 on MS (Murashige and Skoog medium) plates containing different concentrations of NaCl. Seeds were plated on 1/2 strength MS medium

supplemented with 0, 50, 100 and 150 mM of NaCl and were allowed to germinate and grow for two weeks. The photographs were taken after two weeks. Right panel, top: Seedling fresh weight. Right panel, bottom: Seedling root length at different concentrations of NaCl was measured for all genotypes and plotted as % relative to growth on normal (0 mM) MS medium. Three biological replicates were used. Eight to ten seedlings for each genotype per treatment for each biological replicate were included. Student's t-test was performed and significant differences ($p \leq 0.05$) among samples are labeled with different letters. The error bars represent SD. (c) Single mutants of VOZs are not hypersensitive to salt stress. Top: Growth of seedlings of WT, COMP2-4, *voz1-1*, *voz2-1* and *voz2-2* on MS plates containing different concentrations of salt. Seeds were plated on half-strength MS medium supplemented with 0, 50, 100 or 150 mM of NaCl and were allowed to germinate and grow for two weeks. The photographs were taken after two weeks. Bottom: Seedling root length at different concentrations of NaCl was measured for all genotypes and plotted as % relative to growth on normal (0 mM) MS medium. Three biological replicates were used. For each genotype, eight to ten seedlings per treatment and for each biological replicate were used. Student's t-test was performed and significant differences ($p \leq 0.05$) among samples are labeled with different letters. The error bars represent SD.

2.6.2. VOZs Activate the Expression of Salt-Responsive Genes

Prior to analyzing the transcript levels of different salt-responsive genes, we first quantified *VOZ1* and *VOZ2* transcripts in WT seedlings grown on medium supplemented with 0, 50, 100 and 150 mM NaCl. Alterations in transcript levels of *VOZ1* and *VOZ2* under salt stress were observed. The expression of both *VOZ1* and *VOZ2* was significantly enhanced with increasing concentration of NaCl (Figure 8). For example, at 100 mM NaCl, ~2.5- and 2.0-fold increases in transcript levels of *VOZ1* and *VOZ2*, respectively, were observed. However, 150 mM NaCl reduced the salt-induced elevation of transcript levels of *VOZs*, probably due to severe growth inhibition at this concentration. To gain further insights into the role of VOZs in salt stress, the expression level of salt-responsive genes under both "enrichment of salt-responsive genes" and the GO category of "response to salt stimulus" was compared in WT, *DKO* and COMP2-4 RNA-Seq data. The majority of salt-responsive genes were represented in *DKO* and COMP2-4 datasets and their expression profiles were opposite to each other (Figure S5 and Table S3a,b). Motif analysis of upstream (-1000 bp) regions of these genes indicated significant enrichment ($p < 0.0001$) for VOZ binding sites, viz., G-box (*ACGTG*) and NAC bind sites (*CGT[GA]*) and LS-7 (*ACGT*) (Figure S6a,b (lower panels) and Table S3). Arabidopsis genes (At1g16850 (transmembrane protein), At5g59310 (LTP4), At2g37760 (AKR4c8), At5g59820 (ZAT12), At4g23600 (COR13), At5g24770 (VSP2), At1g10585 (bHLH DNA-binding superfamily protein), At2g43510 (ATTI1) and At4g37990 (ATCAD8)), which are closest to genes involved in salt tolerance in other plant species [18–28] and also contain either of VOZ-binding motifs in their promoter (Table S3a,b), were selected as representatives to analyze their expression under control and salt stress conditions. The expression pattern of these nine genes was analyzed by RT-qPCR. Under control conditions, the expression levels of all nine genes were significantly higher in *DKO* as compared to WT or COMP2-4 in 30-day-old plants (Additional File 1). However, in 15-day-old seedlings there was no up-regulation of salt-responsive genes in the *DKO* in untreated seedlings, suggesting differential regulation of these genes by VOZs depending on the developmental stage of the plants. Upon exposure to salt (100 mM), expression of these genes was highly induced in the WT seedlings (Figure 9). However, loss of both VOZs (*DKO*) caused a significant reduction in salt induction of these genes, suggesting that VOZs are essential for the increased expression of these salt-responsive genes under salt stress. The expression levels of three genes, viz., At1g16850, At5g59310 and At3g04720, were partially restored to WT level under salt stress in the COMP2-4 line (Figure S4). Similar to our results, expression of several cold-responsive genes that are highly expressed in the DKO were not restored in the *VOZ2* complemented line [13,14].

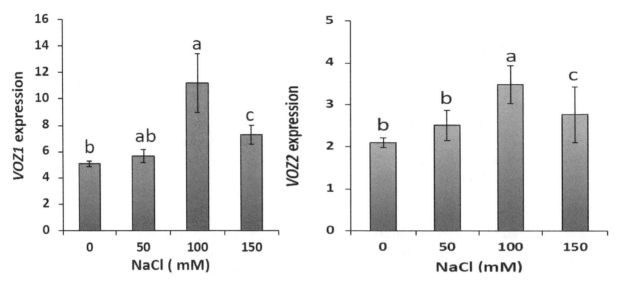

Figure 8. Expression of VOZs in response to salt stress. Expression of *VOZ1* (top panel) and *VOZ2* (bottom panel) in 10-day-old seedlings of WT seedlings grown on 1/2 MS medium supplemented with 0, 50, 100 or 150 mM NaCl was determined by RT-qPCR. The expression of *VOZs* was normalized with *ACTIN2*.

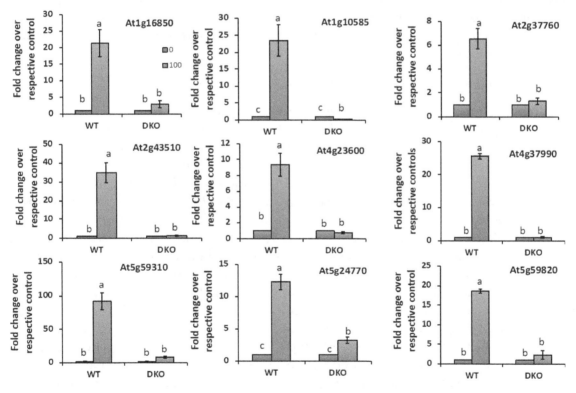

Figure 9. VOZs positively regulate the expression of salt-responsive genes. Expression of salt-responsive genes in 10-day-old seedlings of WT and DKO lines on 1/2 MS medium supplemented with 0 or 100 mM NaCl was determined by RT-qPCR. The expression level of salt-responsive genes was normalized with *ACTIN2*. Fold change in expression level relative to their respective controls (0 mM) is presented. 0 mM values were considered as 1. Three biological replicates were used. Student's *t*-test was performed and significant differences ($p \leq 0.05$) among samples are labeled with different letters. The error bars represent SD.

The majority of the salt-responsive genes contain *cis*-elements in their promoter regions to which known TFs bind. These include *CACGTG, CACG[G/A]C, CATGTG, VCGCGB* and *MCGTGT* that bind G_box bHLH, N_box_bHLH, Nac_box_NAC, and *CAMTA* TFs, respectively. To understand the regulation of these salt-responsive DE genes by VOZs, POBO analysis was carried out for the enrichment of these *cis*-elements in the upstream regions of salt-responsive genes. A significant enrichment ($p < 0.0001$) for *RSREs* (*VCGCGB* and *MCGTGT*) was observed in the upstream region (-1000 bp) of the DE genes (Figure S7). Further, enrichment for *CACGTG*, but not *CACGGC* and *CATGTG*, was found in the promoter regions of the salt stress-responsive genes (Figure S6a,b—top and middle panels). Significantly, enrichment of VOZ binding consensus motifs (*CGT[GA]*, *ACGTG* and *ACGT*) was also observed (Figure S6a,b—bottom panels). Only two genes (viz. At2g37760 and At5g59820) have a canonical binding site ($ACGT_{GATTCAC}ACGC$) for VOZs in their promoter regions. These results suggest that the regulation of salt-responsive genes by VOZs is accomplished via certain *cis*-elements (*CGT[GA]*, *ACGTG* and *ACGT*) within the consensus motifs of VOZs.

3. Discussion

Previous studies with *voz* double mutant reported several developmental defects, such as smaller plants, impaired root growth, delayed flowering, round lamina of juvenile leaves, reduced trichome number on abaxial side, and siliques with aborted seeds [11,14]. Expression of either *VOZ1* or *VOZ2* under their native promoter or expression of *VOZ2* under *CaMV35S* promoter in *DKO* completely rescued the phenotype [11,13]. In this study, we observed another phenotype. Thirty-day-old seedlings of *DKO* plants grown under day neutral conditions at 21 °C exhibited vein-clearing phenotype in older leaves (Figure 1a). This vein-clearing phenotype became more apparent with the age of the plants. Single mutants of *VOZs* did not exhibit this developmental phenotype. Over-expression of *VOZ2* in *DKO* rescued the phenotype (Figure 1A), suggesting that VOZ1 and VOZ2 are functionally redundant in vein-clearing phenotype.

3.1. VOZs are Involved in Regulation of Many Stress-Responsive Genes

Our global transcriptome analysis using RNA-Seq revealed that a significant number of genes that are involved in diverse stress responses are regulated either directly or indirectly by VOZs (Figures 2a and 6, Additional File 1, sheet 1,2). Previously Nakai et al. [14] and more recently Kumar et al. [12] compared the expression of genes in WT and *DKO* using microarrays. Our study significantly differs from these in several ways. Here, we used next-generation sequencing that significantly increases the depth of transcriptome analysis and avoids some problems associated with microarrays. More importantly, the use of a complemented line in which double mutant phenotypes are rescued allowed us to identify the genes that are regulated specifically by VOZs (Figure 1a,b and Additional File 1 sheet 2). Despite using RNAseq, our study revealed a smaller number of DE genes as compared to the previous studies [12,14]. This difference in the number of DE genes and limited overlap between DE genes among different studies could be due to the difference in the age and developmental stage of plants used (30-day-plants in this study vs. 14-day-old-seedlings in previous studies) and/or due to the differences in methodologies (microarrays vs RNA-Seq). In fact, developmental regulation of expression levels of TFs has been previously reported [6,29,30]. Nevertheless, our study identified a new set of 94 genes that are regulated by VOZs (Additional File 1) when compared with Nakai et al. [14] and Kumar et al. [12]. Interestingly, GO enrichment of biological process using DE genes in *DKO* from Nakai et al. [14], we found enrichment of only two biological processes, viz., farnesyl diphosphate metabolic process and sequiterpenoid biosynthetic process. This is in contrast to this study wherein we observed enrichment for 38 GO terms that are consistent with reported functions of VOZs (Figure 6a, Additional File 3). In addition, GO enrichment analysis suggested a new role for VOZs in salt stress, which was confirmed experimentally. Reproducibility among replicates, full or partial restoration of expression of ~98% of DE genes in COMP2-4 to WT level (Figure 2b) and RT-qPCR validation of expression of a number (22 genes) of randomly selected DE genes (Figure 3, Figure S5) indicates that the identified DE genes in this study are bona fide direct or indirect targets of VOZs. Enrichment of DE genes in multiple abiotic stress-responses indicates that VOZs play a major role in crosstalk between multiple stress signal transduction pathways (Figure 6b and Additional File 4). GO analysis of the DE genes indicated high enrichment of GO terms associated with diverse processes that are critical for plant responses to biotic stresses, such as bacteria and fungi, and abiotic stresses including drought, cold, salt and oxidative stress (Figure 6a). Enrichment of genes involved in response to hormones such as abscisic acid (ABA) and jasmonic acid (JA) was also observed (Figure S8 and Additional File 5). Together, these results suggest that VOZs could be integrators of a variety of stress responses. Consistent with these results, VOZs were reported to play an important role in multiple stress responses [13,14].

3.2. Genes with Binding Motifs for VOZs are Both Up- and Down-Regulated

Electrophoretic mobility shift assays showed that a VOZ protein binds to $GCGTN_{x7}ACGC$ sequence in vitro in the V-PPase gene (*AVP1*) promoter [8]. The palindromic sequence-binding site was considered as canonical VOZ TF binding site. We analyzed it to see if the promoter region (−2000 bp upstream to TSS) of DE genes is enriched for this motif, but found no significant enrichment. Only one (DE gene AT5G16360) has two VOZ binding sites (−1590, −1604 bp) in its upstream region. We followed our analysis by screening the promoter regions of DE genes with two suboptimal ($GCGTN_{x7}ACG\textbf{T}$ and $GCGTN_{x8}ACGC$) and other motifs ($GCGTN_{x7}AAGC$, $GCTTN_{x7}ACGC$, $ACGTN_{x7}ACGC$) that are reported to bind VOZs [8,12]. This analysis resulted in the identification of another DE gene (AT2G14247) containing $GCGTN_{x7}ACG\textbf{T}$ motif (−336 bp) and none containing $GCGTN_{x8}ACGC$ in their promoter regions. Recently, Kumar et al. [12] using systematic evolution of ligands by exponential enrichment (SELEX) assay and electrophoretic mobility shift assay (EMSA) not only confirmed $GCGT_{GTGATAC}ACGC$ as VOZ2 binding site but also revealed additional binding sites. Based on this study, we scanned the upstream region of all the DE genes and detected six additional genes (At1g23150, At2g17840, At2g22860, At2g37760, At4g00780, At5g59820) that have these new VOZ-binding sites. In addition to these *cis*-elements, other studies also identified binding of VOZs to alternate palindromic NAC-binding sequences (palNAC-BS) that are similar to other NAC proteins that are responsive to abiotic stress [10,14]. Analysis of DE genes showed that >90% contain *CGT[GA]*, 58% contain *ACGTG* and 83% contain *ACGT* elements and these motifs are significantly enriched in their promoter regions (Figure 4, Table S2). Both up- and down-regulated genes showed enrichment for VOZ-binding sites (Figure 5). Significant enrichment of VOZ-binding motifs in DE genes indicates that VOZs likely regulate the expression of those genes directly by binding to these *cis*-elements.

3.3. VOZs Likely Regulate Expression of Some Genes Indirectly

In the promoter regions of some DE genes we did not find any of the VOZ-binding sites and these genes are likely regulated indirectly by other TFs. We found enrichment of four TF families (bHLH, C2H2, NAC and MYB-related) in DE genes (Figure 4) and TFs in three of these families were up-regulated (Figure S3). Many members of these TF families have multiple binding sites for VOZs in their promoter regions and exhibited expression levels similar to WT in the complemented line (Table S2; Additional File 1). For example, members of bHLH, NAC and C2H2 are up-regulated in *DKO* while they are down-regulated in *COMP2-4* line. In contrast, members of a MYB-related family were down-regulated in *DKO* and up-regulated in *COMP2-4*. It is possible that these TFs may regulate the expression of DE genes that do not contain canonical VOZ binding motif [12]. Together, these data indicate a complex network of regulation of expression of TFs by VOZs.

Recent studies identified *RSRE* element *VCGCGB* as the core element that is enriched in a majority of early-activated genes under stresses [31]. As this element is identical to the binding site of TFs signal responsive/calmodulin-binding transcription activators (SRS/CAMTAs) (*VCGCGB*), many studies showed SRs, in general, and SR1/CAMTA3, in particular, in regulation of multiple biotic and abiotic stress responses [5,6,32–34]. Significantly, SRs regulate genes involved in abiotic stress response, particularly through *MCGTGT* element. Our analysis of the promoter region of all the DE genes indicated that a significant enrichment of the *VCGCGB* and *MCGTGT* elements, suggesting that VOZs might regulate the abiotic stress responses through SRs (Figure S7). Further, the fact that the majority of these genes are misregulated in *DKO* and are implicated in various stress signaling pathways, also suggests an important role for SRs in VOZ-mediated regulation. In support of this, it has been shown that VOZs and CAMTAs interact with the *AVP1* promoter and regulate its expression [8,35]. POBO analysis indicated the enrichment of *RSRE* motif in the promoter regions of the up-regulated DE genes (Figure S7). As shown in Figure 6, significant enrichment of GO term for "responses to salt stress" and "water deprivation" was observed only in the up-regulated DE genes (Additional File 3). Further, enrichment for *VCGCGB* and *MCGTGT* in DE genes and enrichment for GO terms "water deprivation" and "cellular response to osmotic" suggest that VOZs could be regulating drought response genes through utilization of *MCGTGT* and *VCGCGB* by SRs (Figure 6, Figure S7, and Additional File 3). In fact, a significant alteration in cold and drought-responsive genes expression was observed in *DKO* line even under non-stress conditions [14]. One possibility is that VOZs form heterodimers with CAMTAs/SRs or other TFs in regulating some genes. The fact that VOZs and CAMTAs bind to the *AVP1* promoter lends supports to this. However, thus far direct interaction of VOZs and CAMTAs has not been reported. Recently, it has been shown that VOZs interacts with CONSTANS, another TF, in regulating flowering [12].

3.4. VOZ Confers Salt Tolerance by Activating the Expression of Salt-Responsive Genes

Double mutant line (*voz1-1 voz2-1*) were more sensitive to salt stress in terms of seed germination rate, seedling growth and root growth when compared with the WT, COMP2-4 and single mutant lines. Thus, our results suggest that (a) *VOZ1* and *VOZ2* have redundant functions in salt tolerance and (b) *VOZs* act as positive regulators of plants response to salt stress. This positive regulation of salt stress by VOZs is similar to that observed under biotic stress and differs from that of the cold and drought stress response, where it functions as a negative regulator [14]. Previously, Nakai et al. [13,14] have shown that *DKO* was significantly tolerant to cold and drought whereas it is sensitive to bacterial and fungal pathogens. They further reported that the over-expression of *VOZ2* confers tolerance to freezing and drought but curtails tolerance to biotic stresses. Taken together these results suggest that VOZs have opposing functions under salt stress as compared to cold and drought stresses. To further understand the regulation (direct versus indirect) by VOZs, salt-responsive genes were identified and subjected to POBO analysis for enrichment of *RSRE (VCGCGB)*, *NAC (CGT[GA])*, *G-box (ACGTG)* and *ACGT* in their upstream region. This analysis revealed significant enrichment for *RSRE (VCGCGB)*, *CGT[GA])*, *ACGTG*, *ACGT* and *MCGTGT* (Figure 5 and Figure S7). Hence, it is possible that some of these genes could be direct targets of VOZs i.e., they bind to these motifs to regulate expression. Alternatively, other TFs such as CAMTAs/SRs could also participate along with VOZs in this regulation as discussed above.

In summary, our results showed that a large number of genes associated with biotic and abiotic stress responses are regulated by VOZs. Most of these genes are likely direct targets as they contain one or more type of VOZ-binding sites in their promoter region. Analysis of DE genes suggested a new role for VOZs in salt stress. We experimentally showed that VOZs function as positive regulators of salt tolerance. The model in Figure 10 summarizes the role of VOZs in salt stress response. Plants in response to salt stress activate expression of VOZs. This activation of VOZs, in turn, regulates the expression level of salt-responsive genes either directly or indirectly thereby conferring salt tolerant phenotype. The absence of VOZs in *DKO* significantly curtails salt-induced activation of the salt-responsive genes leading to hypersensitive phenotype.

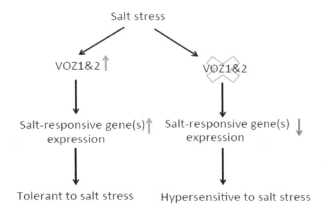

Figure 10. A proposed model for the role of VOZs in salt stress response (see text for details). Green and red arrows indicate the increased and decreased expression levels, respectively.

4. Materials and Methods

4.1. Plant Materials and Growth Conditions

All experiments were performed with *Arabidopsis thaliana* Columbia-0 ecotype. Seeds of single (*voz1-1, voz2-1*) and double mutants (*DKO; voz1-1 voz2-2*) of *VOZ1* and *VOZ2* used in this study were characterized previously [11]. The complemented line (COMP-4) was generated by transforming *DKO* with *VOZ2* cDNA under *CaMV35S* promoter. Plants were grown in soil in a growth chamber at 21 °C, 60% relative humidity under 12/12 h light/dark conditions.

4.2. Salt Stress Treatment

To study the effect of salt stress on seed germination and seedling growth, surface sterilized seeds of wild type (WT) and mutants were sown on half-strength MS medium (containing 0.5 mg/L MES and 1% sucrose) pH 5.7 and supplemented with 0, 50, 100 or 150 mM NaCl. The surface-sterilized seeds of all the lines were stratified at 4 °C in dark for 5 d prior to sowing on plates. The plates with seeds were allowed to germinate under long day condition (16 h light/8 h dark) at 22 °C. Germination rate and fresh weight of seedling and root growth was determined by recording the number of seeds that exhibited emergence of radicle, weight of the seedlings and length of the roots after two weeks of growth, respectively. All experiments were performed three times with a minimum of three replicates.

4.3. RNA-Seq

Total RNA from WT, *DKO* and COMP2-4 genotypes was isolated using miRNAeasy kit (Qiagen, Germantown, MD, USA#217004). Traces of genomic DNA were removed using on column DNAse digestion. RNA-Seq was performed essentially as described previously in Prasad et al. [6].

4.4. Mapping of the Reads and Identification of Differentially Expressed (DE) Genes

The reads were aligned to the TAIR 10 version of the Arabidopsis genome, and DE gene list was generated using the criteria as described earlier [6]. VENNY (http://bioinfogp.cnb.csic.es/tools/venny/) a web-based tool was used for identification of common genes in one or more datasets. Heatmap of DE genes was generated using log2 transformed expression values of each gene using Heatmapper [36]. Box-and-whisker plot of DE genes was generated using the log2 transformed expression values in WT, *DKO* and *COMP2-4* with JMP Pro, version 13, statistical software (SAS, Cary, NC, USA).

4.5. Bioinformatics Analysis of DE Genes for VOZs Binding Motifs

Identification of DE genes containing VOZ binding motifs $GCGTN_{x7}ACGC$, $ACGTG$, $CGT[GA]$ and $ACGT$ in their promoter was carried out using "Patmatch" (Version 1.1) tool (www.arabidopsis.org).

With this tool, we identified motifs on both strands of upstream sequences (-500 bp) preceding the TSS in TAIR10 database. Both up- and down-regulated DE genes were included as input for scoring both type and number of VOZs binding motifs.

4.6. GO Term Enrichment Analysis

GO term enrichment analysis was performed using GeneCodis [37]. Single enrichment analysis with TAIR GO annotations was performed using the hypergeometric test with Benjamin-Hochberg false discovery rate (FDR) correction with a significance of $p \leq 0.05$. The DE genes that are up- or down-regulated were analyzed separately.

4.7. Identification of TFs, Abiotic Stress and Hormone-Responsive Genes in DE List

To identify various TFs in the DE genes, a list of all TFs was obtained from Plant TF Database (version 3.0) (http://planttfdb.cbi.pku.edu.cn) [38] and all DE genes were queried against the total TF list. TAIR 10 ID of all TF genes was used as input for identifying the DE genes encoding the TFs and classifying them based on the similarity with Total TF family list. The TFs and the genes responsive to various abiotic stress conditions were obtained from STIFB (Stress Responsive TF Database) (http://caps.ncbs.res.in/stifdb2/). Promoters of the genes that contained *cis*-element for binding of the TFs that are involved in abiotic stress response were retrieved for the analysis. DE genes were queried against the list of the genes for a specific abiotic stress. Further, on the basis of overlap of locus ID (TAIR ID) between the lists of genes, they were further categorized into different subsets. Similarly, plant hormone biosynthesis and signaling genes in DE list were identified by comparing the DE genes with that of genes list of each individual hormone available from the Arabidopsis Hormone Database 2.0 (http://ahd.cbi.pku.edu.cn).

4.8. Promoter Analysis for Enrichment of Cis-Elements

To identify the *cis*-elements in promoters, either 500 or 1000 bp sequence upstream of the transcription start site was extracted from TAIR using an online tool for bulk sequence retrieval. For the estimation of the enrichment for particular *cis*-elements, promoter sequences (-500 or -1000 bp) were used as input for POBO analysis [39]. The upstream sequences of -500 or -1000 bp of the genes in the data set were used as an input into the web portal and analyzed for *cis*-element/motif against *Arabidopsis thaliana* background (clean). The following parameters were used for this analysis: number of promoters in cluster is equivalent to number of input sequences; number of pseudoclusters to generate =1000. For statistical significance, a linked GraphPad application calculates a two-tailed p-value using generated t-value and degrees of freedom for determination of the statistical differences between the input sequences and the background.

4.9. Validation of DE Genes Using RT-qPCR Analysis

Primers for validation of DE genes using Real-time qPCR (RT-qPCR) were designed using Primer Quest web tool (http://www.idtdna.com/Primerquest/Home/Index) from IDT, Coralville, IA, USA (Additional File 6). DE genes were randomly selected and analyzed for their expression levels using RT-qPCR. cDNA from 30-day-old plants was prepared and expression of each gene in all genotypes was estimated essentially as described [6]. For each genotype, cDNA from two independent biological replicates was used. Three technical replicates were used for each sample. *ROC5 (CYCLOPHILIN)* was used as a reference gene. Fold change in expression was calculated and plotted with respect to WT. The expression level in WT for each gene is considered as 1.

4.10. RT-qPCR Analysis of Salt-Responsive Genes

Ten-day-old control and salt-treated seedlings of different genotypes were used for extraction of total RNA. A quantity of 1 µg of RNA was used for the preparation of cDNA using Superscript III reverse transcriptase system as described in Prasad et al. [6]. The cDNA was diluted 6 times and 1.5 µL per reaction was used as a template. Expression analysis was performed using RT-qPCR as described above. The data obtained were normalized with *ACTIN2* and fold change in the expression level was calculated relative to their respective control, i.e., 0 mM NaCl. The expression level in control was considered as 1. A minimum of 3 technical replicates and 3 biological replicates were used for each experiment.

Author Contributions: A.S.N.R. conceived and directed the project. K.V.S.K.P. and A.S.N.R. designed the experiments. K.V.S.K.P. performed RNA-Seq, all bioinformatics analysis with the DE gene list and all experiments pertinent to salt stress and RT-qPCR analysis of gene expression. DX analyzed RNA-Seq data and generated DE genes list. K.V.S.K.P. and A.S.N.R. wrote the manuscript. DX read and commented on the manuscript.

Abbreviation: *DKO (voz1-1 voz2-1), COMP2-4 (CaMV35S: VOZ2: OCS)*; DE (Differentially expressed genes).

References

1.   Cheeseman, J.M. The evolution of halophytes, glycophytes and crops, and its implications for food security under saline conditions. *New Phytol.* **2015**, *206*, 557–570. [CrossRef] [PubMed]

2.   Zhu, J.K. Abiotic stress signaling and responses in plants. *Cell* **2016**, *167*, 313–324. [CrossRef] [PubMed]

3.   Reddy, A.S.; Ali, G.S.; Celesnik, H.; Day, I.S. Coping with stresses: Roles of calcium- and calcium/calmodulin-regulated gene expression. *Plant Cell* **2011**, *23*, 2010–2032. [CrossRef] [PubMed]

4.   Ohama, N.; Sato, H.; Shinozaki, K.; Yamaguchi-Shinozaki, K. Transcriptional Regulatory Network of Plant Heat Stress Response. *Trends Plant Sci.* **2017**, *22*, 53–65. [CrossRef] [PubMed]

5.   Kim, Y.S.; An, C.; Park, S.; Gilmour, S.J.; Wang, L.; Renna, L.; Brandizzi, F.; Grumet, R.; Thomashow, M.F. CAMTA-Mediated Regulation of Salicylic Acid Immunity Pathway Genes in Arabidopsis Exposed to Low Temperature and Pathogen Infection. *Plant Cell* **2017**, *29*, 2465–2477. [CrossRef] [PubMed]

6.   Prasad, K.V.S.K.; Abdel-Hameed, A.A.E.; Xing, D.; Reddy, A.S.N. Global gene expression analysis using RNA-seq uncovered a new role for SR1/CAMTA3 transcription factor in salt stress. *Sci. Rep.* **2016**, *6*, 27021. [CrossRef] [PubMed]

7.   Khan, S.A.; Li, M.Z.; Wang, S.M.; Yin, H.J. Revisiting the Role of Plant Transcription Factors in the Battle against Abiotic Stress. *Int. J. Mol. Sci.* **2018**, *19*, 1634. [CrossRef] [PubMed]

8.   Mitsuda, N.; Hisabori, T.; Takeyasu, K.; Sato, M.H. VOZ; isolation and characterization of novel vascular plant transcription factors with a one-zinc finger from Arabidopsis thaliana. *Plant Cell Physiol.* **2004**, *45*, 845–854. [CrossRef] [PubMed]

9.   Yasui, Y.; Mukougawa, K.; Uemoto, M.; Yokofuji, A.; Suzuri, R.; Nishitani, A.; Kohchi, T. The phytochrome-interacting vascular plant one-zinc finger1 and VOZ2 redundantly regulate flowering in Arabidopsis. *Plant Cell* **2012**, *24*, 3248–3263. [CrossRef] [PubMed]

10.   Jensen, M.K.; Kjaersgaard, T.; Nielsen, M.M.; Galberg, P.; Petersen, K.; O'Shea, C.; Skriver, K. The Arabidopsis thaliana NAC transcription factor family: Structure-function relationships and determinants of ANAC019 stress signalling. *Biochem. J.* **2010**, *426*, 183–196. [CrossRef] [PubMed]

11.   Celesnik, H.; Ali, G.S.; Robison, F.M.; Reddy, A.S. Arabidopsis thaliana VOZ (Vascular plant One-Zinc finger) transcription factors are required for proper regulation of flowering time. *Biol. Open* **2013**, *2*, 424–431. [CrossRef] [PubMed]

12.   Kumar, S.; Choudhary, P.; Gupta, M.; Nath, U. VASCULAR PLANT ONE-ZINC FINGER1 (VOZ1) and VOZ2 Interact with CONSTANS and Promote Photoperiodic Flowering Transition. *Plant Physiol.* **2018**, *176*, 2917–2930. [CrossRef] [PubMed]

13.   Nakai, Y.; Fujiwara, S.; Kubo, Y.; Sato, M.H. Overexpression of VOZ2 confers biotic stress tolerance but decreases abiotic stress resistance in Arabidopsis. *Plant Signal. Behav.* **2013**, *8*, e23358. [CrossRef] [PubMed]

14.   Nakai, Y.; Nakahira, Y.; Sumida, H.; Takebayashi, K.; Nagasawa, Y.; Yamasaki, K.; Akiyama, M.; Ohme-Takagi, M.; Fujiwara, S.; Shiina, T.; et al. Vascular plant one-zinc-finger protein 1/2 transcription factors regulate abiotic and biotic stress responses in Arabidopsis. *Plant J.* **2013**, *73*, 761–775. [CrossRef] [PubMed]

15.   Koguchi, M.; Yamasaki, K.; Hirano, T.; Sato, M.H. Vascular plant one-zinc-finger protein 2 is localized both to the nucleus and stress granules under heat stress in Arabidopsis. *Plant Signal. Behav.* **2017**, *12*, e1295907. [CrossRef] [PubMed]

16. Song, C.; Lee, J.; Kim, T.; Hong, J.C.; Lim, C.O. VOZ1, a transcriptional repressor of DREB2C, mediates heat stress responses in Arabidopsis. *Planta* **2018**, *247*, 1439–1448. [CrossRef] [PubMed]

17. Yasui, Y.; Kohchi, T. VASCULAR PLANT ONE-ZINC FINGER1 and VOZ2 repress the FLOWERING LOCUS C clade members to control flowering time in Arabidopsis. *Biosci. Biotechnol. Biochem.* **2014**, *78*, 1850–1855. [CrossRef] [PubMed]

18. Oberschall, A.; Deák, M.; Török, K.; Sass, L.; Vass, I.; Kovács, I.; Fehér, A.; Dudits, D.; Horváth, G.V. A novel aldose/aldehyde reductase protects transgenic plants against lipid peroxidation under chemical and drought stresses. *Plant J.* **2000**, *24*, 437–446. [CrossRef] [PubMed]

19. Simpson, P.J.; Tantitadapitak, C.; Reed, A.M.; Mather, O.C.; Bunce, C.M.; White, S.A.; Ride, J.P. Characterization of Two Novel Aldo-Keto Reductases from Arabidopsis: Expression Patterns, Broad Substrate Specificity, and an Open Active-Site Structure Suggest a Role in Toxicant Metabolism Following Stress. *J. Mol. Biol.* **2009**, *392*, 465–480. [CrossRef] [PubMed]

20. Zhou, J.; Li, F.; Wang, J.L.; Ma, Y.; Chong, K.; Xu, Y.Y. Basic helix-loop-helix transcription factor from wild rice (OrbHLH2) improves tolerance to salt- and osmotic stress in Arabidopsis. *J. Plant Physiol.* **2009**, *166*, 1296–1306. [CrossRef] [PubMed]

21. Safi, H.; Saibi, W.; Alaoui, M.M.; Hmyene, A.; Masmoudi, K.; Hanin, M.; Brini, F. A wheat lipid transfer protein (TdLTP4) promotes tolerance to abiotic and biotic stress in Arabidopsis thaliana. *Plant Physiol. Biochem.* **2015**, *89*, 64–75. [CrossRef] [PubMed]

22. Kim, I.J.L.; Yun, B.W.; Jamil, M. GA Mediated OsZAT-12 Expression Improves Salt Resistance of Rice. *Int. J. Agric. Biol.* **2016**, *18*, 330–336.

23. Le, C.T.; Brumbarova, T.; Ivanov, R.; Stoof, C.; Weber, E.; Mohrbacher, J.; Fink-Straube, C.; Bauer, P. ZINC FINGER OF ARABIDOPSIS THALIANA12 (ZAT12) Interacts with FER-LIKE IRON DEFICIENCY-INDUCED TRANSCRIPTION FACTOR (FIT) Linking Iron Deficiency and Oxidative Stress Responses. *Plant Physiol.* **2016**, *170*, 540–557. [CrossRef] [PubMed]

24. Vogel, J.T.; Zarka, D.G.; Van Buskirk, H.A.; Fowler, S.G.; Thomashow, M.F. Roles of the CBF2 and ZAT12 transcription factors in configuring the low temperature transcriptome of Arabidopsis. *Plant J.* **2005**, *41*, 195–211. [CrossRef] [PubMed]

25. Gong, Z.; Koiwa, H.; Cushman, M.A.; Ray, A.; Bufford, D.; Kore-eda, S.; Matsumoto, T.K.; Zhu, J.; Cushman, J.C.; Bressan, R.A.; et al. Genes that are uniquely stress regulated in salt overly sensitive (sos) mutants. *Plant Physiol.* **2001**, *126*, 363–375. [CrossRef] [PubMed]

26. Shan, L.; Li, C.L.; Chen, F.; Zhao, S.Y.; Xia, G.M. A Bowman-Birk type protease inhibitor is involved in the tolerance to salt stress in wheat. *Plant Cell Environ.* **2008**, *31*, 1128–1137. [CrossRef] [PubMed]

27. Li, R.; Wang, W.J.; Wang, W.G.; Li, F.S.; Wang, Q.W.; Xu, Y.; Wang, S.H. Overexpression of a cysteine proteinase inhibitor gene from Jatropha curcas confers enhanced tolerance to salinity stress. *Electron. J. Biotechnol.* **2015**, *18*, 368–375. [CrossRef]

28. Srinivasan, T.; Kumar, K.R.; Kirti, P.B. Constitutive expression of a trypsin protease inhibitor confers multiple stress tolerance in transgenic tobacco. *Plant Cell Physiol.* **2009**, *50*, 541–553. [CrossRef] [PubMed]

29. Reddy, A.S.; Reddy, V.S.; Golovkin, M. A calmodulin binding protein from Arabidopsis is induced by ethylene and contains a DNA-binding motif. *Biochem. Biophys. Res. Commun.* **2000**, *279*, 762–769. [CrossRef] [PubMed]

30. Yang, T.B.; Poovaiah, B.W. A calmodulin-binding/CGCG box DNA-binding protein family involved in multiple signaling pathways in plants. *J. Boil. Chem.* **2002**, *277*, 45049–45058. [CrossRef] [PubMed]

31. Walley, J.W.; Coughlan, S.; Hudson, M.E.; Covington, M.F.; Kaspi, R.; Banu, G.; Harmer, S.L.; Dehesh, K. Mechanical stress induces biotic and abiotic stress responses via a novel cis-element. *PLoS Genet.* **2007**, *3*, e172. [CrossRef] [PubMed]

32. Du, L.; Ali, G.S.; Simons, K.A.; Hou, J.; Yang, T.; Reddy, A.S.; Poovaiah, B.W. Ca(2+)/calmodulin regulates salicylic-acid-mediated plant immunity. *Nature* **2009**, *457*, 1154–1158. [CrossRef] [PubMed]

33. Laluk, K.; Prasad, K.V.; Savchenko, T.; Celesnik, H.; Dehesh, K.; Levy, M.; Mitchell-Olds, T.; Reddy, A.S. The calmodulin-binding transcription factor SIGNAL RESPONSIVE1 is a novel regulator of glucosinolate metabolism and herbivory tolerance in Arabidopsis. *Plant Cell Physiol.* **2012**, *53*, 2008–2015. [CrossRef] [PubMed]

34. Galon, Y.; Nave, R.; Boyce, J.M.; Nachmias, D.; Knight, M.R.; Fromm, H. Calmodulin-binding transcription activator (CAMTA) 3 mediates biotic defense responses in Arabidopsis. *FEBS Lett.* **2008**, *582*, 943–948. [CrossRef] [PubMed]

35. Mitsuda, N.; Isono, T.; Sato, M.H. Arabidopsis CAMTA family proteins enhance V-PPase expression in pollen. *Plant Cell Physiol.* **2003**, *44*, 975–981. [CrossRef] [PubMed]

36. Babicki, S.; Arndt, D.; Marcu, A.; Liang, Y.; Grant, J.R.; Maciejewski, A.; Wishart, D.S. Heatmapper: Web-enabled heat mapping for all. *Nucleic Acids Res.* **2016**, *44*, W147–W153. [CrossRef] [PubMed]

37. Tabas-Madrid, D.; Nogales-Cadenas, R.; Pascual-Montano, A. GeneCodis3: A non-redundant and modular enrichment analysis tool for functional genomics. *Nucleic Acids Res.* **2012**, *40*, W478–W483. [CrossRef] [PubMed]

38. Jin, J.; Zhang, H.; Kong, L.; Gao, G.; Luo, J. PlantTFDB 3.0: A portal for the functional and evolutionary study of plant transcription factors. *Nucleic Acids Res.* **2014**, *42*, D1182–D1187. [CrossRef] [PubMed]

39. Kankainen, M.; Holm, L. POBO, transcription factor binding site verification with bootstrapping. *Nucleic Acids Res.* **2004**, *32*, W222–W229. [CrossRef] [PubMed]

6

SNF1-Related Protein Kinases SnRK2.4 and SnRK2.10 Modulate ROS Homeostasis in Plant Response to Salt Stress

Katarzyna Patrycja Szymańska *, Lidia Polkowska-Kowalczyk, Małgorzata Lichocka, Justyna Maszkowska and Grażyna Dobrowolska *

Institute of Biochemistry and Biophysics, Polish Academy of Sciences, Pawińskiego 5a, 02-106 Warsaw, Poland; lidekp@ibb.waw.pl (L.P.-K.); mlichocka@ibb.waw.pl (M.L.); j.maszkowska@ibb.waw.pl (J.M.)
* Correspondence: kszymanska@ibb.waw.pl (K.P.S.); dobrowol@ibb.waw.pl (G.D.)

Abstract: In response to salinity and various other environmental stresses, plants accumulate reactive oxygen species (ROS). The ROS produced at very early stages of the stress response act as signaling molecules activating defense mechanisms, whereas those produced at later stages in an uncontrolled way are detrimental to plant cells by damaging lipids, DNA, and proteins. Multiple systems are involved in ROS generation and also in ROS scavenging. Their level and activity are tightly controlled to ensure ROS homeostasis and protect the plant against the negative effects of the environment. The signaling pathways responsible for maintaining ROS homeostasis in abiotic stress conditions remain largely unknown. Here, we show that in *Arabidopsis thaliana*, two abscisic acid- (ABA)-non-activated SNF1-releted protein kinases 2 (SnRK2) kinases, SnRK2.4 and SnRK2.10, are involved in the regulation of ROS homeostasis in response to salinity. They regulate the expression of several genes responsible for ROS generation at early stages of the stress response as well as those responsible for their removal. Moreover, the SnRK2.4 regulate catalase levels and its activity and the level of ascorbate in seedlings exposed to salt stress.

Keywords: antioxidant enzymes; *Arabidopsis thaliana*; ascorbate cycle; hydrogen peroxide; reactive oxygen species; salinity; SnRK2

1. Introduction

Plants growing in nature are exposed to ever changing environmental conditions. They experience various abiotic stresses, such as drought, temperature extremes, and salinity. Salinity and drought are among the most detrimental factors limiting plant growth and development. Salinity causes ion-related stress, limitations in nutrient uptake as well as osmotic stress.

A secondary effect of salt stress and several other stresses is the accumulation of reactive oxygen species (ROS) in plant cells. Various ROS, such as singlet oxygen ($^1O_2$), superoxide radical (O_2^-), hydroxyl radical ($-OH$), and hydrogen peroxide (H_2O_2), are produced at low levels in chloroplasts, mitochondria, peroxisomes, and the apoplast during plant growth in optimal conditions [1,2]. They are involved in the regulation of plant growth and development, acting as signaling molecules. Upon stress, however, ROS play a double role [1,3–5]. ROS production at a low level is needed at the first stages of the stress response for induction of the plant defense, e.g., activation of signaling cascades, expression of stress response genes encoding enzymes involved in the synthesis of osmoprotectants, and some enzymes responsible for ROS scavenging [6–8]. At the later stages, ROS that accumulate in a non-controlled way have a widespread toxic effect, causing peroxidation of lipids, and damaging proteins and DNA, eventually leading to cell death. In response to stress, ROS are produced by diverse enzymes, e.g., NADPH oxidases, glycolate oxidases, oxalate oxidase, xanthine oxidase, and some peroxidases [9,10]. In *Arabidopsis thaliana* subjected to salinity stress, mainly two NADPH oxidases, respiratory burst oxidases, AtRbohD and AtRbohF, are involved in ROS production. They generate O_2^- free radicals in the apoplastic space by transferring electrons from NADPH to O_2. Then, the O_2^- is dismutated to H_2O_2 by superoxide dismutase (SOD) and the H_2O_2 molecules diffuse to adjacent cells, where they can play a role of signaling molecules, inducing plant defense, or they cause oxidative stress and cell damage. To achieve ROS homeostasis, which is required for efficient defense against the negative effects of harmful environmental conditions, plants have evolved several systems for ROS removal, both enzymatic, such as catalases (CATs), SODs, and various peroxidases (PRXs), and non-enzymatic, i.e., the ascorbate-glutathione cycle [9,11–16]. The enzymes involved in the ROS production and removal are encoded by multiple genes and are strictly regulated in response to stress, at the transcriptional, protein, and activity levels.

There are several reports showing that kinases from the SNF1-releted protein kinases 2 (SnRK2s) family are major regulators of the plant response to osmotic stress (drought, salinity). SnRK2s are plant-specific kinases activated in response to osmotic stress and some of them additionally in response to abscisic acid (ABA). They have been found in every plant species analyzed [17]. The SnRK2s are classified into three groups based on their phylogeny. The classification correlates well with their response to ABA: Group 1 comprises ABA-non-activated kinases; in group 2, are kinases weakly activated or non-activated by ABA (depending on plant species); and group 3 contains kinases strongly activated by ABA [18,19]. SnRK2s play a crucial role in the induction of defense mechanisms against drought [20,21] and salinity [22–24] via ABA-dependent and ABA-independent pathways. So far, the role of ABA-dependent SnRK2s (SnRK2.2, SnRK2.3, and SnRK2.6) has been mostly studied and found to be crucial for ABA signaling [25,26]. In response to drought, they regulate stomatal closure in an ABA-dependent manner by phosphorylating several ion channels [27–29], expression of stress-response genes by phosphorylating transcription factors activated in response to ABA [30–32], and the activity of aquaporins [33]. SnRK2.6 is involved in the ABA-dependent ROS production indispensable for stomatal closure [34], possibly by phosphorylating the NADPH oxidase, RbohF [35,36].

Much less is known about the involvement of SnRK2s in response to salinity, even though these kinases are known to be strongly activated in response to this stress. It has been shown that ABA-non-responsive kinases (belonging to group 1) are activated rapidly and transiently in response to salt stress [23,37,38]. McLoughlin et al. [23] showed that two kinases from group 1, SnRK2.4 and SnRK2.10, are fully active within seconds in roots of Arabidopsis plants after treatment with NaCl. Both kinases were found to be required for plant tolerance to salinity stress by regulating root growth and architecture [23]. Moreover, it has been shown that SnRK2s from group 1 influence the plant tolerance to salt stress via regulation of mRNA decay. They phosphorylate VARICOSE (VCS), a protein regulating mRNA decapping [24]. Very recently, using the phosphoproteomic approach, several potential ABA-non-activated SnRK2s' targets that phosphorylated in response to salinity have been found [39]. Among them there were several proteins, e.g., RNA- and DNA- binding proteins, protein kinases, phosphatases, and dehydrins, Early Responsive to Dehydration 10 (ERD10) and ERD14, whose phosphorylation likely affects the plant tolerance to salt stress. It has also been suggested that kinases from this group could be involved in the regulation of tolerance to salt stress via regulation of oxidative stress generated in response to salinity. Diédhiou et al. [22] showed that transgenic rice overexpressing Stress-Activated Protein Kinase 4 (SAPK4), the rice ABA-non-activated SnRK2, exhibited improved tolerance to salt stress. Their results indicated that SAPK4 regulates Na^+ and Cl^- accumulation and ROS homeostasis; the enhanced level of the kinase caused up-regulation of the *CatA* gene encoding catalase. Additionally, it has been shown that SnRK2.4 positively regulates the accumulation of ROS in response to stress induced by cadmium ions [40].

These data suggest that at least some members of the group 1 of the SnRK2 family are likely to be involved in the regulation of plant tolerance to osmotic stress by controlling the ROS level. The aim of the present study was to establish the role of SnRK2.4 and SnRK2.10 in the regulation of the ROS homeostasis in response to salt stress in *Arabidopsis thaliana*.

2. Results

2.1. SnRK2.4 and SnRK2.10 Kinases Are Involved in H_2O_2 Accumulation in Response to Salt Stress

To check whether SnRK2.4 and/or SnRK2.10 are involved in ROS accumulation in the early response to salt stress, we compared the accumulation of H_2O_2 in leaves of four-week-old plants of the *snrk2.4* and *snrk2.10* knockout mutants and wild type Col-0 plants exposed to 150 mM NaCl for various time using a luminol-based assay. The mutant lines accumulated significantly less H_2O_2 than the wild type (wt) plants did in response to salinity (Figure 1A). The maximal level of H_2O_2 in Col-0 leaves was observed 30 min after the stressor application (over a three-fold increase in respect to the control level), whereas in the *snrk2.4* and *snrk2.10* mutants, the maximal H_2O_2 accumulation was only two-fold and occurred at 60 min. Notably, the level of H_2O_2 in control conditions was by ca. 30% lower in three out of four mutants tested relative to wt. The lower initial H_2O_2 content combined with the smaller increase resulted in the mutant plants having less than half of the H_2O_2 level found in the wt at the peak of the response to salt stress. Since the two independent lines of both mutants showed similar behavior for further studies, we decided to use only one line of each mutant, *snrk2.4-1* and *snrk2.10-1* (later referred to as *snrk2.4* and *snrk2.10* mutants), previously well characterized [23,39,40]. To verify the observed differences between the wt plants and the *snrk2* mutants in ROS accumulation in response to salt stress, we monitored their level in Arabidopsis roots using the fluorescent dye, dichlorofluorescin diacetate (H_2DCFDA). We analyzed the accumulation of ROS in roots of five-day-old seedlings of wt plants, and the *snrk2.4* and *snrk2.10* mutants (Figure 1B,C) exposed to 250 mM NaCl for 15 min. Similarly to what was observed for Arabidopsis leaves exposed to the salt treatment, in roots of the *snrk2* mutant lines, the basal level of ROS was lower than in the roots of the wt plants and the ROS accumulation after the stress application was lower in those mutants in comparison with the one observed for wt seedlings. Thus, the obtained results strongly suggested the role of SnRK2.4 and SnRK2.10 in the regulation of ROS accumulation at the early stages of plant response to salinity.

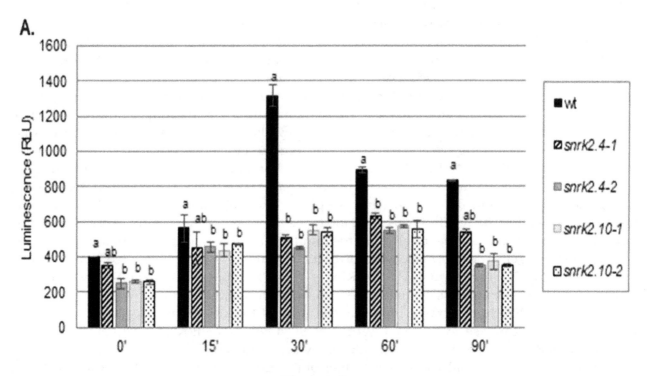

Figure 1. *Cont.*

B.

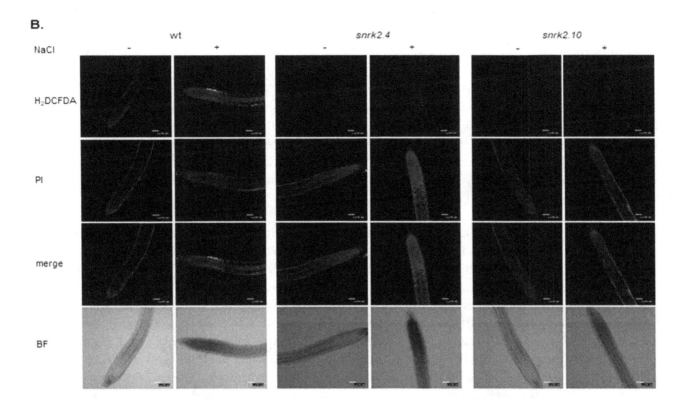

C.

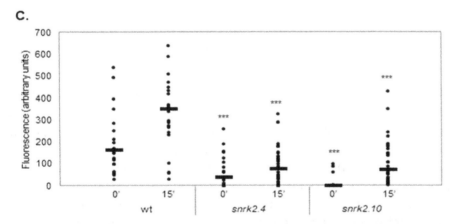

Figure 1. SnRK2.4 and SnRK2.10 affect ROS level in plants subjected to salt stress. (**A**). Leaves of wt plants and *snrk2.4* and *snrk2.10* mutant lines were subjected to 150 mM NaCl for the indicated time and H_2O_2 was determined using a luminol-based assay. Letters represent statistical differences in respect to the wt plants where a means no significant difference, and b means a significant difference [one way analysis of variance (ANOVA). Error bars represent standard deviation (SD). Three independent biological replicates, each with four samples per data point were performed. Results of all combined experiments are shown. B and C. Roots of five-day-old Arabidopsis seedlings (wt plants and *snrk2.4* and *snrk2.10* mutant lines) were stained with propidium iodide (PI; 20 µg/mL) and 2′,7′-Dichlorofluorescin diacetate (H_2DCFDA; 30 µg/mL) and then treated for 15 min with 250 mM NaCl in $\frac{1}{2}$ MS (+) or $\frac{1}{2}$ MS only (−). (**B**) The production of ROS was monitored by imaging of H_2DCFDA fluorescence in the roots using confocal microscopy; BF – bright field image, scale bars =50 µm (**C**) Fluorescence intensity of H_2DCFDA was calculated from well-defined region of interest (4000 µm^2) in the root meristematic zone on each single confocal section; stars represent statistically significant differences in respect to the wt plants (Mann-Whitney *U* test) where *** $p < 0.0001$; results represent data collected from at least 30 seedlings/line/conditions where each dot represents the sample value and a dash represents the median of measurements.

2.2. SnRK2.4 and SnRK2.10 Regulate Expression of Genes Involved in ROS Generation in Response to Salinity

To determine the mechanism by which SnRK2.4 and SnRK2.10 affect the H_2O_2 homeostasis in salt-stressed plants, we investigated their impact on the enzymes involved in ROS production and scavenging. Since mitogen-activated protein kinase (MAPK, MPK) cascades regulate the ROS homeostasis by controlling the expression of genes encoding enzymes involved in ROS production [41,42], scavenging, as well as genes involved in ROS signaling [43], we studied the expression of several genes playing a role in the regulation of ROS levels. As the first approach, we analyzed the expression of genes whose products are responsible for ROS generation. Several cellular ROS generating enzymes, especially NADPH oxidases and apoplastic peroxidases, are involved in the plant response to environmental stresses [12].

Since, in the Arabidopsis, in response to salinity ROS, are generated mainly by the RbohD and RbohF oxidases [8,44,45], we analyzed transcript levels of *RbohD* and *RbohF*. For the studies, we used two-week-old Arabidopsis seedlings of the *snrk2.4* and *snrk2.10* mutant lines and wt plants treated with 150 mM NaCl up to 24 h. At the first stage of the response to salt stress (1 h), the transcript levels of *RbohD* and *RbohF* were significantly lower in the *snrk2.4* mutant than in the wt plants (Figure 2A,B) in agreement with the lower level of H_2O_2 found in this mutant compared to the wt. At the later time points, this difference was reversed and the expression of *RbohD* and *RbohF* became higher in the mutant than in the wt plants. Notably, the two genes assayed showed markedly different kinetics and extents of activation during the salt treatment. The expression pattern of *RbohD* and *RbohF* in the *snrk2.10* mutant in response to the stress differed significantly from the one observed for the *snrk2.4* mutant. *RbohD* expression was hardly affected by the mutation (it was lower by only 10–15% than in the wt plants throughout the experiment), while the expression of *RbohF* was enhanced markedly at the early stages (up to 3 h) of the response and then fell to slightly below that in the wt. This behavior was opposite to that shown by the *snrk2.4* mutant.

Additionally, we analyzed the expression of genes encoding two apoplastic peroxidases (PRX), PRX33 and PRX34, known to play an important role in the oxidative burst in response to biotic stresses [46,47]. The expression of both *PRXs* was induced in the wt plants in response to salinity stress, indicating their involvement in an abiotic stress response as well (Figure 2C,D). Notably, their expression patterns were affected in a complex manner in both *snrk2* mutants indicating an involvement of SnRK2.4 and SnRK2.10 in their regulation upon salt stress. As for the two *Rboh* genes described above, also here, the two mutant lines showed contrasting responses. In the *snrk2.4* mutant, the *PRX33* gene showed a delayed induction compared to the wt, and at 24 h, its transcript level was double that in the wt. The induction of the *PRX34* gene was enhanced several fold at 3 h and 6 h and was only slightly higher at 24 h relative to the wt. In the *snrk2.10* mutant, *PRX33* showed about a two-fold higher expression level than in the wt at 1 h, nearly identical ones at 3 and 6 h, and again much higher (over three-fold) at 24 h. One should note that also in control conditions, the expression of *PRX33* was markedly up-regulated in the *snrk2.10* mutant. The *SnRK2.10* mutation had a negligible effect on the expression of the other *PRX* gene studied, *PRX34*.

These results indicate that both kinases have an impact on the expression of *RbohD/F* and *PRX33/34* upon salt stress and that their roles are markedly different—they cannot substitute each other in this respect.

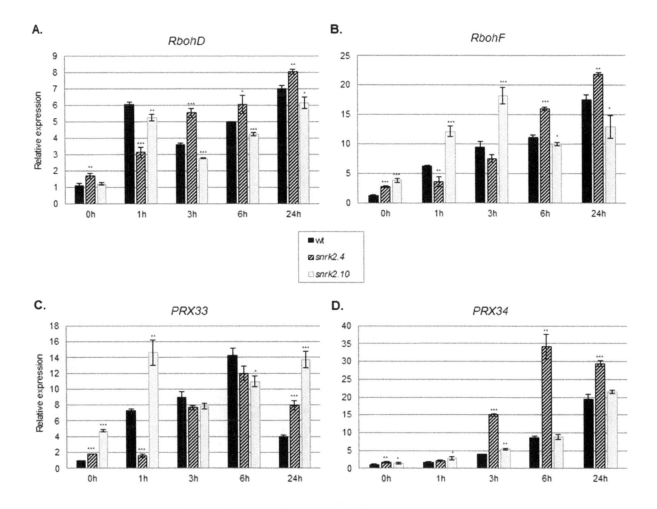

Figure 2. SnRK2.4 and SnRK2.10 affect the expression of genes involved in ROS homeostasis during response to salt stress. Expression (mRNA level) of (**A**). *RbohD—respiratory burst oxidase homolog protein D*; (**B**) *RbohF—respiratory burst oxidase homolog protein F*; (**C**) *PRX33—peroxidase 33*; and (**D**) *PRX34—peroxidase 34* was determined by RT-qPCR in wt plants and *snrk2* mutant lines subjected to treatment with 150 mM NaCl at times indicated (h); error bars represent SD; stars represent statistically significant differences in comparison with the wt plants (Student *t*-test) where * $p < 0.05$; ** $p < 0.001$; *** $p < 0.0001$. At least two independent biological replicates of the experiment were performed. Results of one representative experiment are shown.

2.3. SnRK2.4 and SnRK2.10 Are Involved in Regulation of ROS Scavenging in Response to Salt Stress

Plants have evolved several ROS scavenging pathways, both enzymatic and non-enzymatic. We analyzed the involvement of SnRK2.4 and SnRK2.10 in the regulation of some of them upon salinity stress by comparing the responses of two-week-old *snrk2.4* and *snrk2.10* mutants and wt seedlings exposed to 150 mM NaCl.

2.3.1. SnRK2s Affect CAT1 Gene Expression, Catalase Level, and Activity

Among the most prominent ROS scavenging enzymes are catalases. The Arabidopsis genome encodes three catalases—CAT1, CAT2, and CAT3. Expression of all of them is induced by salinity, however, the strongest changes were observed for the *CAT1* gene [48], therefore, we focused our studies on this gene. Our analysis revealed significant differences in the pattern of *CAT1* expression in response to salt stress between the *snrk2.4* and *snrk2.10* seedlings and the wt ones (Figure 3A).

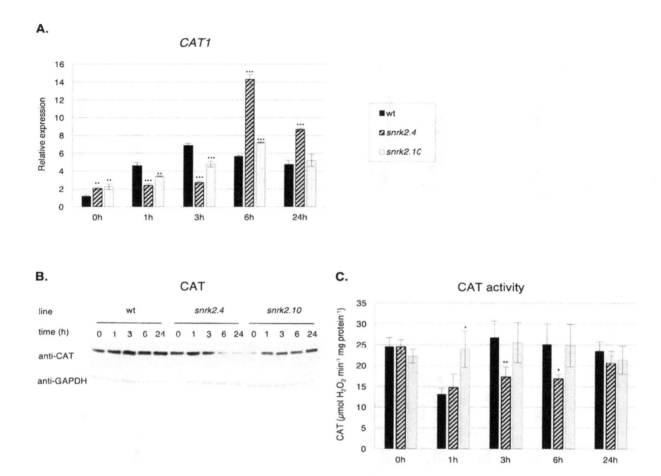

Figure 3. SnRK2.4 and SnRK2.10 modulate catalase (CAT) on multiple levels during response to salt stress. Wild type and *snrk2* mutants' seedlings were subjected to treatment with 150 mM NaCl for times indicated. *CAT1* expression was determined by RT-qPCR (**A**), total catalase protein was determined by immunoblot analysis (**B**), and total catalase activity assay was performed (**C**); error bars represent SD; stars represent statistically significant differences in comparison with the wt plants (Student *t*-test) where * $p < 0.05$; ** $p < 0.001$; *** $p < 0.0001$. After exposure, membranes were stripped and reused for glyceraldehyde 3-phosphate dehydrogenase (GAPDH) detection as a loading control. At least two independent biological replicates of the experiment were performed, each with four samples per data point. Results of one representative experiment are shown.

At the first stages of the response (up to 3 h), the *CAT1* transcript level was significantly lower in both mutants than in the wt, whereas at the later stages, the reverse was true, especially in *snrk2.4*. This suggests that at first, the *CAT1* expression is positively regulated by SnRK2.4 and SnRK2.10, whereas at the later stages of the stress response, the SnRK2s, especially SnRK2.4, exert an inhibitory action. Notably, the *CAT1* transcript level was up-regulated ca. two-fold in the both mutants in control conditions. In that, the effect of a lack of either SnRK2s resembled the situation observed earlier for *PRX33* and *RbohD/RbohF* genes. Unexpectedly, the differences in *CAT1* transcript levels were not reflected by the amount of catalase protein (Figure 3B) or activity (Figure 3C). The catalase protein level was in fact lower in the *snrk2* mutants exposed to salinity stress than in wt plants (Figure 3B); the lowest level was observed in the *snrk2.4* mutant after salt treatment. These data apparently indicate discrepancies between the transcript and protein levels. However, the immunoblot analysis was performed using antibodies recognizing all three isoforms of catalase—CAT1, CAT2, and CAT3. It should be mentioned that in Arabidopsis rosettes, *CAT2* and *CAT3* transcripts are much more abundant than that for *CAT1* [49,50]. Possibly the same is true for the corresponding proteins, therefore, the changes in the amount of the least abundant CAT1 were likely obscured by changes in the other two catalase isoenzymes. The latter, however, suggests that CAT2 and CAT3 could be also under the control of SnRK2s.

The catalase activity, reflecting the combined activity of the three isoenzymes, showed a different pattern for each line studied (Figure 3C); two mutants differed substantially from each other and also from the wt. Strikingly, at 1 h of salt treatment, there was a substantial drop of the catalase activity in the *snrk2.4* and the wt, whereas in the *snrk2.10* mutant, the activity basically did not change.

Taken together, our results show that during the plant response to salinity, SnRK2.4 and SnRK2.10 (especially SnRK2.4) regulate catalase at various levels, including gene expression, catalase protein level, and probably also its enzymatic activity.

2.3.2. SnRK2.4 and SnRK2.10 Regulate the Ascorbate Cycle

Ascorbate is a major antioxidant in plants. To check whether SnRK2.4 and SnRK2.10 play a role in the regulation of the ascorbate cycle in plants subjected to salinity, we compared the expression of genes and their protein products involved in the ascorbate cycle (ascorbate peroxidases, APXs, and dehydroascorbate reductase 1, DHAR1) as well as the APX activity in the *snrk2.4* and *snrk2.10* mutants and wt seedlings exposed to salt stress (Figure 4A–D). Expression of all the genes encoding cytoplasmic *APXs* (*APX1*, *APX2*, and *APX6*) and *DHAR1* was highly induced in response to salinity and that response was markedly and in a complex manner affected in both *snrk2* mutants. Notably, in control conditions, their expression was significantly higher in both mutants than in the wt.

APX1 expression was induced rapidly in response to salinity in wt seedlings, reached a maximum at 1 h and declined below control level at 6 h and 24 h, whereas in the *snrk2.4* mutant, it increased gradually between 3 h and 24 h; in *snrk2.10*, the pattern was similar except for the presence of an early (1 h) peak of expression.

A different situation was observed for the *APX2* gene. It underwent progressive very strong induction in wt plants (1000-fold increase at 24 h of salt treatment), and a similarly progressive, but much less pronounced, induction in the mutants.

APX6 expression was rapidly induced up to six-fold in response to salinity in the wt to reach the maximum at 3 h followed by a slight drop to four-times value at time 0. In contrast, in the *snrk2.4* and *snrk2.10* mutants, where the *APX6* expression was only slightly (*snrk2.10*) or not at all (*snrk2.4*) induced upon salt application, and dropped below the initial level at 24 h in both mutants.

Immunoblot analysis (with antibodies recognizing all three APXs) showed a slight decrease of the APX amount during salt stress in the wt and substantial accumulation at the later stages (6 h and 24 h) in the *snrk2* mutants, especially in the *snrk2.4* (Figure 4E). This accumulation roughly parallels the transcript pattern of *APX1* and *APX2* and suggests that the SnRK2.4 and SnRK2.10 kinases are negative regulators of APX accumulation at the later, but not the early stages of the plant response to salinity.

DHAR1 expression was highly and progressively induced in wt seedlings exposed to salt stress, and to a lesser extent also in the mutants. Notably, in the latter, its level fell between 6 h and 24 h of salt stress. DHAR1 protein showed strong and rapid accumulation in the wt line in response to the stress, slightly lower in the *snrk2.10* mutant, whereas in the *snrk2.4* mutant, it was barely detectable in control conditions and then grew gradually until the end of the treatment (Figure 4F), but it was significantly lower than in the wt and the *snrk2.10* mutant plants. These results suggest that the two SnRK2s, especially SnRK2.4, positively regulate DHAR1 accumulation.

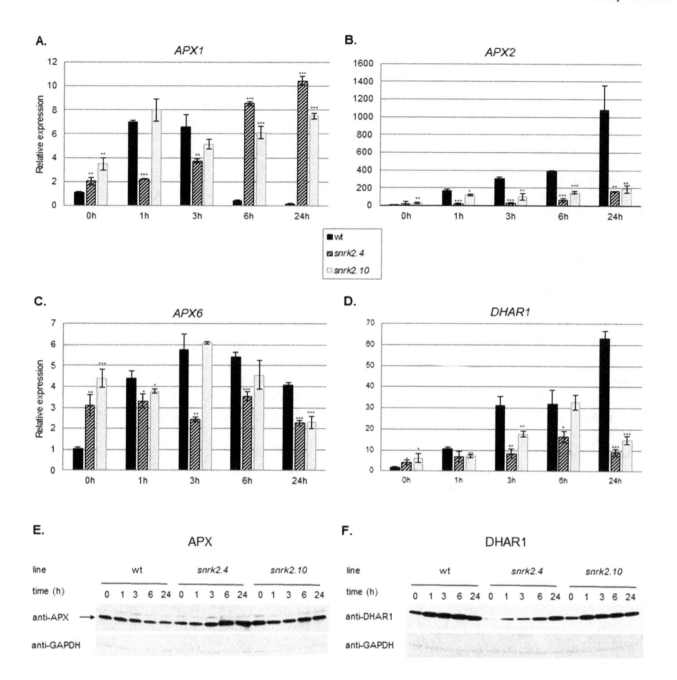

Figure 4. SnRK2.4 and SnRK2.10 regulate enzymes of the ascorbate cycle during response to salt stress. Wild type and *snrk2* mutant seedlings were subjected to treatment with 150 mM NaCl for times indicated. Expression of (**A**) *APX1—Ascorbate Peroxidase 1*, (**B**) *APX2—Ascorbate Peroxidase 2*, (**C**) *APX6—Ascorbate Peroxidase 6*, and (**D**) *DHAR1—Dehydroascorbate Reductase 1* was monitored by RT-qPCR; error bars represent SD; stars represent statistically significant differences in comparison with the wt plants (Student *t*-test) where * $p < 0.05$; ** $p < 0.001$; *** $p < 0.0001$. Total protein level of (**E**) APX and (**F**) DHAR1 was monitored with immunoblot analysis; after exposure, membranes were stripped and reused for GAPDH detection as a loading control. At least two independent biological replicates of the experiment were performed. Results of one representative experiment are shown.

The above results indicate that SnRK2.4, and to lesser extent also SnRK2.10, regulate the level of the enzymes of the ascorbate cycle during the plant response to salinity.

Next, we compared the total ascorbate (Asc) content, APX activity, and the ascorbate/dehydroascorbate (Asc/DHAsc) ratio in the *snrk2.4* and *snrk2.10* mutants and wt seedlings subjected to salinity (Figure 5).

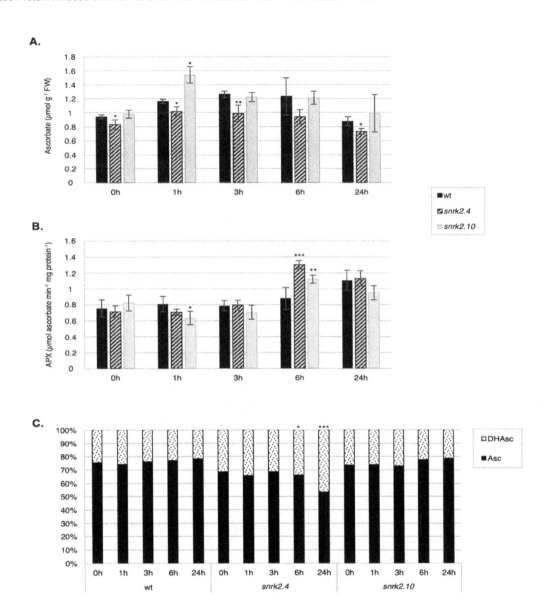

Figure 5. SnRK2.4 and SnRK2.10 regulate ascorbate cycle during response to salt stress. Wild type and *snrk2* mutant seedlings were subjected to treatment with 150 mM NaCl for times indicated and (**A**) ascorbate content, (**B**) ascorbate peroxidase (APX) activity, and (**C**) ascorbate redox status were monitored. Asc—ascorbate, DHAsc—dehydroascorbate; error bars represent SD; stars represent statistically significant differences from wt plants (Student *t*-test for Asc and APX activity, Chi-square test for Asc/DHAsc ratio) where * $p < 0.05$; ** $p < 0.001$; *** $p < 0.0001$. At least two independent biological replicates of the experiment were performed, each with four samples per data point. Results of one representative experiment are shown.

The total Asc content in wt plants was slightly increased in response to salt stress (from 0.9 µmol g^{-1} FW in control conditions to 1.3 µmol g^{-1} FW after 6 h of the treatment) (Figure 5A). A similar pattern was observed for the *snrk2.10* mutant, but with stronger Asc accumulation after 1 h of treatment (from 1 µmol g^{-1} FW in control conditions to 1.55 µmol g^{-1} FW after 1 h and 1.2 µmol g^{-1} FW after 3 h and 6 h of salt stress), while in the *snrk2.4* mutant, the Asc content only slightly increased in response to salt stress (0.83–1.0 µmol g^{-1} FW).

The ratio between oxidized (DHAsc) and the reduced form of Asc was virtually identical in the wt and the *snrk2.10* mutant and did not change upon salt stress (Figure 5C). In the *snrk2.4* mutant line, the fraction of the reduced form of Asc was lower than that in the other lines at all time points of the treatment and additionally showed a significant decrease after 6 h and 24 h of the stress (from 69% in control to 53% after 24 h), suggesting an increased APX activity. Indeed, measurements of the APX activity confirmed this conjecture, albeit the activity pattern did not exactly match the DHAsc/Asc pattern (Figure 5B). Generally, the APX activity reflected the APX protein level (compare Figure 4E).

3. Discussion

Salinity imposes ion and osmotic stresses on plant cells and leads to accumulation of ROS. The understanding of the signaling pathways controlling redox homeostasis during salt stress remains limited. A majority of the data concerning this subject pertains to the plant responses to biotic stresses. Several kinases regulating ROS production in response to pathogen infection have been identified. Some of them act as positive regulators of ROS generation via direct or indirect regulation of RbohD and/or RbohF activity or of their transcript accumulation (e.g., MPK3/6, Flagellin-sensitive 2, EF-Tu receptor, Brassinosteroid insensitive 1 associated receptor kinase 1, Botrytis-induced kinase 1, CPK5), some positively regulate PRX activity (like ZmMPK7), whereas others inhibit ROS accumulation (e.g., MPK4, AtCPK28) [51–60]. Much less is known regarding the protein kinases involved in ROS production or scavenging in response to abiotic stresses. In response to salinity, ROS are generated in different cellular compartments: Chloroplasts (by the photosynthetic electron transport), mitochondria (by the respiratory electron transport), peroxisomes, and, in the apoplast, by the action of oxidases present in the plasma membrane. The ROS generated in the various cellular compartments cross talk with each other. Recently it has been shown that, similarly to animals, the ROS-induced ROS release (RIRR) process (e.g., ROS produced in one cellular organelle or compartment induce ROS production in another one) takes place also in plants [61]. An RIRR-generated ROS wave leads to ROS amplification and signal transduction to neighboring compartments and cells. Therefore, enzymes localized to the cytoplasm or nucleus (as is in the case of the SnRK2s studied here) can indirectly affect the ROS level also in other subcellular compartments.

3.1. Role of SnRK2.4/SnRK2.10 in ROS Accumulation in Response to Salinity Stress

Previously published results have indicated that ABA-non-responsive SnRK2s (from group 1) regulate ROS levels in response to abiotic stresses. Diédhiou et al. [22] showed that an ABA-non-activated SnRK2, SAPK4, regulates ROS homeostasis in rice in response to salt stress, and Kulik et al. [40] that SnRK2.4 and SnRK2.10 are involved in positive regulation of H_2O_2 accumulation in Arabidopsis roots in the early response to heavy metal stress.

Surprisingly, studies performed on Arabidopsis, the snrk2.2/3/6 triple, the septuple, and the decuple mutants defective in several SnRK2s did not show any correlation between SnRK2s level and H_2O_2 accumulation in response to osmotic stress (polyethylene glycol (PEG) treatment of Arabidopsis seedlings) [62]. However, first, the measurement was done 12 h after PEG addition. Moreover, it seems likely that individual SnRK2s might differently affect ROS production and scavenging and their defects in the multiple snrk2 mutants could effectively cancel out, resulting in no net change in H_2O_2 accumulation. Finally, the role of various SnRK2s might be different in response to ABA, PEG, and salinity.

Our present studies focused on the role of two ABA-non-responsive SnRK2 kinases, SnRK2.4 and SnRK2.10, in the regulation of ROS homeostasis in Arabidopsis exposed to salinity stress. Since these kinases localize to the cytoplasm (SnRK2.10) and the cytoplasm and nucleus (SnRK2.4), we studied those enzymes involved in ROS homeostasis that in principle could be regulated by cytoplasmic and nuclear kinases: Directly by phosphorylation or indirectly at the transcriptional level. In respect to ROS production, we studied two plasma membrane NADPH oxidases, RbohD and RbohF, and two apoplastic peroxidases, PRX33 and PRX34. It has been established that RbohD and RbohF are regulated at the activity and expression levels. Their activity is tightly controlled by phosphorylation and by Ca^{2+} binding. Several kinases capable of phosphorylating RbohD/F have been identified, but the role of these phosphorylations is not fully clear. Drerup et al. [63] showed that Calcineurin B-like protein 1/9 (CBL1/9)-CIPK26 (from CBL-interacting protein kinase 26) complexes phosphorylate and activate RbohF. Similarly, Han et al. [36] presented that CIPK11 and CIPK26 phosphorylate RbohF, constituting alternative paths for RbohF activation, whereas Kimura et al. [64] suggest that the binding of CIPK26 to RbohF decreases ROS production. It has also been shown that RbohD/F are regulated also by Ca^{2+}-independent kinases, like MPK8, which inhibits RbohD activity in response to wounding [65]. Some data indicate an involvement of SnRK2s in the regulation of NADPH oxidase activity. OST1/SnRK2.6, an ABA-dependent kinase, regulates the ROS level required for the stomatal closure [34]. Sirichandra et al. [35] showed that OST1 phosphorylates RbohF in vitro and suggested that this phosphorylation plays a role in its activation and possibly in the regulation of stomatal movement in response to ABA. Recently, Han et al. [36] showed that OST1 together with CIPKs is involved in RbohF activation.

Our present results revealed that SnRK2.4 and SnRK2.10 positively regulate the ROS production at the early stages of the response to salinity; we observed significantly lower H_2O_2 levels in the *snrk2.4* and *snrk2.10* mutants than in wt plants salt-treated for up to 90 min. These results suggest that SnRK2.4 and SnRK2.10 might phosphorylate RbohD and/or RbohF in response to salinity and thereby regulate the ROS level. The phosphorylation of RbohD/F by SnRK2.4/10 is highly plausible since the substrate specificities of SnRK2s, CIPKs, and calcium-dependent protein kinases (CDPKs) are quite similar; all of them belong to the CDPK-SnRK superfamily [66]. The SnRK2.4 and SnRK210 kinases are activated rapidly in response to salinity, within seconds after the stressor application. Therefore, it is likely that they are involved in the earliest events of the response to salt stress, i.e., activation of the Rbohs and production of ROS responsible for triggering the defense mechanisms. However, this hypothesis needs further studies.

Our results pointed out to a role of SnRK2.4 and SnRK2.10 in the regulation of *RbohD* and *RbohF* expression. In *Arabidopsis thaliana* seedlings, the expression of *RbohD* and *RbohF* is induced in response to salinity [14,44,45,67]. At the first stages of the response (in our case, the treatment with NaCl for 1 h), the transcript levels of *RbohD* and *RbohF* were significantly lower in the *snrk2.4* mutant than in wt plants, which is in agreement with the lower level of H_2O_2 found in the mutants. Later, during the salt treatment, this correlation was no longer sustained and the *RbohD/F* expression in the mutant became elevated above the level observed for the wt plants. The effect of disruption of the *SnRK2.10* gene was more complex, as it has little effect on induction of *RbohD* expression, but actually enhanced that of *RbohF*.

Additionally, we analyzed the expression of genes encoding apoplastic peroxidases, *PRX33* and *PRX34*, whose involvement in the response to salt stress has not been considered so far. Expression of both *PRXs* was induced in response to salinity stress, which indicates their role in the abiotic stress response, and this induction was apparently regulated by SnRK2s. Expression of *PRX33* was significantly lower in the *snrk2.4* mutant early in the response to salinity, but became highly elevated relative to the wt plants after prolonged salt treatment. For *PRX34*, the effect of the *snrk2.4* mutation was visible only at later stages of the response and manifested as a several fold enhancement of the induction. This suggests an inhibitory role of SnRK2.4 on the salt-induced *PRX34* expression.

As for the *Rboh* genes, also here, SnRK2.10 turned out to act differently to SnRK2.4. The *snrk2.10* mutation had virtually no effect on *PRX34* expression, but greatly stimulated the expression of *PRX33* in control conditions and also at early and late response to salt stress.

These data indicate that even though in both the *snrk2.4* and *snrk2.10* mutants, the level of H_2O_2 produced early in response to salt stress is significantly lower than in wt plants, the mechanisms of the regulation of ROS accumulation by the two kinases seem to be different. It should be stressed here, that unlike SnRK2.4, SnRK2.10 does not localize to the nucleus, therefore, the different modes of regulation of gene expression by these kinases are not surprising (Figure 6).

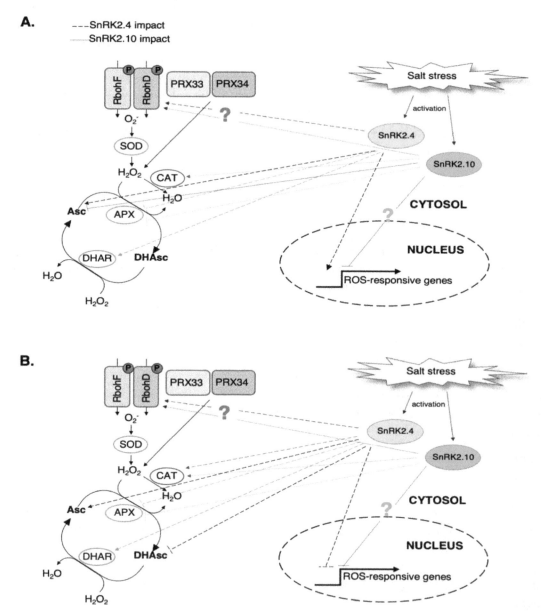

Figure 6. Possible roles of SnRK2.4 and SnRK2.10 in the regulation of the ROS homeostasis in Arabidopsis seedlings exposed to the salt stress. Proposed role of the SnRK2s in (**A**) early response and (**B**) late response to the salt stress. In response to salinity, SnRK2.4 along with SnRK2.10 regulate the ROS production/accumulation as well as ROS scavenging at the transcription as well as protein and/or activity levels. Detailed description in the text; dash lines—SnRK2.4 impact; dotted lines—SnRK2.10 impact; green question mark—probably indirect regulation; red question mark—plausible direct regulation by phosphorylation.

The expression of genes encoding ROS producing enzymes in the *snrk2.4* and *snrk2.10* (especially *snrk2.4*) is very different at early and late stages of the response. We propose that at the later stages the expression of genes studied might be regulated by other signaling pathways, for example, MAPK cascade(s), which are involved in controlling ROS homeostasis. Those pathways might be triggered to compensate for the low ROS level in the *snrk2s* mutants. In response to several stimuli MAPK cascade(s) control *RbohD, RbohF, PRX33*, and *PRX34* expression [52,68]. MPK3/MPK6 phosphorylate and thus activate the ERF6 transcription factor, whose targets are *RbohD* and *PRX33* in response to fungal pathogen [69–71]. Moreover, in *Nicotiana benthamiana* during the ETI (from effector-triggered immunity) and Elicitin-(INF1)-triggered PTI (from pattern-triggered immunity), salicylic acid induced protein kinase (SIPK, orthologue of Arabidopsis MPK6) phosphorylates four

W-box binding transcription factors (WRKYs), which are responsible for the expression of *RBOHB* (ortholog of Arabidopsis *RbohD*), and positively regulates the *RBOHB* transcript level [72]. Since MPK6 and SIPK are activated in response to salinity and water deficits [37,73] and ERF6 is involved in response to the water limitation [74], it seems likely that the MPK6 pathway, and presumably some others, could overcompensate for the low expression of *RbohD* and *PRX33* at early stages of the response in the *snrk2.4* mutant.

3.2. Involvement of SnRK2.4/SnRK2.10 in ROS Removal under Salinity Stress Conditions

Data on the signaling pathways and protein kinases involved in the regulation of the antioxidant systems engaged in ROS scavenging in response to salinity or osmotic stress are scarce. It has been shown that GSK3 kinase (ASKα) regulates salt stress tolerance of Arabidopsis by phosphorylation and activation of glucose-6-phosphate dehydrogenase, an enzyme important for maintaining the cellular redox balance [75]. Zong et al. [76] have shown that ectopic expression of *ZmMPK7* in *Nicotiana tabaccum* enhances peroxidase activity, which results in lower accumulation of H_2O_2 in response to osmotic stress. It has been reported that several protein kinases of MAPK cascades as well as CIPK are involved in the regulation of expression and/or activity of CAT1 [43,48,77].

We analyzed here the impact of SnRK2.4 and SnRK2.10 on several enzymes involved in ROS scavenging (CATs, APXs, and DHAR1). A comparison of the changes in the *CAT1* transcript level in the *snrk2.4*, *snrk2.10*, and wt plants exposed to salinity indicated that SnRK2.4, and to a lesser extent also SnRK2.10, positively regulate the expression of *CAT1* during the first stages of the stress response. However, similar to what was observed for *RbohD*, at the later stages of the response, the impact of the *snrk2* mutations became just the opposite, which results in a higher *CAT1* expression than in the wt plants. We conjecture that this effect was not due to a direct regulation of *CAT1* expression by SnRK2.4/SnRK2.10, but rather because of reflected activation of some other signaling pathway(s) in order to compensate for the low *CAT1* expression. One such pathway might be again the MPK6 cascade, known to mediate *CAT1* expression and H_2O_2 production [77].

Besides the regulation of the *CAT1* expression, SnRK2.4 and, less markedly, also SnRK2.10 positively regulated catalase protein accumulation and activity during salt stress, as they were both significantly lower in the *snrk2* mutants exposed to the stress than in the wt. The discrepancy between the enhanced expression of *CAT1* later in the response and the lower catalase protein level and activity indicates that in response to salinity, the SnRK2s affect not only CAT1, but most likely also the CAT2 and CAT3 levels, in opposing directions. Since CAT2 and CAT3 are more abundant than CAT1 in Arabidopsis seedlings, it is likely that the catalase activity, which we measured, represented mainly the activity of CAT2 and CAT3. It is plausible that SnRK2s regulate also the expression of *CAT2* and/or *CAT3* genes. We observed lower catalase activity in the *snrk2.4* mutant exposed for up to 6 h to the salt stress. It is not clear whether SnRK2s modulate the enzyme(s) specific activity or only the catalases' protein level or through the phosphorylation they impact the targeting of catalases into the peroxisomes (their final destination). Phosphorylation of catalases by SnRK2s is quite feasible, since it has been shown that salt overly sensitive 2 (SOS2), a kinase belonging to the SnRK3 subfamily, interacts with CAT2 and CAT3 and possibly influences H_2O_2 accumulation in response to salinity [78]. SOS2 localizes to the plasma membrane and cytoplasm and its substrate specificity is nearly the same as that of the SnRK2's.

The available information on the regulation of the ascorbate cycle in response to abiotic stresses and the protein kinases is very limited. It has been suggested that in response to strong light, the SnRK2.6/OST1 kinase activates *APX2* expression [79]. Pitzschke and Hirt [80] have speculated that MAPK cascades could be involved in ascorbate cycle regulation based on transcriptomic analysis performed on *mapk* mutants and wt plants. They revealed that several genes encoding enzymes involved in ascorbate biosynthesis and metabolism were differentially expressed compared to wt plants [43], but those results have not been confirmed so far.

Our data regarding the expression of genes encoding selected enzymes of the ascorbate cycle, APX (APX1, 2, 6) and DHAR1, indicate that SnRK2.4 and SnRK2.10 also modulate the ascorbate cycle. The two kinases had a significant impact on *APXs'* expression upon salt stress, but the different *APX* genes were regulated differently. The effect of the *snrk2* mutations on the *APX1* expression profile was similar to those observed for *RbohD* or *PRX33*. At present, we do not know how the *APX2* and *APX6* up-regulation by SnRK2s affects the overall APX activity. The combined APX protein level in the mutants and wt plants exposed to the stress correlated well with the level of expression of *APX1*, but not of *APX2* or *APX6*. These data indicate that most likely APX1 has the largest share in the overall cytoplasmic APX pool, and importantly, SnRK2.4 and SnRK2.10 play a role in its regulation. Furthermore, another enzyme of the ascorbate cycle that regenerates DHAsc to Asc, DHAR1, was strongly up-regulated at both the transcript and protein levels by SnRK2.4 and slightly less by SnRK2.10 in plants exposed to salinity.

The total Asc level and the Asc/DHAsc ratio were significantly lower in the *snrk2.4* mutant in comparison with the wt plants in response to salinity, which suggests that SnRK2.4 kinase positively regulates Asc accumulation. In agreement with these data, the APX activity was higher in *snrk2.4* than in the two other lines studied. Taken together, these data indicate that SnRK2.4 plays a substantial role in the regulation of the ascorbate cycle in response to salt stress, by direct or indirect regulation of APX and DHAR1 (Figure 6).

Discussing our results, one issue should be pointed out—the role of the circadian clock in the regulation of the redox homeostasis. Circadian clocks regulate the plant growth and development as well as responses to multiple environmental cues, both biotic as well as abiotic [81,82]. Numerous genes involved in the response to osmotic stress and ABA signaling have been identified as circadian clock-dependent, including *SnRK2.6* and several genes encoding stress-responsive transcription factors (for review see [83]). It has been shown that the redox homeostasis (ROS production, scavenging, and expression of ROS-responsive genes) is tuned with diurnal and circadian rhythms, for example, the expression of *CAT1* and *CAT3* is highest at noon, whereas *CAT2* is highest at dawn [84]. On the other hand, in the feedback response, ROS signals affect clock responses [84,85]. Since our results show that SnRK2.4 and SnRK2.10 regulate the ROS level, it is highly likely that the kinases have some impact on the circadian clock. We also conjecture that expression of SnRK2.4 and/or SnRK2.10 might be regulated by a circadian rhythm. It should be stressed at this point that the expression of some genes studied by us might be affected not only by the salt, but also, to some extent, by the diurnal/circadian rhythms.

It has been reported that the transcriptomic analysis using a circadian-guided network approach might be used for identification of the genes involved in the early sensing of mild drought [86], indicating again a close relation between the stress responses and the circadian clock. Salt stress and dehydration signaling pathways have several common elements, including SnRK2s. Importantly, dehydration accompanies salt stress and it has been shown that not only ABA-activated SnRK2s, but also SnRK2, which are not activated in response to ABA, e.g., SnRK2.10 are involved in the plant response to a water deficit [39]. Therefore, when analyzing the plant response to salinity stress, as well as all other stresses, one should be aware of circadian/diurnal rhythms, which play a role in tuning those responses [84–86].

Regulation of the plant response to salt stress by SnRK2s is complex, and our knowledge on this subject is very limited. To provide the full picture, presenting the role of SnRK2s in the salt stress response, additional extensive work is required, e.g., the elucidation of the interplay between SnRK2s, ROS, circadian clock, and various signaling pathways.

A model summarizing our knowledge on the involvement of SnRK2.4 and SnRK2.10 in the plant response to salt stress is presented in Figure 7. In response to salinity stress, the kinases regulate root growth and architecture [23], mRNA decay (by phosphorylation of VCS) [24], have an impact on dehydrin ERD14 localization and likely interactions with plant membranes [39], and on ROS homeostasis (the results described here).

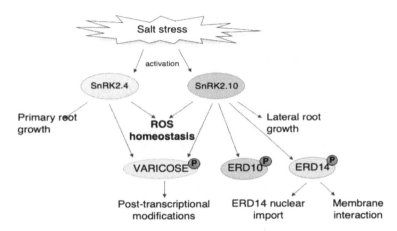

Figure 7. Schematic model illustrating the role of SnRK2.4 and SnRK2.10 in Arabidopsis' response to salt stress. SnRK2.4 and SnRK2.10 modulate root growth under the salinity conditions. Moreover, in response to salt stress, the ABA-non-activated SnRK2s phosphorylate VARICOSE (VCS), a protein participating in mRNA decay, and two dehydrins, Early Responsive to Dehydration 10 (ERD10) and ERD14. Our results presented here revealed that SnRK2.4 and SnRK2.10 regulate the ROS homeostasis in the response to salinity.

In conclusion, our data described here show that SnRK2.4 along with SnRK2.10 positively regulate the first ROS wave that transduces the salt stress signal. The kinases regulate ROS accumulation as well as ROS scavenging, by modulating the catalase level and the ascorbate cycle (Figure 6). These results suggest that the two studied SnRK2s are involved in the fine tuning of the ROS level and thus contribute to the regulation of ROS homeostasis required for the plant acclimation to unfavorable environmental conditions.

4. Materials and Methods

4.1. Plant Material, Growth, and Treatment Conditions

The following *Arabidopsis thaliana* lines were used: Arabidopsis Col-0 ecotype ("wild type"; wt); homozygous T-DNA insertion lines *snrk2.4-1* (SALK_080588), *snrk2.4-2* (SALK_146522), and *snrk2.10-1* (WiscDsLox233E9) kindly provided by Prof. C. Testerink (University of Amsterdam, The Netherlands), and *snrk2.10-2* (SAIL_698_C05) from the Nottingham Arabidopsis Stock Center (NASC). Seedlings were grown in a sterile hydroponic culture as described Kulik et al. [40] for two weeks.

For luminol-based H_2O_2 determination, plants were grown for four weeks on Jiffy pods (Jiffy-7, Jiffy Group) in a growth chamber under 8 h of light /16 h dark conditions at 21 °C/18 °C.

For ROS production measurements with H_2DCFDA, Arabidopsis seedling were grown on $\frac{1}{2}$ MS plates supplemented with 0.8% agar for five days in a growth chamber under 8 h of light/16 h dark conditions at 21 °C/18 °C.

Two- or four-week-old plants were treated with 150 mM NaCl for the indicated time (as described in the results section; stress was applied 2 h after the light was turn on), harvested by sieving, and immediately frozen in liquid nitrogen. Plant material was kept at −80 °C until further analysis.

4.2. Determination of H_2O_2

Luminol-based assay for H_2O_2 was performed according to Rasul et al. [87]. Discs of 2-mm diameter were excised from leaves of four-week-old plants using a cork borer, from 5 leaves per sample per condition, and placed into assay vials with 200 μL of MQ water, sealed with parafilm, and incubated at RT overnight. Next, stress conditions were applied (either 150 mM NaCl or MQ water as a control) and 4 μL of luminol solution [3 mM luminol dissolved in dimethyl sulfoxide (DMSO); final concentration 60 μM] was added at appropriate time points. Vials were gently mixed and luminescence was measured using a luminometer for a total time of 120 s. The measurements were performed at selected time points up to 90 min post treatment. Statistical analysis was performed using one way analysis of variance (ANOVA).

ROS detection with H_2DCFDA was performed as described previously by Kulik et al. [40] and Srivastava et al. [88] with minor modifications. Staining of five-day-old Arabidopsis seedlings roots with PI and H_2DCFDA was performed before treatment with 250 mM NaCl in $\frac{1}{2}$ MS or $\frac{1}{2}$ MS only. Single confocal sections were collected with a $20\times$ (NA 0.75) Plan Fluor multiimmersion objective mounted on an inverted epifluorescence TE 2000E microscope (Nikon, Tokyo, Japan) coupled with an EZ-C1 confocal laser-scanning head (Nikon). H_2DCFDA fluorescence was excited with blue light at 488 nm emitted by a Sapphire 488 nm laser (Coherent, Santa Clara, CA, USA) and detected with a 515/30-nm band-pass-filter and rendered in false green, PI fluorescence was excited with green light at 543 nm emitted by a 1 mW He-Ne laser (Melles Griot, Carlsbad, CA, USA) and detected with a 610 nm long-pass filter and rendered in false magenta. All confocal parameters (laser power, gain, etc.) and conditions were the same during the experiment. EZ-C1 FreeViewer software was used to quantify the fluorescence intensity from the 4000 μm^2 area of the root meristematic zone in each Arabidopsis seedling. Each experimental variant was repeated at least twice with a total of 30 single images collected. Statistical analysis was performed using the Mann-Whitney U test.

4.3. RNA Extraction and RT-qPCR Analysis

Total RNA was extracted with TRI Reagent® according to the manufacturer's protocol (MRC). Approximately 150–200 mg of frozen ground plant material was used. DNA contamination was removed from the obtained RNA using a RapidOut DNA Removal kit (Thermo Scientific, Waltham, MA, USA). cDNA was synthesized from 4 µg of purified RNA using an Enhanced Avian HS RT-PCR Kit (Sigma-Aldrich, St. Louis, MO, USA) following the manufacturer's protocol. RT-qPCR was performed on 50 ng of the cDNA using LightCycler® 480 SYBR Green I Master Mix (Roche, Basel, Switzerland) and a Roche LightCycler® 480 machine. Relative transcript levels were calculated according to Livak and Schmittgen [89] with *UBQ10* (AT4G05320) and *UBC* (AT5G25760) as reference genes [90,91]. Statistical analysis was performed using Student *t*-test. All primers used in this study are listed in Table S1.

4.4. Protein Extraction and Immunoblot Analysis

Total proteins were extracted from frozen plant samples in two volumes of extraction buffer: 100 mM HEPES, pH7.5; 5 mM EDTA; 5 mM EGTA; 10 mM DTT; 1 mM Na_3VO_4; 10 mM NaF; 50 mM β-glycerophosphate; 10 mM pyridoxal 5-phosphate; 10% glycerol; and 1 × Complete protease inhibitors (EDTA-free, Roche) on a rotator for 30 min at 4 °C and then centrifuged at 12,000 rpm for 30 min at 4 °C. Protein concentration in the supernatant was measured using a Bradford Protein Assay. The extracts were used immediately or flash-frozen and kept at −80 °C for further analysis. The immunoblot blot analysis was based on a standard procedure described by Sambrook [92]. Protein samples (7–15 µg) were separated on 12% SDS-polyacrylamide gels and transferred to Immobilon®® P membrane by electroblotting in transfer buffer, TB (25 mM Tris base, 192 mM glycine), overnight at 18 V. Transferred proteins were visualized by staining the membranes with Ponceau S (2% Ponceau S in 3% trichloroacetic acid). Immunodetection with anti-APX rabbit IgG (AS08 368, Agrisera, Vännäs, Sweden), anti-CAT rabbit IgG (AS09 501, Agrisera), and anti-DHAR1 rabbit IgG (AS11 1746, Agrisera) was performed as described in the manufacturer's protocols. Anti-glyceraldehyde-3-phosphate dehydrogenase (GAPDH) rabbit IgG (raised against the CYDDIKAAIKEESEG peptide of GAPDH; BioGenes, Berlin, Germany) was used as described previously in Wawer et al. [93]. Secondary anti-rabbit antibodies (alkaline phosphatase (AP) conjugated—AS09 607, Agrisera; horseradish peroxidase (HRP) conjugated—AS09 602, Agrisera) were visualized using appropriate substrates—5-bromo-4-chloro-3-indolyl-phosphate/nitroblue tetrazolium (BCIP/NBT, Roche) for AP, and—ECL detection reagent (Pierce™ ECL Western Blotting Substrate, Thermo Scientific) for HRP according to the manufacturer's protocol. Membranes were reused for

GAPDH protein detection used as a loading control for Western blots. Stripping of the membranes was performed according to Abcam online protocols.

For APX and CAT activity assays (see further), proteins were extracted from frozen plant samples (0.5 g FW) with 1 mL of ice-cold 50 mM sodium phosphate buffer, pH 7.5 or 100 mM potassium phosphate buffer, pH 7.0, respectively, containing 1 mM polyethylene glycol, 1 mM phenylmethylsulfonyl fluoride, 8% (w/v) polyvinylpolypyrolydone, and 0.01% (v/v) Triton X-100, according to Venisse et al. [94].

4.5. Determination of Ascorbate and Ascorbate/Dehydroascorbate Ratio

Ascorbate (Asc) and dehydroascorbate (DHAsc) was determined using a modified bipyridyl method described in detail by Polkowska-Kowalczyk et al. [95]. Statistical analysis was performed using the Student t-test and Chi-square test.

4.6. Determination of APX and CAT Activity

Ascorbate peroxidase (APX, EC 1.11.1.11) activity was assayed as described previously in Polkowska-Kowalczyk et al. [95]. Enzyme activity was expressed as μmol of oxidized ascorbate per min per mg of protein.

Catalase (CAT, EC 1.11.1.6) activity was assayed at 25 °C following the decomposition of H_2O_2 at 240 nm (extinction coefficient 0.036 mM^{-1} cm^{-1}) according to a modified method of Aebi [96]. The reaction mixture contained 50 μL of plant extract in 1 mL 50 mM potassium phosphate buffer (pH 7.0) and 9.8 mM H_2O_2. Enzyme activity was expressed as μmol H_2O_2 decomposed per min per mg of protein.

Statistical analysis was performed using the Student t-test.

Author Contributions: Conceptualization, K.P.S. and G.D.; Methodology, K.P.S. and L.P.-K.; Formal Analysis, K.P.S.; Investigation, K.P.S., L.P.-K. and M.L.; Resources, K.P.S. and J.M.; Visualization K.P.S.; Writing—Original Draft Preparation, G.D. and K.P.S.; Writing—Review & Editing K.P.S., L.P.-K., J.M. and G.D.; Funding Acquisition K.P.S.

Acknowledgments: We kindly thank Christa Testerink for providing the *snrk2* knockout mutants seeds.

Abbreviations

APX	Ascorbate peroxidase
Asc	Ascorbate
CAT	Catalase
CDPK	Calcium-dependent protein kinase
CIPK	CBL-interacting protein kinase
DAsc	Dehydroascorbate
DHAR	Dehydroascorbate reductase
H_2O_2	Hydrogen peroxide
MPK	Mitogen activated protein kinase
O_2^-	Superoxide radical
PRX	Peroxidase
Rboh	Respiratory burst oxidase homologue
ROS	Reactive oxygen species
SnRK2	SNF-1 Related Protein Kinases type 2

References

1. Mittler, R. ROS Are Good. *Trends Plant Sci.* **2017**, *22*, 11–19. [CrossRef] [PubMed]
2. Mhamdi, A.; Van Breusegem, F. Reactive Oxygen Species in Plant Development. *Development* **2018**, *145*, dev164376. [CrossRef] [PubMed]
3. Foyer, C.H.; Ruban, A.V.; Noctor, G. Viewing Oxidative Stress through the Lens of Oxidative Signalling Rather than Damage. *Biochem. J.* **2017**, *474*, 877–883. [CrossRef] [PubMed]
4. Waszczak, C.; Carmody, M.; Kangasjärvi, J. Reactive Oxygen Species in Plant Signaling. *Annu. Rev. Plant Biol.* **2018**, *69*, 209–236. [CrossRef] [PubMed]
5. Noctor, G.; Reichheld, J.P.; Foyer, C.H. ROS-Related Redox Regulation and Signaling in Plants. *Semin. Cell Dev. Biol.* **2018**, *80*, 3–12. [CrossRef]
6. Foyer, C.H.; Noctor, G. Redox Homeostasis and Antioxidant Signaling: A Metabolic Interface between Stress Perception and Physiological Responses. *Plant Cell Online* **2005**, *17*, 1866–1875. [CrossRef] [PubMed]
7. Sofo, A.; Scopa, A.; Nuzzaci, M.; Vitti, A. Ascorbate Peroxidase and Catalase Activities and Their Genetic Regulation in Plants Subjected to Drought and Salinity Stresses. *Int. J. Mol. Sci.* **2015**, *16*, 13561–13578. [CrossRef] [PubMed]
8. Ben Rejeb, K.; Lefebvre-De Vos, D.; Le Disquet, I.; Leprince, A.S.; Bordenave, M.; Maldiney, R.; Jdey, A.; Abdelly, C.; Savouré, A. Hydrogen Peroxide Produced by NADPH Oxidases Increases Proline Accumulation during Salt or Mannitol Stress in *Arabidopsis thaliana*. *New Phytol.* **2015**, *208*, 1138–1148. [CrossRef] [PubMed]
9. Mittler, R. Oxidative Stress, Antioxidants and Stress Tolerance. *Trends Plant Sci.* **2002**, *7*, 405–410. [CrossRef]
10. Baxter, A.; Mittler, R.; Suzuki, N. ROS as Key Players in Plant Stress Signalling. *J. Exp. Bot.* **2014**, *65*, 1229–1240. [CrossRef]
11. Mittler, R.; Vanderauwera, S.; Gollery, M.; Van Breusegem, F. Reactive Oxygen Gene Network of Plants. *Trends Plant Sci.* **2004**, *9*, 490–498. [CrossRef] [PubMed]
12. Miller, G.; Suzuki, N.; Ciftci-Yilmaz, S.; Mittler, R. Reactive Oxygen Species Homeostasis and Signalling during Drought and Salinity Stresses. *Plant Cell Environ.* **2010**, *33*, 453–467. [CrossRef] [PubMed]
13. Choudhury, S.; Panda, P.; Sahoo, L.; Panda, S.K. Reactive Oxygen Species Signaling in Plants under Abiotic Stress. *Plant Signal. Behav.* **2013**, *8*, e23681. [CrossRef] [PubMed]
14. Hossain, M.S.; Dietz, K.-J. Tuning of Redox Regulatory Mechanisms, Reactive Oxygen Species and Redox Homeostasis under Salinity Stress. *Front. Plant Sci.* **2016**, *7*, 548. [CrossRef] [PubMed]
15. Inupakutika, M.A.; Sengupta, S.; Devireddy, A.R.; Azad, R.K.; Mittler, R. The Evolution of Reactive Oxygen Species Metabolism. *J. Exp. Bot.* **2016**, *67*, 5933–5943. [CrossRef] [PubMed]
16. Czarnocka, W.; Karpiński, S. Friend or Foe? Reactive Oxygen Species Production, Scavenging and Signaling in Plant Response to Environmental Stresses. *Free Radic. Biol. Med.* **2018**, *122*, 4–20. [CrossRef] [PubMed]
17. Kulik, A.; Wawer, I.; Krzywińska, E.; Bucholc, M.; Dobrowolska, G. SnRK2 Protein Kinases-Key Regulators of Plant Response to Abiotic Stresses. *OMICS* **2011**, *15*, 859–872. [CrossRef] [PubMed]
18. Boudsocq, M.; Barbier-Brygoo, H.; Laurière, C. Identification of Nine Sucrose Nonfermenting 1-Related Protein Kinases 2 Activated by Hyperosmotic and Saline Stresses in *Arabidopsis thaliana*. *J. Biol. Chem.* **2004**, *279*, 41758–41766. [CrossRef]
19. Kobayashi, Y.; Yamamoto, S.; Minami, H.; Kagaya, Y.; Hattori, T. Differential Activation of the Rice Sucrose Nonfermenting1-Related Protein Kinase2 Family by Hyperosmotic Stress and Abscisic Acid. *Plant Cell* **2004**, *16*, 1163–1177. [CrossRef]
20. Fujii, H.; Zhu, J.K. *Arabidopsis* Mutant Deficient in 3 Abscisic Acid-Activated Protein Kinases Reveals Critical Roles in Growth, Reproduction, and Stress. *Proc. Natl. Acad. Sci. USA* **2009**, *106*, 8380–8385. [CrossRef]
21. Fujita, Y.; Nakashima, K.; Yoshida, T.; Katagiri, T.; Kidokoro, S.; Kanamori, N.; Umezawa, T.; Fujita, M.; Maruyama, K.; Ishiyama, K.; et al. Three SnRK2 Protein Kinases Are the Main Positive Regulators of Abscisic Acid Signaling in Response to Water Stress in Arabidopsis. *Plant Cell Physiol.* **2009**, *50*, 2123–2132. [CrossRef] [PubMed]
22. Diédhiou, C.J.; Popova, O.V.; Dietz, K.J.; Golldack, D. The SNF1-Type Serine-Threonine Protein Kinase SAPK4 Regulates Stress-Responsive Gene Expression in Rice. *BMC Plant Biol.* **2008**, *8*, 49. [CrossRef] [PubMed]

23. McLoughlin, F.; Galvan-Ampudia, C.S.; Julkowska, M.M.; Caarls, L.; Van Der Does, D.; Laurière, C.; Munnik, T.; Haring, M.A.; Testerink, C. The Snf1-Related Protein Kinases SnRK2.4 and SnRK2.10 Are Involved in Maintenance of Root System Architecture during Salt Stress. *Plant J.* **2012**, *72*, 436–449. [CrossRef] [PubMed]

24. Soma, F.; Mogami, J.; Yoshida, T.; Abekura, M.; Takahashi, F.; Kidokoro, S.; Mizoi, J.; Shinozaki, K.; Yamaguchi-Shinozaki, K. ABA-Unresponsive SnRK2 Protein Kinases Regulate MRNA Decay under Osmotic Stress in Plants. *Nat. Plants* **2017**, *3*, 16204. [CrossRef] [PubMed]

25. Hubbard, K.E.; Nishimura, N.; Hitomi, K.; Getzoff, E.D.; Schroeder, J.I. Early Abscisic Acid Signal Transduction Mechanisms: Newly Discovered Components and Newly Emerging Questions. *Genes Dev.* **2010**, *24*, 1695–1708. [CrossRef] [PubMed]

26. Umezawa, T.; Nakashima, K.; Miyakawa, T.; Kuromori, T.; Tanokura, M.; Shinozaki, K.; Yamaguchi-Shinozaki, K. Molecular Basis of the Core Regulatory Network in ABA Responses: Sensing, Signaling and Transport. *Plant Cell Physiol.* **2010**, *51*, 1821–1839. [CrossRef] [PubMed]

27. Geiger, D.; Scherzer, S.; Mumm, P.; Stange, A.; Marten, I.; Bauer, H.; Ache, P.; Matschi, S.; Liese, A.; Al-Rasheid, K.A.S.; et al. Activity of Guard Cell Anion Channel SLAC1 Is Controlled by Drought-Stress Signaling Kinase-Phosphatase Pair. *Proc. Natl. Acad. Sci. USA* **2009**, *106*, 21425–21430. [CrossRef] [PubMed]

28. Lee, S.C.; Lan, W.; Buchanan, B.B.; Luan, S. A Protein Kinase-Phosphatase Pair Interacts with an Ion Channel to Regulate ABA Signaling in Plant Guard Cells. *Proc. Natl. Acad. Sci. USA* **2009**, *106*, 21419–21424. [CrossRef] [PubMed]

29. Sato, A.; Sato, Y.; Fukao, Y.; Fujiwara, M.; Umezawa, T.; Shinozaki, K.; Hibi, T.; Taniguchi, M.; Miyake, H.; Goto, D.B.; et al. Threonine at Position 306 of the KAT1 Potassium Channel Is Essential for Channel Activity and Is a Target Site for ABA-Activated SnRK2/OST1/SnRK2.6 Protein Kinase. *Biochem. J.* **2009**, *424*, 439–448. [CrossRef]

30. Furihata, T.; Maruyama, K.; Fujita, Y.; Umezawa, T.; Yoshida, R.; Shinozaki, K.; Yamaguchi-Shinozaki, K. Abscisic Acid-Dependent Multisite Phosphorylation Regulates the Activity of a Transcription Activator AREB1. *Proc. Natl. Acad. Sci. USA* **2006**, *103*, 1988–1993. [CrossRef]

31. Wang, P.; Xue, L.; Batelli, G.; Lee, S.; Hou, Y.-J.; Van Oosten, M.J.; Zhang, H.; Tao, W.A.; Zhu, J.-K. Quantitative Phosphoproteomics Identifies SnRK2 Protein Kinase Substrates and Reveals the Effectors of Abscisic Acid Action. *Proc. Natl. Acad. Sci. USA* **2013**, *110*, 11205–11210. [CrossRef]

32. Umezawa, T.; Sugiyama, N.; Takahashi, F.; Anderson, J.C.; Ishihama, Y.; Peck, S.C.; Shinozaki, K. Genetics and Phosphoproteomics Reveal a Protein Phosphorylation Network in the Abscisic Acid Signaling Pathway in *Arabidopsis thaliana*. *Sci. Signal.* **2013**, *6*, rs8. [CrossRef]

33. Grondin, A.; Rodrigues, O.; Verdoucq, L.; Merlot, S.; Leonhardt, N.; Maurel, C. Aquaporins Contribute to ABA-Triggered Stomatal Closure through OST1-Mediated Phosphorylation. *Plant Cell* **2015**, *27*, 1945–1954. [CrossRef]

34. Mustilli, A.-C.; Merlot, S.; Vavasseur, A.; Fenzi, F.; Giraudat, J. *Arabidopsis* OST1 Protein Kinase Mediates the Regulation of Stomatal Aperture by Abscisic Acid and Acts Upstream of Reactive Oxygen Species Production. *Plant Cell* **2002**, *14*, 3089–3099. [CrossRef]

35. Sirichandra, C.; Gu, D.; Hu, H.C.; Davanture, M.; Lee, S.; Djaoui, M.; Valot, B.; Zivy, M.; Leung, J.; Merlot, S.; et al. Phosphorylation of the *Arabidopsis* AtrbohF NADPH Oxidase by OST1 Protein Kinase. *FEBS Lett.* **2009**, *583*, 2982–2986. [CrossRef]

36. Han, J.-P.; Köster, P.; Drerup, M.M.; Scholz, M.; Li, S.; Edel, K.H.; Hashimoto, K.; Kuchitsu, K.; Hippler, M.; Kudla, J. Fine Tuning of RBOHF Activity Is Achieved by Differential Phosphorylation and Ca^{2+} Binding. *New Phytol.* **2018**. [CrossRef]

37. Mikołajczyk, M.; Awotunde, O.S.; Muszyńska, G.; Klessig, D.F.; Dobrowolska, G. Osmotic Stress Induces Rapid Activation of a Salicylic Acid-Induced Protein Kinase and a Homolog of Protein Kinase ASK1 in Tobacco Cells. *Plant Cell* **2000**, *12*, 165–178. [CrossRef]

38. Burza, A.M.; Pekala, I.; Sikora, J.; Siedlecki, P.; Małagocki, P.; Bucholc, M.; Koper, L.; Zielenkiewicz, P.; Dadlez, M.; Dobrowolska, G. *Nicotiana tabacum* Osmotic Stress-Activated Kinase Is Regulated by Phosphorylation on Ser-154 and Ser-158 in the Kinase Activation Loop. *J. Biol. Chem.* **2006**, *281*, 34299–34311. [CrossRef]

39. Maszkowska, J.; Dębski, J.; Kulik, A.; Kistowski, M.; Bucholc, M.; Lichocka, M.; Klimecka, M.; Sztatelman, O.; Szymańska, K.P.; Dadlez, M.; et al. Phosphoproteomic Analysis Reveals That Dehydrins ERD10 and ERD14 Are Phosphorylated by SNF1-Related Protein Kinase 2.10 in Response to Osmotic Stress. *Plant Cell Environ.* **2018**. [CrossRef]

40. Kulik, A.; Anielska-Mazur, A.; Bucholc, M.; Koen, E.; Szymanska, K.; Zmienko, A.; Krzywinska, E.; Wawer, I.; McLoughlin, F.; Ruszkowski, D.; et al. SNF1-Related Protein Kinases Type 2 Are Involved in Plant Responses to Cadmium Stress. *Plant Physiol.* **2012**, *160*, 868–883. [CrossRef]

41. Nakagami, H.; Soukupová, H.; Schikora, A.; Zárský, V.; Hirt, H. A Mitogen-Activated Protein Kinase Kinase Kinase Mediates Reactive Oxygen Species Homeostasis in Arabidopsis. *J. Biol. Chem.* **2006**, *281*, 38697–38704. [CrossRef]

42. Yang, L.; Ye, C.; Zhao, Y.; Cheng, X.; Wang, Y.; Jiang, Y.Q.; Yang, B. An Oilseed Rape WRKY-Type Transcription Factor Regulates ROS Accumulation and Leaf Senescence in Nicotiana Benthamiana and *Arabidopsis* through Modulating Transcription of RbohD and RbohF. *Planta* **2018**, *247*, 1323–1338. [CrossRef]

43. Pitzschke, A.; Djamei, A.; Bitton, F.; Hirt, H. A Major Role of the MEKK1-MKK1/2-MPK4 Pathway in ROS Signalling. *Mol. Plant* **2009**, *2*, 120–137. [CrossRef]

44. Ben Rejeb, K.; Benzarti, M.; Debez, A.; Bailly, C.; Savouré, A.; Abdelly, C. NADPH Oxidase-Dependent H_2O_2 Production Is Required for Salt-Induced Antioxidant Defense in *Arabidopsis thaliana*. *J. Plant Physiol.* **2015**, *174*, 5–15. [CrossRef]

45. Ma, L.; Zhang, H.; Sun, L.; Jiao, Y.; Zhang, G.; Miao, C.; Hao, F. NADPH Oxidase AtrbohD and AtrbohF Function in ROS-Dependent Regulation of Na^+/K^+ Homeostasis in *Arabidopsis* under Salt Stress. *J. Exp. Bot.* **2012**, *63*, 305–317. [CrossRef]

46. O'Brien, J.A.; Daudi, A.; Finch, P.; Butt, V.S.; Whitelegge, J.P.; Souda, P.; Ausubel, F.M.; Bolwell, G.P. A Peroxidase-Dependent Apoplastic Oxidative Burst in Cultured *Arabidopsis* Cells Functions in MAMP-Elicited Defense. *Plant Physiol.* **2012**, *158*, 2013–2027. [CrossRef]

47. Daudi, A.; Cheng, Z.; O'Brien, J.A.; Mammarella, N.; Khan, S.; Ausubel, F.M.; Bolwell, G.P. The Apoplastic Oxidative Burst Peroxidase in *Arabidopsis* Is a Major Component of Pattern-Triggered Immunity. *Plant Cell* **2012**, *24*, 275–287. [CrossRef]

48. Xing, Y.; Jia, W.; Zhang, J. AtMEK1 Mediates Stress-Induced Gene Expression of CAT1 Catalase by Triggering H_2O_2 Production in Arabidopsis. *J. Exp. Bot.* **2007**, *58*, 2969–2981. [CrossRef]

49. Frugoli, J.A.; Zhong, H.H.; Nuccio, M.L.; McCourt, P.; McPeek, M.A.; Thomas, T.L.; McClung, C.R. Catalase Is Encoded by a Multigene Family in *Arabidopsis thaliana* (L.) Heynh. *Plant Physiol.* **1996**, *112*, 327–336. [CrossRef]

50. Mhamdi, A.; Queval, G.; Chaouch, S.; Vanderauwera, S.; Van Breusegem, F.; Noctor, G. Catalase Function in Plants: A Focus on *Arabidopsis* Mutants as Stress-Mimic Models. *J. Exp. Bot.* **2010**, *61*, 4197–4220. [CrossRef]

51. Gao, M.; Liu, J.; Bi, D.; Zhang, Z.; Cheng, F.; Chen, S.; Zhang, Y. MEKK1, MKK1/MKK2 and MPK4 Function Together in a Mitogen-Activated Protein Kinase Cascade to Regulate Innate Immunity in Plants. *Cell Res.* **2008**, *18*, 1190–1198. [CrossRef]

52. Asai, S.; Ohta, K.; Yoshioka, H. MAPK Signaling Regulates Nitric Oxide and NADPH Oxidase-Dependent Oxidative Bursts in Nicotiana Benthamiana. *Plant Cell Online* **2008**, *20*, 1390–1406. [CrossRef]

53. Dubiella, U.; Seybold, H.; Durian, G.; Komander, E.; Lassig, R.; Witte, C.-P.; Schulze, W.X.; Romeis, T. Calcium-Dependent Protein Kinase/NADPH Oxidase Activation Circuit Is Required for Rapid Defense Signal Propagation. *Proc. Natl. Acad. Sci. USA* **2013**, *110*, 8744–8749. [CrossRef]

54. Kadota, Y.; Sklenar, J.; Derbyshire, P.; Stransfeld, L.; Asai, S.; Ntoukakis, V.; Jones, J.D.; Shirasu, K.; Menke, F.; Jones, A.; et al. Direct Regulation of the NADPH Oxidase RBOHD by the PRR-Associated Kinase BIK1 during Plant Immunity. *Mol. Cell* **2014**, *54*, 43–55. [CrossRef]

55. Li, L.; Li, M.; Yu, L.; Zhou, Z.; Liang, X.; Liu, Z.; Cai, G.; Gao, L.; Zhang, X.; Wang, Y.; et al. The FLS2-Associated Kinase BIK1 Directly Phosphorylates the NADPH Oxidase RbohD to Control Plant Immunity. *Cell Host Microbe* **2014**, *15*, 329–338. [CrossRef]

56. Monaghan, J.; Matschi, S.; Shorinola, O.; Rovenich, H.; Matei, A.; Segonzac, C.; Malinovsky, F.G.G.; Rathjen, J.P.P.; Maclean, D.; Romeis, T.; et al. The Calcium-Dependent Protein Kinase CPK28 Buffers Plant Immunity and Regulates BIK1 Turnover. *Cell Host Microbe* **2014**, *16*, 605–615. [CrossRef]

57. Monaghan, J.; Matschi, S.; Romeis, T.; Zipfel, C. The Calcium-Dependent Protein Kinase CPK28 Negatively Regulates the BIK1-Mediated PAMP-Induced Calcium Burst. *Plant Signal. Behav.* **2015**, *10*, e1018497. [CrossRef]

58. Liu, Y.; He, C. A Review of Redox Signaling and the Control of MAP Kinase Pathway in Plants. *Redox Biol.* **2017**, *11*, 192–204. [CrossRef]

59. Kawasaki, T.; Yamada, K.; Yoshimura, S.; Yamaguchi, K. Chitin Receptor-Mediated Activation of MAP Kinases and ROS Production in Rice and Arabidopsis. *Plant Signal. Behav.* **2017**, *12*, e1361076. [CrossRef]

60. Zhang, M.; Chiang, Y.H.; Toruño, T.Y.; Lee, D.H.; Ma, M.; Liang, X.; Lal, N.K.; Lemos, M.; Lu, Y.J.; Ma, S.; et al. The MAP4 Kinase SIK1 Ensures Robust Extracellular ROS Burst and Antibacterial Immunity in Plants. *Cell Host Microbe* **2018**, *24*, 379–391.e5. [CrossRef]

61. Zandalinas, S.I.; Mittler, R. ROS-Induced ROS Release in Plant and Animal Cells. *Free Radic. Biol. Med.* **2018**, *122*, 21–27. [CrossRef]

62. Fujii, H.; Verslues, P.E.; Zhu, J.-K. *Arabidopsis* Decuple Mutant Reveals the Importance of SnRK2 Kinases in Osmotic Stress Responses in Vivo. *Proc. Natl. Acad. Sci. USA* **2011**, *108*, 1717–1722. [CrossRef]

63. Drerup, M.M.; Schlücking, K.; Hashimoto, K.; Manishankar, P.; Steinhorst, L.; Kuchitsu, K.; Kudla, J. The Calcineurin B-like Calcium Sensors CBL1 and CBL9 Together with Their Interacting Protein Kinase CIPK26 Regulate the *Arabidopsis* NADPH Oxidase RBOHF. *Mol. Plant* **2013**, *6*, 559–569. [CrossRef]

64. Kimura, S.; Kawarazaki, T.; Nibori, H.; Michikawa, M.; Imai, A.; Kaya, H.; Kuchitsu, K. The CBL-Interacting Protein Kinase CIPK26 Is a Novel Interactor of *Arabidopsis* NADPH Oxidase AtRbohF That Negatively Modulates Its ROS-Producing Activity in a Heterologous Expression System. *J. Biochem.* **2013**, *153*, 191–195. [CrossRef]

65. Takahashi, F.; Mizoguchi, T.; Yoshida, R.; Ichimura, K.; Shinozaki, K. Calmodulin-Dependent Activation of MAP Kinase for ROS Homeostasis in Arabidopsis. *Mol. Cell* **2011**, *41*, 649–660. [CrossRef]

66. Hrabak, E.M.; Chan, C.W.; Gribskov, M.; Harper, J.; Choi, J.; Halford, N.; Kudla, J.; Luan, S.; Nimmo, H.; Sussman, M.; et al. The *Arabidopsis* CDPK-SnRK Superfamily of Protein Kinases. *Plant Physiol.* **2003**, *132*, 666–680. [CrossRef]

67. Xie, Y.J.; Xu, S.; Han, B.; Wu, M.Z.; Yuan, X.X.; Han, Y.; Gu, Q.; Xu, D.K.; Yang, Q.; Shen, W.B. Evidence of *Arabidopsis* Salt Acclimation Induced by Up-Regulation of HY1 and the Regulatory Role of RbohD-Derived Reactive Oxygen Species Synthesis. *Plant J.* **2011**, *66*, 280–292. [CrossRef]

68. Arnaud, D.; Lee, S.; Takebayashi, Y.; Choi, D.; Choi, J.; Sakakibara, H.; Hwang, I. Cytokinin-Mediated Regulation of Reactive Oxygen Species Homeostasis Modulates Stomatal Immunity in Arabidopsis. *Plant Cell* **2017**, *29*, 543–559. [CrossRef]

69. Meng, X.; Xu, J.; He, Y.; Yang, K.-Y.; Mordorski, B.; Liu, Y.; Zhang, S. Phosphorylation of an ERF Transcription Factor by *Arabidopsis* MPK3/MPK6 Regulates Plant Defense Gene Induction and Fungal Resistance. *Plant Cell* **2013**, *25*, 1126–1142. [CrossRef]

70. Sewelam, N.; Kazan, K.; Thomas-Hall, S.R.; Kidd, B.N.; Manners, J.M.; Schenk, P.M. Ethylene Response Factor 6 Is a Regulator of Reactive Oxygen Species Signaling in Arabidopsis. *PLoS ONE* **2013**, *8*, e70289. [CrossRef]

71. Wang, P.; Du, Y.; Zhao, X.; Miao, Y.; Song, C.-P. The MPK6-ERF6-ROS-Responsive Cis-Acting Element7/GCC Box Complex Modulates Oxidative Gene Transcription and the Oxidative Response in Arabidopsis. *Plant Physiol.* **2013**, *161*, 1392–1408. [CrossRef]

72. Adachi, H.; Nakano, T.; Miyagawa, N.; Ishihama, N.; Yoshioka, M.; Katou, Y.; Yaeno, T.; Shirasu, K.; Yoshioka, H. WRKY Transcription Factors Phosphorylated by MAPK Regulate a Plant Immune NADPH Oxidase in Nicotiana Benthamiana. *Plant Cell* **2015**, *27*, 2645–2663. [CrossRef]

73. Ichimura, K.; Mizoguchi, T.; Yoshida, R.; Yuasa, T.; Shinozaki, K. Various Abiotic Stresses Rapidly Activate *Arabidopsis* MAP Kinases ATMPK4 and ATMPK6. *Plant J.* **2000**, *24*, 655–665. [CrossRef]

74. Dubois, M.; Skirycz, A.; Claeys, H.; Maleux, K.; Dhondt, S.; De Bodt, S.; Vanden Bossche, R.; De Milde, L.; Yoshizumi, T.; Matsui, M.; et al. Ethylene Response FACTOR6 Acts as a Central Regulator of Leaf Growth under Water-Limiting Conditions in Arabidopsis. *Plant Physiol.* **2013**, *162*, 319–332. [CrossRef]

75. Dal Santo, S.; Stampfl, H.; Krasensky, J.; Kempa, S.; Gibon, Y.; Petutschnig, E.; Rozhon, W.; Heuck, A.; Clausen, T.; Jonak, C. Stress-Induced GSK3 Regulates the Redox Stress Response by Phosphorylating Glucose-6-Phosphate Dehydrogenase in Arabidopsis. *Plant Cell* **2012**, *24*, 3380–3392. [CrossRef]

76. Zong, X.J.; Li, D.P.D.Q.; Gu, L.K.; Li, D.P.D.Q.; Liu, L.X.; Hu, X.L. Abscisic Acid and Hydrogen Peroxide Induce a Novel Maize Group C MAP Kinase Gene, ZmMPK7, Which Is Responsible for the Removal of Reactive Oxygen Species. *Planta* **2009**, *229*, 485–495. [CrossRef]

77. Xing, Y.; Jia, W.; Zhang, J. AtMKK1 Mediates ABA-Induced CAT1 Expression and H2O2 production via AtMPK6-Coupled Signaling in Arabidopsis. *Plant J.* **2008**, *54*, 440–451. [CrossRef]

78. Verslues, P.E.; Batelli, G.; Grillo, S.; Agius, F.; Kim, Y.-S.; Zhu, J.-K.; Agarwal, M.; Katiyar-Agarwal, S.; Zhu, J.-K. Interaction of SOS2 with Nucleoside Diphosphate Kinase 2 and Catalases Reveals a Point of Connection between Salt Stress and H2O2 Signaling in *Arabidopsis thaliana*. *Mol. Cell. Biol.* **2007**, *27*, 7771–7780. [CrossRef]

79. Galvez-Valdivieso, G.; Fryer, M.J.; Lawson, T.; Slattery, K.; Truman, W.; Smirnoff, N.; Asami, T.; Davies, W.J.; Jones, A.M.; Baker, N.R.; et al. The High Light Response in *Arabidopsis* Involves ABA Signaling between Vascular and Bundle Sheath Cells. *Plant Cell Online* **2009**, *21*, 2143–2162. [CrossRef]

80. Pitzschke, A.; Hirt, H. Disentangling the Complexity of Mitogen-Activated Protein Kinases and Reactive Oxygen Species Signaling. *Plant Physiol.* **2009**, *149*, 606–615. [CrossRef]

81. Hotta, C.T.; Gardner, M.J.; Hubbard, K.E.; Baek, S.J.; Dalchau, N.; Suhita, D.; Dodd, A.N.; Webb, A.A.R. Modulation of Environmental Responses of Plants by Circadian Clocks. *Plant Cell Environ.* **2007**, *33*, 333–349. [CrossRef]

82. Bhardwaj, V.; Meier, S.; Petersen, L.N.; Ingle, R.A.; Roden, L.C. Defence Responses of *Arabidopsis thaliana* to Infection by Pseudomonas Syringae Are Regulated by the Circadian Clock. *PLoS ONE* **2011**, *6*, e26968. [CrossRef] [PubMed]

83. Seung, D.; Risopatron, J.P.M.; Jones, B.J.; Marc, J. Circadian Clock-Dependent Gating in ABA Signalling Networks. *Protoplasma* **2012**, *249*, 445–457. [CrossRef] [PubMed]

84. Lai, A.G.; Doherty, C.J.; Mueller-Roeber, B.; Kay, S.A.; Schippers, J.H.M.; Dijkwel, P.P. Circadian Clock-Associated 1 Regulates ROS Homeostasis and Oxidative Stress Responses. *Proc. Natl. Acad. Sci. USA* **2012**, *109*, 17129–17134. [CrossRef] [PubMed]

85. Li, Z.; Bonaldi, K.; Uribe, F.; Pruneda-Paz, J.L. A Localized Pseudomonas Syringae Infection Triggers Systemic Clock Responses in Arabidopsis. *Curr. Biol.* **2018**, *28*, 630–639.e4. [CrossRef] [PubMed]

86. Greenham, K.; Guadagno, C.R.; Gehan, M.A.; Mockler, T.C.; Weinig, C.; Ewers, B.E.; McClung, C.R. Temporal Network Analysis Identifies Early Physiological and Transcriptomic Indicators of Mild Drought in Brassica Rapa. *Elife* **2017**, *6*, e29655. [CrossRef] [PubMed]

87. Rasul, S.; Dubreuil-Maurizi, C.; Lamotte, O.; Koen, E.; Poinssot, B.; Alcaraz, G.; Wendehenne, D.; Jeandroz, S. Nitric Oxide Production Mediates Oligogalacturonide-Triggered Immunity and Resistance to Botrytis Cinerea in *Arabidopsis thaliana*. *Plant Cell Environ.* **2012**, *35*, 1483–1499. [CrossRef]

88. Srivastava, A.K.; Sablok, G.; Hackenberg, M.; Deshpande, U.; Suprasanna, P. Thiourea Priming Enhances Salt Tolerance through Co-Ordinated Regulation of MicroRNAs and Hormones in Brassica Juncea. *Sci. Rep.* **2017**, *7*, 1–15. [CrossRef] [PubMed]

89. Livak, K.J.; Schmittgen, T.D. Analysis of Relative Gene Expression Data Using Real-Time Quantitative PCR and the 2-$\Delta\Delta$CT Method. *Methods* **2001**, *25*, 402–408. [CrossRef]

90. Czechowski, T. Genome-Wide Identification and Testing of Superior Reference Genes for Transcript Normalization in Arabidopsis. *Plant Physiol.* **2005**, *139*, 5–17. [CrossRef]

91. Remans, T.; Smeets, K.; Opdenakker, K.; Mathijsen, D.; Vangronsveld, J.; Cuypers, A. Normalisation of Real-Time RT-PCR Gene Expression Measurements in *Arabidopsis thaliana* Exposed to Increased Metal Concentrations. *Planta* **2008**, *227*, 1343–1349. [CrossRef] [PubMed]

92. Sambrook, J.; Fritsch, E.F.; Maniatis, T. *Molecular Cloning: A Laboratory Manual*; Cold Spring Harbor Laboratory Press: Cold Spring Harbor, NY, USA, 1989.

93. Wawer, I.; Bucholc, M.; Astier, J.; Anielska-Mazur, A.; Dahan, J.; Kulik, A.; Wysłouch-Cieszynska, A.; Zaręba-Kozioł, M.; Krzywinska, E.; Dadlez, M.; et al. Regulation of *Nicotiana tabacum* Osmotic Stress-Activated Protein Kinase and Its Cellular Partner GAPDH by Nitric Oxide in Response to Salinity. *Biochem. J.* **2010**, *429*, 73–83. [CrossRef] [PubMed]

94. Venisse, J.S.; Gullner, G.; Brisset, M.N. Evidence for the Involvement of an Oxidative Stress in the Initiation of Infection of Pear by Erwinia Amylovora. *Plant Physiol.* **2001**, *125*, 2164–2172. [CrossRef] [PubMed]

95. Polkowska-Kowalczyk, L.; Wielgat, B.; Maciejewska, U. Changes in the Antioxidant Status in Leaves of Solanum Species in Response to Elicitor from Phytophthora Infestans. *J. Plant Physiol.* **2007**, *164*, 1268–1277. [CrossRef] [PubMed]

96. Aebi, H. Catalase in Vitro. *Methods Enzymol.* **1984**, *105*, 121–126. [PubMed]

iTRAQ-Based Protein Profiling and Biochemical Analysis of Two Contrasting Rice Genotypes Revealed their Differential Responses to Salt Stress

Sajid Hussain [1,†], Chunquan Zhu [1,†], Zhigang Bai [1], Jie Huang [1], Lianfeng Zhu [1], Xiaochuang Cao [1], Satyabrata Nanda [1], Saddam Hussain [2], Aamir Riaz [1], Qingduo Liang [1], Liping Wang [1], Yefeng Li [1], Qianyu Jin [1,*] and Junhua Zhang [1,*]

[1]   State Key Laboratory of Rice Biology, China National Rice Research Institute, Hangzhou 310006, Zhejiang, China; sajid_2077uaf@yahoo.com (S.H.); zhuchunquan@caas.cn (C.Z.); baizg1989@163.com (Z.B.); huangjie67179484@163.com (J.H.); zlfnj@163.com (L.Z.); caoxiaochuang@126.com (X.C.); sbn.satyananda@gmail.com (S.N.); aamirriaz33@gmail.com (A.R.); 15550883578@163.com (Q.L.); 664948431@163.com (L.W.); m13067998118@163.com (Y.L.)

[2]   Department of Agronomy, University of Agriculture Faisalabad, Punjab 38000, Pakistan; sadamhussainuaf@gmail.com

*   Correspondence: jinqianyu@caas.cn (Q.J.); zhangjunhua@caas.cn (J.Z.)

†   Equal Contribution.

Abstract: Salt stress is one of the key abiotic stresses causing huge productivity losses in rice. In addition, the differential sensitivity to salinity of different rice genotypes during different growth stages is a major issue in mitigating salt stress in rice. Further, information on quantitative proteomics in rice addressing such an issue is scarce. In the present study, an isobaric tags for relative and absolute quantitation (iTRAQ)-based comparative protein quantification was carried out to investigate the salinity-responsive proteins and related biochemical features of two contrasting rice genotypes—Nipponbare (NPBA, *japonica*) and Liangyoupeijiu (LYP9, *indica*), at the maximum tillering stage. The rice genotypes were exposed to four levels of salinity: 0 (control; CK), 1.5 (low salt stress; LS), 4.5 (moderate salt stress; MS), and 7.5 g of NaCl/kg dry soil (high salt stress, HS). The iTRAQ protein profiling under different salinity conditions identified a total of 5340 proteins with 1% FDR in both rice genotypes. In LYP9, comparisons of LS, MS, and HS compared with CK revealed the up-regulation of 28, 368, and 491 proteins, respectively. On the other hand, in NPBA, 239 and 337 proteins were differentially upregulated in LS and MS compared with CK, respectively. Functional characterization by KEGG and COG, along with the GO enrichment results, suggests that the differentially expressed proteins are mainly involved in regulation of salt stress responses, oxidation-reduction responses, photosynthesis, and carbohydrate metabolism. Biochemical analysis of the rice genotypes revealed that the Na^+ and Cl^- uptake from soil to the leaves via the roots was increased with increasing salt stress levels in both rice genotypes. Further, increasing the salinity levels resulted in increased cell membrane injury in both rice cultivars, however more severely in NPBA. Moreover, the rice root activity was found to be higher in LYP9 roots compared with NPBA under salt stress conditions, suggesting the positive role of rice root activity in mitigating salinity. Overall, the results from the study add further insights into the differential proteome dynamics in two contrasting rice genotypes with respect to salt tolerance, and imply the candidature of LYP9 to be a greater salt tolerant genotype over NPBA.

Keywords: Salt stress; *Oryza sativa*; proteomics; iTRAQ quantification; cell membrane injury; root activity

1. Introduction

To satisfy the food demands of a population of more than nine billion people by 2050, the world's food productivity needs to be increased by 50% above current production [1,2]. The current growth trends of the major food crops, including wheat, rice, maize, and soybean, suggest that crop production will not be sufficient to meet these ever-rising food demands [3]. Further, the occurrence of abiotic stresses owing to climate change is one of the major reasons for the productivity gap [4]. Soil salinity is considered to be a major problem in the productivity of rice (*Oryza sativa* L.) worldwide [4]. Rice is highly sensitive to salt stress; however, the range of sensitivity varies with rice ecotypes, genotypes, and growth stages [5,6]. Salt tolerance in rice is correlated with variations in the translocation of sodium (Na^+) and chloride (Cl^-) ions in the aboveground plant organs, including the shoot and panicles [7–12]. Salinity affects rice physiology and growth by causing osmotic stress, nutrient imbalance, ionic toxicity, oxidative damage, alteration of metabolic processes, reduced cell division, genotoxicity, decline of growth and yield, and even the death of the plant [8,9,13–17]. In rice, salinity tolerance is usually achieved as a result of a cocktail of physiological and genetic reprogramming, including selective ion uptake and exclusion, preferential compartmentation of Na^+, alternation in stomatal closure, reactive oxygen species (ROS) signaling, and expression of salt-stress responsive genes and transcription factors [18–22].

Alterations in physiological and biochemical processes lead to changes in the protein pool in plants. In recent times, proteomic analysis has emerged as a significant molecular technique for the profiling and identification of proteins expressed in response to various abiotic stresses [23]. Isobaric tags for relative and absolute quantitation (iTRAQ)-based protein profiling and analysis has been performed in several crops, including rice [24], maize [25], wheat [26], tomatoes [27], and cotton [23], in response to abiotic stresses. Differential protein expressions in the areal tissues of rice subjected to salt stress have been reported by a few studies [28–30]. However, most of these studies have employed the 2D gel electrophoresis method to quantify the protein dynamics in rice. The 2D gel electrophoresis technique lacks efficiency in identifying the low abundant proteins, including extreme-acidic or basic proteins, proteins with molecular weights <15 kDa or >150 kDa, and hydrophobic proteins [23]. Furthermore, most of these works have been performed using the *japonica* rice genotype "Nipponbare" as the plant material. Therefore, in this study, we explored the proteomic dynamics of rice under salt stress in both *japonica* (Nipponbare, NPBA) and *indica* (Liangyoupeijiu, LYP9) rice genotypes by employing an iTRAQ-based proteomic study.

In the current study, the iTRAQ-based proteomic technique was used to identify the differentially expressed proteins in two rice genotypes of contrasting salt tolerance levels. The *indica* rice LYP9 has a higher salt tolerance level than the *japonica* rice NPBA [13]. Therefore, the proteomic analysis was performed with the aim of elucidating and comparing the effects of salt stress in these rice genotypes. Further, the physiological responses, such as cell membrane injury (CMI) and rice root activity of the NPBA and LYP9 genotypes, were assessed in response to various salt stress levels at the maximum tillering stage. Additionally, the Na^+ and Cl^- uptake from soil to leaf via root under the subjected salt stress levels were determined in both rice genotypes. The results from this study will help us to achieve better insights into the salt stress resistance mechanisms in rice.

2. Results

2.1. Na⁺ and Cl⁻ in the Soil

The soil Na^+ concentrations for LYP9 rice were recorded to be 0.17, 0.95, 1.7, and 2.0 mg·g^{-1} for the control (no salt stress, CK), low salt stress (LS), moderate salt stress (MS), and high salt stress (HS) treatments, respectively. In NPBA, the soil Na^+ was recorded to be 0.18, 1.0, 1.6, and 2.15 mg·g^{-1} for the CK, LS, MS, and HS treatments, respectively. The Na^+ concentration was found to be the highest in the HS treatment for NPBA rice, as most of the rice seedlings died under the HS condition before attaining the maximum tillering stage. Furthermore, the soil Na^+ concentration was lower for the LYP9 rice than the NPBA rice (Table 1). On the other hand, the soil Cl^- concentrations were recorded to be 0.04, 0.59, 2.17, and 2.43 $mg \cdot g^{-1}$ for the CK, LS, MS, and HS treatments in LYP9 rice, respectively. In NPBA, the soil Cl^- was found to be 0.01, 0.66, 1.64, and 3.03 $mg \cdot g^{-1}$ for the CK, LS, MS, and HS treatments, respectively.

Table 1. Differential Na^+ and Cl^- uptake from soil to leaf via root in LYP9 and NPBA under different salt stress levels at rice maximum tillering stage.

Cultivars	Treatments	Na^+ (mg/g)		Cl^- (mg/g)		Na^+ (mg/g)	Cl^- (mg/g)
		Root	Leaf	Root	Leaf	Soil	Soil
LYP9	CK	0.7 ± 0.05d	0.2 ± 0.03c	0.5 ± 0.3d	6.8 ± 0.4d	0.2 ± 0.01d	0.04 ± 0.01e
	LS	1.1 ± 0.03bc	0.5 ± 0.08b	1.6 ± 0.7cd	12.4 ± 1.7bcd	1.0 ± 0.03c	0.6 ± 0.06de
	MS	1.5 ± 0.04b	0.8 ± 0.10a	7.9 ± 1.3ab	17.6 ± 2.4ab	1.7 ± 0.04b	2.2 ± 0.06bc
	HS	1.6 ± 0.09a	0.9 ± 0.11a	9.5 ± 1.6a	19.1 ± 2.9a	2.0 ± 0.07a	2.4 ± 0.33b
NPBA	CK	0.7 ± 0.03d	0.15 ± 0.01cd	1.0 ± 0.3d	9.9 ± 0.9cd	0.2 ± 0.01d	0.01 ± 0.01e
	LS	1.0 ± 0.07c	0.3 ± 0.01c	3.8 ± 0.4c	14.7 ± 2.8abc	1.0 ± 0.02c	0.7 ± 0.08e
	MS	1.3 ± 0.07b	0.9 ± 0.02a	6.3 ± 0.6b	18.7 ± 2.4ab	1.6 ± 0.12b	1.6 ± 0.035d
	HS	-	-	-	-	2.2 ± 0.07a	3.0 ± 0.23a

Values are denoted as mean ± SE ($n = 3$). Values followed by different letters denote significant difference ($p \leq 0.05$) according to LSD test. Abbreviations: control (no salt stress, CK), low salt stress (LS), moderate salt stress (MS), and high salt stress (HS), Liangyoupeijiu (LYP9), Nipponbare (NPBA). The similar lettering within rice genotype shows the significant and different lettering mean non-significance within treatment levels.

2.2. Na^+ and Cl^- in the Rice Plants

The concentration of Na^+ was found to increase in rice in proportion to rice growth. At the time of rice transplanting, the Na^+ concentration in the LYP9 and NPBA roots was 0.44 and 0.37 $mg \cdot g^{-1}$, respectively. However, at the maximum tillering stage, Na^+ concentrations in rice roots was increased in both rice genotypes, with the increase in subjected salt stress levels. In LYP9 rice, LS, MS, and HS levels of salt stress resulted in the increase of Na^+ concentrations in rice roots amounting to 67.2%, 126.9%, and 138.8%, respectively, as compared with the CK treatment. Similarly, in NPBA rice, Na^+ concentration in the roots was increased by 42.9% for LS and 128.6% for MS as compared with the CK treatment. However, the NPBA rice could not survive under HS salinity conditions. These results indicated that the uptake of Na^+ is higher in rice in the maximum tillering stage as compared to the seedling stage (Table 1). Similar proportions were observed for Na^+ concentration in rice leaves, where the Na^+ concentrations were found to be increased by 163.2%, 305.3%, and 357.9% under LS, MS, and HS conditions, respectively, as compared with the CK condition in LYP9 rice, and by 86.7% and 480% under LS and MS conditions, respectively, as compared with the CK condition in NPBA rice (Table 1). The Na^+ uptake from root to shoot was found to be higher in LYP9 than NPBA. These results suggest that LYP9 has an enhanced ability to uptake Na^+ in the plant parts than compared to NPBA, which might aid in improved salt tolerance in LYP9 compared with NPBA. Likewise, at the maximum tillering stage, the Cl^- uptake by the rice roots and leaves was increased with the increase in the salt stress levels (Table 1). Moreover, these increases in the Cl^- ion uptakes were found to be higher in LYP9 leaves and roots than those of NPBA.

2.3. Cell Membrane Injury (CMI) in Rice Flag Leaves

Evaluations of cell membrane injury (CMI) in both LYP9 and NPBA rice revealed that salt concentrations and CMI are directly proportional, where higher salt concentrations cause severe cell membrane damage. The CMI was found to be higher in the HS condition as compared with MS, LS, and CK conditions in both rice cultivars (Figure 1). CMI was recorded as 5% for CK, 6.7% for LS, 7% for MS, and 15.2% for HS in LYP9. However, CMI in NPBA was recorded as 9.8% for CK, 10.6% for LS, and 11.9% for MS. Compared with the control (CK), the CMI in the LYP9 rice cultivar was increased by 34%, 40%, and 204% under LS, MS, and HS, respectively. On the other hand, CMI was increased by 8.1% (LS), and 21.4% (MS) in the NPBA rice, whilst rice seedlings died under HS conditions before reaching the maximum tillering stage in this genotype of rice (Figure 2). These results strongly suggest that salt stress negatively affects the cell membrane stability, and cell membrane integrity was found to be higher in LYP9 as compared with NPBA. Collectively, these results indicated that LYP9 is more tolerant to salt stress than NPBA.

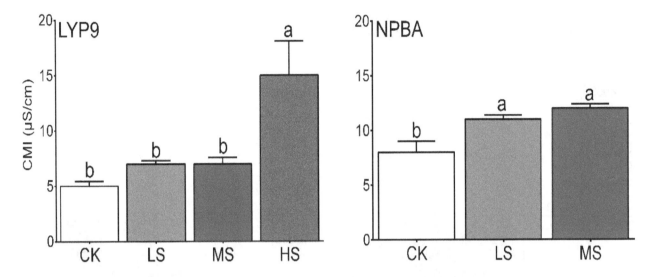

Figure 1. Evaluation of cell membrane injury under the subjected salt stress in LYP9 and NPBA. Bars denoted mean values ± SE ($n = 3$). Values followed by different letters denote significant difference ($p \leq 0.05$) according to LSD test. The similar lettering within rice genotype shows the significant and different lettering mean non-significance within treatment levels.

CK-LYP9   LS-LYP9   MS-LYP9   HS-LYP9   CK-NPBA   LS-NPBA   MS-NPBA   HS-NPBA

Figure 2. Effects of different levels of salt stress on the rice growth at the early stage in both LYP9 and NPBA.

2.4. Rice Root Activity

High root activity is an indicator of resistance against stress [31]. Rice root activity was increased by 2.1% for LS, 50.2% for MS, and 173.7% for HS as compared with CK in LYP9. In the case of NPBA, the rice root activity was decreased by 3.3% for LS, while it increased by 111.4% for MS, as compared to CK. In this study, the rice root activity was higher in LYP9 compared with NPBA under various salt stress levels, inferring the role of root activity in salt tolerance (Figure 3).

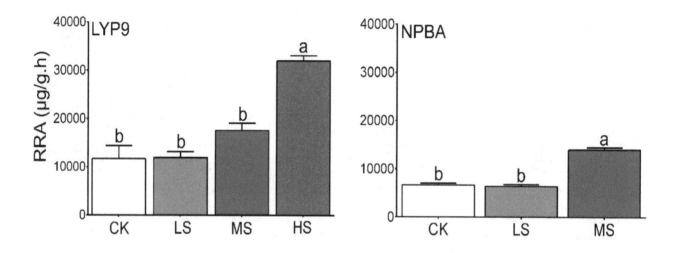

Figure 3. Rice root activity under different salt stress in LYP9 and NPBA. Bars denoted mean values ± SE (*n* = 3). Bars denoted mean values ±SE (*n* = 3). Values followed by different letters denote significant difference (*p* ≤ 05) according to LSD test. The similar lettering within rice genotype shows the significant and different lettering mean non-significance within treatment levels.

2.5. iTRAQ-Based Protein Identification at the Rice Maximum Tillering Stage

Quantitative proteomic analysis of three leaf samples (CK, LS, and MS) from NPBA rice and four leaf samples (CK, LS, MS, and HS) from LYP9 rice were performed using the iTRAQ method. In total, 5340 proteins were identified with 1% FDR (Table 2). In LYP9, 28, 368, and 491 proteins were found to be up-regulated under LS, MS, and HS treatments, respectively, as compared with the CK treatment. On the other hand, in NPBA, 239 and 337 up-regulated proteins were detected under the LS and MS treatments as compared with the CK treatment (Table 3). The longest length of enriched peptides was 7 to 18, with the mass error below 0.025 to 1.00 and with a high performing Pearson correlation coefficient with repeated samples, showing a high quality of the mass spectroscopy data and sample preparation. Proteins with a 1.2 fold change and Q-value of >0.05 were considered as differentially expressed proteins.

Table 2. Overview of the total protein identification in both rice genotypes.

Total Spectra	Spectra	Unique Spectra	Peptides	Unique Peptide
402,823	71,146	53,833	21,741	18,899

Table 3. Differentially expressed proteins in NPBA and LYP9 rice under different salt levels with 1.2 fold change and Q-value > 0.05.

Protein ID	NCBI Accession	Protein Name	NPBA			LYP9	
			LS vs. CK	MS vs. CK	LS vs. CK	MS vs. CK	HS vs. CK
Salt responsive							
tr|B9FWE4|B9FWE4_ORYSJ	gi|222636749	Uncharacterized protein	1.516	1.415	0.906	1.234	1.255
tr|A2Y7R4|A2Y7R4_ORYSI	gi|115465579	Malate dehydrogenase	1.393	2	1.014	1.488	1.573
tr|B8BBS3|B8BBS3_ORYSI	gi|115476908	Os08g0478200 protein	1.389	1.593	0.951	1.402	2.706
tr|A2WT84|A2WT84_ORYSI	gi|115438875	Malate dehydrogenase	1.897	2.835	1.027	1.871	2.006
tr|A0A0P0VS15|A0A0P0VS15_ORYSJ	gi|115450217	Nascent polypeptide-associated complex subunit β (Fragment)	2.523	2.558	1.017	1.594	1.384
tr|A2XA10|A2XA10_ORYSI	gi|46805452	Os02g0768600 protein	1.506	2.225	1.071	2.213	2.212
tr|A0A190X658|A0A190X658_ORYSI	gi|115477769	L-isoaspartate methyltransferase	1.575	2.403	0.901	1.591	1.69
sp|Q43008|SODM_ORYSJ	gi|115463191	Superoxide dismutase	1.775	2.06	1.071	1.534	1.828
sp|Q9FE01|APX2_ORYSJ	gi|115474285	Ascorbate peroxidase	1.308	1.26	0.966	1.227	1.119
sp|Q07661|NDK1_ORYSJ	gi|61679782	Nucleoside diphosphate kinase 1	1.295	1.816	0.909	1.068	1.435
sp|Q5N725|ALFC3_ORYSJ	gi|297598143	Fructose-bisphosphate aldolase 3	1.399	1.639	1.023	1.089	1.532
sp|Q7XDC8|MDHC_ORYSJ	gi|115482534	Malate dehydrogenase	1.37	1.749	1.004	1.284	1.523
tr|A2X753|A2X753_ORYSI	gi|115447273	Os02g0612900 protein	1.441	1.552	1.036	1.506	1.597
tr|A2X7X9|A2X7X9_ORYSI	gi|125540544	Putative uncharacterized protein	1.152	1.502	0.882	1.378	1.25
tr|A0A0P0VTX8|A0A0P0VTX8_ORYSJ	gi|108706531	Os03g0182600 protein	0.852	1.876	0.908	0.949	1.367
tr|E0X6V4|E0X6V4_ORYSJ	gi|306415973	Triosephosphate isomerase	1.003	1.232	1.027	1.107	1.256
tr|A2ZAA7|A2ZAA7_ORYSI	gi|115483468	Nucleoside diphosphate kinase	1.197	2.406	0.882	1.08	1.768
tr|Q9ATR3|Q9ATR3_ORYSA	gi|13249140	Glucanase	1.063	2.009	0.876	0.912	1.39
tr|A2ZIH2|A2ZIH2_ORYSI	gi|115487556	Expressed protein	1.048	1.561	0.957	1.204	1.424
tr|B9FV80|B9FV80_ORYSJ	gi|222636335	Peroxidase	0.888	1.812	0.966	1.298	1.814
tr|B8B893|B8B893_ORYSI	gi|218199240	Plasma membrane ATPase	1.404	1.435	0.906	0.716	0.86
tr|A2XA20|A2XA20_ORYSI	gi|115448935	Proteasome subunit β type	0.862	1.109	0.946	1.067	1.059
tr|A2Y628|A2Y628_ORYSI	gi|125552829	Cysteine proteinase inhibitor	0.96	1.775	1.056	1.438	2.205
tr|Q9ZNZ1|Q9ZNZ1_ORYSA	gi|4097938	Beta-1,3-glucanase	0.795	1.711	1.003	0.932	1.619
tr|A2ZCK1|A2ZCK1_ORYSI	gi|148762354	Alcohol dehydrogenase 2	0.63	0.866	1.055	1.019	1.019
sp|A2XFC7|APX1_ORYSI	gi|158512874	L-ascorbate peroxidase 1	1.216	1.343	0.942	1.21	1.327
tr|A2X822|A2X822_ORYSI	gi|125540587	Glutathione peroxidase	0.717	0.617	0.924	1.77	1.567
tr|A2XFD1|A2XFD1_ORYSI	gi|125543402	Putative uncharacterized protein	1.1	1.543	0.943	1.23	1.554
tr|A2YLI3|A2YLI3_ORYSI	gi|115472191	Os07g0495200 protein	1.159	1.821	0.971	1.505	1.818
tr|B8ADI1|B8ADI1_ORYSI	gi|218187601	NADH-cytochrome b5 reductase	0.771	0.72	1.081	2.182	2.316
tr|A2YSB2|A2YSB2_ORYSI	gi|115475275	Os08g0205400 protein	1.587	2.217	0.908	1.597	1.194
tr|B8AY35|B8AY35_ORYSI	gi|218196772	Fructose-bisphosphate aldolase	0.458	0.214	0.964	0.74	1.505
tr|B8AY17|B8AY17_ORYSI	gi|218196757	Putative uncharacterized protein	0.725	0.849	0.996	1.174	1.649
tr|Q9ZNZ1|Q9ZNZ1_ORYSA	gi|4097938	Beta-1,3-glucanase	0.795	1.711	1.003	0.932	1.619
sp|Q941Z0|NQR1_ORYSJ	gi|115442299	Putative uncharacterized protein	0.686	0.766	0.984	0.931	1.369
tr|A2WWV4|A2WWV4_ORYSI	gi|125528336	Putative uncharacterized protein	0.518	0.55	1.031	1.159	1.304

Table 3. *Cont.*

Protein ID	NCBI Accession	Protein Name	NPBA			LYP9	
			LS vs. CK	MS vs. CK	LS vs. CK	MS vs. CK	HS vs. CK
sp|P93438|METK2_ORYSJ	gi|3024122	S-adenosylmethionine synthase	1.282	1.017	1.013	1.226	1.092
tr|A2XUB9|A2XUB9_ORYI	gi|90265194	B0812A04.3 protein	1.074	1.225	1.215	1.186	1.437
tr|A2ZZZ0|A2ZZZ0_ORYSI	gi|125564321	Putative uncharacterized protein	1.01	0.776	0.921	1.202	1.102
tr|B8AEU4|B8AEU4_ORYSI	gi|218191814	Putative uncharacterized protein	0.954	1.212	0.948	0.908	1.134
tr|A0A0P0VTX8|A0A0P0VTX8_ORYSJ	gi|108706531	Os03g0182600 protein	0.852	1.876	0.908	0.949	1.367
tr|Q688M9|Q688M9_ORYSJ	gi|51854423	putative endo-1,31,4-β-D-glucanase	1.16	1.14	0.992	1.111	1.144
tr|B8ATW7|B8ATW7_ORYSJ	gi|115460338	Os04g0602100 protein	1.386	1.494	1.115	1.448	1.482
sp|Q7FAH2|G3PC2_ORYSJ	gi|115459078	Glyceraldehyde-3-phosphate dehydrogenase 2	0.887	1.03	1.004	0.996	1.196
tr|Q0JG30|Q0JG30_ORYSJ	gi|297598314	Os01g046500 protein	0.95	0.844	0.995	0.799	0.959
tr|Q6L5I4|Q6L5I4_ORYSJ	gi|47900421	Putative aldehyde dehydrogenase	0.735	0.737	0.911	1.167	1.008
sp|A2XW22|DHE2_ORYSI	gi|81686712	Glutamate dehydrogenase 2	1.177	1.142	0.912	0.8	1.184
sp|Q7FAY6|RGP2_ORYSJ	gi|115461086	Amylogenin	1.357	1.021	0.776	0.558	0.683
sp|Q259G4|PMM_ORYSJ	gi|115461390	Phosphomannomutase	0.836	1.19	1.007	0.924	1.216
Photosynthesis related							
tr|A2YWS7|A2YWS7_ORYSJ	gi|115477166	Os08g0504500 protein	1.317	2.22	0.883	1.852	1.77
tr|Q2QWM7|Q2QWM7_ORYSJ	gi|108862278	Os12g0190200 protein	1.053	1.535	0.91	1.363	1.279
tr|B8BCC6|B8BCC6_ORYSI	gi|115477246	Os08g0512500 protein	2.311	3.409	1.022	1.551	1.349
tr|A2XZK1|A2XZK1_ORYSI	gi|125550552	Putative uncharacterized protein	1.246	1.015	1.301	2.61	2.439
tr|B8AAX3|B8AAX3_ORYSI	gi|115440559	Os01g0805300 protein	1.418	2.178	0.986	1.943	1.703
tr|Q0D6V8|Q0D6V8_ORYSJ	gi|297607127	Os07g0435300 protein	2.246	3.387	0.982	2.305	2.053
tr|Q7XHS1|Q7XHS1_ORYSJ	gi|115472141	2Fe-2S iron-sulfur cluster protein-like	1.016	1.414	0.936	1.62	1.548
tr|A2X7M2|A2X7M2_ORYSI	gi|115447507	Os02g0638300 protein	1.096	1.611	1.14	1.73	1.825
tr|B0FFP0|B0FFP0_ORYSJ	gi|115470529	Chloroplast 23 kDa polypeptide of PS II (Fragment)	1.319	1.705	0.997	1.747	1.609
tr|Q7M1U9|Q7M1U9_ORYSA	gi|218186547	Photosystem I 9K protein	1.832	3.172	1.027	2.206	2.383
tr|A0A0P0XF80|A0A0P0XF80_ORYSJ	gi|38636895	Os08g0347500 protein	1.642	2.347	0.926	1.756	1.81
tr|Q7M1Y7|Q7M1Y7_ORYSA	gi|164375543	Photosystem II oxygen-evolving complex protein 2 (Fragment)	1.77	2.373	0.989	2.015	1.756
tr|B8AJX7|B8AJX7_ORYSI	gi|115455221	Serine hydroxymethyltransferase	2.24	2.885	1.078	1.525	1.334
tr|B8AY24|B8AY24_ORYSI	gi|218196765	Putative uncharacterized protein	1.288	1.56	1.026	1.639	1.326
sp|Q6Z2T6|CHLP_ORYSJ	gi|297599916	Geranylgeranyl reductase	0.956	1.173	0.973	0.957	1.174
sp|P0C420|PSBH_ORYSA	gi|11466818	Photosystem II reaction center protein H	0.795	0.694	1.029	0.864	1.14
Oxidation reduction responsive							
tr|A3BVS6|A3BVS6_ORYSJ	gi|125600340	Superoxide dismutase	1.512	1.903	0.96	1.618	1.684
sp|Q6H7E4|TRXM1_ORYSJ	gi|115447527	Putative uncharacterized protein	0.941	1.681	1.049	1.75	2.489
sp|Q9SDD6|PRX2F_ORYSJ	gi|115435844	Peroxiredoxin-2F, mitochondrial	1.363	1.772	1.021	1.642	1.687
tr|B7FAE9|B7FAE9_ORYSJ	gi|215769368	Glutathione peroxidase	0.98	1.347	0.965	1.38	1.176
tr|A2YO43|A2YO43_ORYSI	gi|125550744	Peroxidase	1.232	2.046	0.87	0.633	1.271
tr|Q9FTN6|Q9FTN6_ORYSJ	gi|115434034	Os01g0106300 protein	0.732	1.977	0.788	0.606	1.476

Table 3. *Cont.*

Protein ID	NCBI Accession	Protein Name	NPBA		LYP9		
			LS vs. CK	MS vs. CK	LS vs. CK	MS vs. CK	HS vs. CK
tr|A2X2T0|A2X2T0_ORYSI	gi|55700921	Peroxidase	0.775	1.122	0.913	0.85	1.697
tr|O22440|O22440_ORYSA	gi|115474063	Peroxidase	1.763	2.554	0.963	2.051	1.612
tr|A3A7Y3|A3A7Y3_ORYSJ	gi|125582491	Uncharacterized protein	1.099	1.361	1.101	1.555	2.4
tr|B9FL20|B9FL20_ORYSJ	gi|115464801	Uncharacterized protein	1.175	1.416	0.965	1.159	1.356
tr|Q9AS12|Q9AS12_ORYSJ	gi|115436300	Peroxidase	4.654	5.188	0.78	1.948	2.334
tr|B8ATW7|B8ATW7_ORYSI	gi|115460338	Os04g0602100 protein	1.386	1.494	1.115	1.448	1.482
tr|B9FCM4|B9FCM4_ORYSJ	gi|116309795	OSIGBa0148A10.12 protein	2.208	2.05	1.017	1.365	1.123
tr|Q0JB49|Q0JB49_ORYSJ	gi|115459848	Glutathione peroxidase	1.449	1.435	0.933	1.537	1.271
tr|Q43006|Q43006_ORYSA	gi|20286|emb	Peroxidase	4.58	4.923	1.16	1.421	1.233
tr|Q5Z7J7|Q5Z7J7_ORYSJ	gi|55701041	Peroxidase	5.025	5.222	0.802	2.469	2.659
tr|Q25AK7|Q25AK7_ORYSA	gi|90265065	H0510A06.15 protein	1.326	1.047	0.91	1.209	1.022
tr|Q6K4J4|Q6K4J4_ORYSJ	gi|115479691	Peroxidase	1.23	1.049	0.988	1.16	0.919
tr|A2WJQ7|A2WJQ7_ORYSI	gi|115434036	Os01g0106400 protein	0.884	2.12	0.973	1.313	2.057
sp|P41095|RLA0_ORYSI	gi|115474653	60S acidic ribosomal protein	1.312	1.231	1.073	0.882	0.83
sp|B8AUI3|GLO3_ORYSI	gi|115460650	Peroxisomal (S)-2-hydroxy-acid oxidase GLO3	0.627	0.615	1.305	0.833	0.972
tr|A0A0N7KI36|A0A0N7KI36_ORYSJ	gi|55700967	Peroxidase	0.895	0.817	0.934	1.403	1.041
tr|B8B5W7|B8B5W7_ORYSI	gi|218200254	Peroxidase	1.11	1.51	0.996	2.708	1.966
tr|A2WPA1|A2WPA1_ORYSI	gi|125525683	Peroxidase	1.258	1.625	1.07	2.133	3.577
tr|A2ZAA6|A2ZAA6_ORYSI	gi|115483466	Putative peptide methionine sulfoxide reductase	1.121	1.184	0.902	1.739	1.32
tr|A2XVK6|A2XVK6_ORYSI	gi|125549044	Putative uncharacterized protein	0.844	0.93	0.946	1.321	1.19
tr|B9F688|B9F688_ORYSJ	gi|222624472	Uncharacterized protein	2.063	3.091	1.018	2.407	2.351
tr|B8AU10|B8AU10_ORYSI	gi|218194884	Putative uncharacterized protein	1.206	0.747	1.145	1.386	1.226
tr|Q7F1J9|Q7F1J9_ORYSJ	gi|115477368	Os08g0522400 protein	1.225	1.347	1.072	1.309	1.121
sp|Q6K471|FTRC_ORYSJ	gi|75125055	Ferredoxin-thioredoxin reductase	1.28	2.03	0.905	1.598	1.668
tr|A0A0B4U1V7|A0A0B4U1V7_ORYSA	gi|115467518	Aldehyde dehydrogenase ALDH2b	1.178	1.029	1.016	1.006	1.233
sp|Q6AV34|ARGC_ORYSJ	gi|218193315	Probable N-acetyl-gamma-glutamyl-phosphate reductase	1.046	1.075	0.971	0.966	1.237
tr|Q2QV45|Q2QV45_ORYSJ	gi|115487998	70 kDa heat shock protein	1.415	1.387	1.068	1.391	1.254
sp|Q84VG0|CML7_ORYSJ	gi|115474531	Putative uncharacterized protein	1.351	1.579	0.859	1.35	1.392
tr|A2Y8A8|A2Y8A8_ORYSI	gi|115465902	Os06g0104300 protein	0.877	2.193	1.078	1.336	1.654
tr|A0A0P0X7V0|A0A0P0X7V0_ORYSJ	gi|115472943	Os07g0573800 protein (Fragment)	1.61	1.898	0.916	1.161	1.207
tr|B8BAM3|B8BAM3_ORYSJ	gi|115474739	Os08g0139200 protein	1.096	1.539	0.841	0.9	1.207
sp|Q69TY4|PR2E1_ORYSJ	gi|115469028	Putative uncharacterized protein	1.289	1.212	0.944	1.336	1.361
sp|Q8W3D9|PORB_ORYSJ	gi|75248671	Protochlorophyllide reductase B	0.881	1.621	0.891	1.192	2.065
tr|B8AGN1|B8AGN1_ORYSI	gi|115445869	Os02g0328300 protein	1.63	2.925	0.927	1.814	1.808
tr|B9F604|B9F604_ORYSJ	gi|222625905	Uncharacterized protein	1.474	1.82	1.086	1.62	1.613
tr|Q7F229|Q7F229_ORYSJ	gi|115471449	Os07g0260300 protein	0.924	1.084	0.949	1.375	2.104
tr|A6N0B2|A6N0B2_ORYSI	gi|149391329	Mitochondrial formate dehydrogenase 1 (Fragment)	0.993	1.084	0.931	0.99	1.212

Table 3. *Cont.*

Protein ID	NCBI Accession	Protein Name	NPBA			LYP9	
			LS vs. CK	MS vs. CK	LS vs. CK	MS vs. CK	HS vs. CK
sp|Q10L32|MSRB5_ORYSJ	gi|115453111	Putative uncharacterized protein	1.116	1.471	1.004	1.272	1.479
tr|Q941T6|Q941T6_ORYSJ	gi|15408884	Os01g0847700 protein	1.028	0.871	—	1.416	1.504
tr|B8B2F2|B8B2F2_ORYSI	gi|218198209	Formate dehydrogenase	1.014	1.19	—	0.898	1.278
sp|Q7XPL2|HEM6_ORYSI	gi|75232919	OSIGBa0152L12.9 protein	0.993	1.224	—	0.963	1.253
sp|P0C5D4|PRXQ_ORYSI	gi|115466906	Peroxiredoxin Q, chloroplastic	1.215	1.691	—	1.71	2.077
tr|A0A0P0WR9|A0A0P0WWR9_ORYSJ	gi|300681235	Os06g0472000 protein	1.308	1.269	—	1.532	1.593
tr|A2WL79|A2WL79_ORYSJ	gi|125524611	Peroxidase	0.826	0.852	—	1.182	1.233
sp|P37834|PER1_ORYSJ	gi|115464711	Peroxidase	0.702	0.999	—	0.742	1.764
tr|Q01LB1|Q01LB1_ORYSA	gi|115458104	OSJNBa0072K14.5 protein	1.175	1.243	—	1.124	1.221
sp|P0C0L1|APX6_ORYSJ	gi|115487636	Putative uncharacterized protein	1.127	1.213	—	1.282	1.477
sp|Q7X8R5|TRXM2_ORYSJ	gi|115459582	B1011H02.3 protein	1.557	2.198	—	1.577	3.486
tr|B7E4J4|B7E4J4_ORYSJ	gi|215704355	Putative uncharacterized protein	0.853	1.062	—	0.579	1.105
tr|Q7XV08|Q7XV08_ORYSJ	gi|38567882	OSJNBa0036B21.10 protein	1.159	1.382	—	1.084	1.397
Carbohydrate metabolism							
sp|Q8L7J2|BGL06_ORYSJ	gi|218192323	Beta-glucosidase 6	0.177	0.383	1.004	0.424	1.741
sp|Q76BW5|XTH8_ORYSJ	gi|115475445	Xyloglucan endotransglycosylase/hydrolase protein 8	0.953	2.101	0.939	1.074	1.369
tr|Q01JC3|Q01JC3_ORYSA	gi|116310134	Malate dehydrogenase	0.795	0.74	0.995	0.591	0.889
tr|Q0DCB1|Q0DCB1_ORYSJ	gi|115467998	Os06g0356700 protein	0.849	1.073	0.912	1.227	2.764
tr|Q10CU4|Q10CU4_ORYSJ	gi|115455353	GH family 3 N terminal domain containing protein, expressed	0.72	2.799	0.66	0.663	2.234
tr|Q9ZNZ1|Q9ZNZ1_ORYSA	gi|4097938	Beta-1,3-glucanase	0.795	1.711	1.003	0.932	1.619
tr|H2KWT0|H2KWT0_ORYSJ	gi|108630034	HIPL1 protein, putative, expressed	1.106	2.099	0.908	1.231	2.014
tr|B8AIS2|B8AIS2_ORYSI	gi|218191593	Putative uncharacterized protein	0.773	0.837	0.875	1.437	1.452
sp|Q0INM3|BGA15_ORYSJ	gi|115488372	Beta-galactosidase 15	1.348	1.602	0.924	1.187	1.56
tr|B9FWS5|B9FWS5_ORYSJ	gi|222636880	Uncharacterized protein	0.838	1.177	1.122	0.866	1.141
tr|Q0IJG3|Q0IJG3_ORYSJ	gi|297599314	Os01g0946500 protein	0.95	0.844	0.995	0.799	0.959
tr|Q0IOQ9|Q0IOQ9_ORYSJ	gi|115479865	Os09g0487600 protein	0.829	1.294	0.88	1.272	1.594
tr|A2XM08|A2XM08_ORYSI	gi|115455349	GH family 3 N terminal domain containing protein, expressed	0.859	1.124	0.866	0.78	1.387
sp|Q10NX8|BGAL6_ORYSJ	gi|152013362	Beta-galactosidase 6	1.063	1.684	0.938	1.413	1.814
tr|B8AII1|B8AII1_ORYSI	gi|218190145	Putative uncharacterized protein	0.904	1.55	0.954	1.167	1.478
tr|Q01IH0|Q01IH0_ORYSA	gi|116310092	H0502G05.3 protein	0.728	0.783	0.894	1.001	1.193
tr|Q01JK3|Q01JK3_ORYSA	gi|116310050	Aldose 1-epimerase	0.823	1.226	0.939	1.515	1.515
tr|B8BHM7|B8BHM7_ORYSI	gi|10140702	Alpha-galactosidase	0.73	1.32	1.006	1.331	1.544
tr|A229V6|A229V6_ORYSI	gi|125532825	Uncharacterized protein	1.101	2.017	0.876	1.276	1.194
tr|Q0DTS9|Q0DTS9_ORYSJ	gi|297600575	Os03g0227400 protein (Fragment)	1.049	1.226	0.804	1.06	1.306
tr|A2XME9|A2XME9_ORYSI	gi|115455637	Malate dehydrogenase	1.143	1.261	1.151	1.502	1.548
tr|Q6Z8F4|Q6Z8F4_ORYSJ	gi|115448091	Phosphoribulokinase		1.318	1.066	1.144	1.249

Table 3. *Cont.*

Protein ID	NCBI Accession	Protein Name	NPBA		LYP9		
			LS vs. CK	MS vs. CK	LS vs. CK	MS vs. CK	HS vs. CK
tr|A2YIJ5|A2YIJ5_ORYSI	gi|50509727	Os07g0168600 protein	0.779	0.93	0.952	1.118	1.317
sp|Q75I93|BGL07_ORYSJ	gi|115454825	Beta-glucosidase	1.201	1.066	0.95	1.276	1.58
tr|Q7XIV4|Q7XIV4_ORYSJ	gi|115474081	Alpha-galactosidase	0.786	1.367	0.919	1.115	1.491
tr|A3A285|A3A285_ORYSJ	gi|115443693	Uncharacterized protein	0.83	1.101	0.843	1.151	1.262
tr|A0A0P0XVT5|A0A0P0XVT5_ORYSJ	gi|297610712	Alpha-galactosidase (Fragment)	0.72	1.115	0.848	1.16	1.321
tr|B7F946|B7F946_ORYSJ	gi|297605789	Os06g0356800 protein	0.681	1.016	0.7	1.159	2.984
Stress responsive							
tr|Q9AQU0|Q9AQU0_ORYSJ	gi|13486733	Peptidyl-prolyl cis-trans isomerase	1.249	1.825	0.965	1.578	1.772
tr|Q8GTB0|Q8GTB0_ORYSJ	gi|27476086	Putative heat shock 70 KD protein, mitochondrial	1.294	1.354	0.92	1.027	1.208
tr|Q84S20|Q84S20_ORYSJ	gi|28971968	CHP-rich zinc finger protein-like	2.605	2.416	0.869	1.374	1.439
tr|Q5JKK9|Q5JKK9_ORYSJ	gi|115442153	Os01g0940700 protein	1.897	3.948	0.955	1.04	0.959
sp|Q75HQ0|BIP4_ORYSJ	gi|115464027	Heat shock 70 kDa protein BIP4	10	10	1.058	0.8	0.72
tr|Q53NM9|Q53NM9_ORYSJ	gi|115486793	DnaK-type molecular chaperone hsp70-rice	1.87	1.487	1.009	0.821	0.793
tr|Q10NA9|Q10NA9_ORYSJ	gi|115452223	70 kDa heat shock protein	2.198	1.668	1.086	0.907	0.816
sp|Q5VRY1|HSP18_ORYSJ	gi|115434946	17.5 kDa heat shock protein	1.413	3.508	1.045	1.084	1.043
tr|Q6YUA7|Q6YUA7_ORYSJ	gi|115476792	Os08g0464000 protein	1.323	1.3	1.041	0.866	1.034
tr|A2YK26|A2YK26_ORYSI	gi|115471453	Os07g0262200 protein	1.096	1.252	0.994	0.995	1.314
tr|B9FK56|B9FK56_ORYSI	gi|222631026	Uncharacterized protein	1.028	1.106	1.031	1.314	1.25
tr|A2Z3L9|A2Z3L9_ORYSI	gi|115480445	Os09g0541700 protein	1.1	1.218	0.999	1.144	1.342
tr|O82143|O82143_ORYSJ	gi|115451853	26S proteasome regulatory particle	1.138	1.146	0.981	1.293	1.427
tr|Q5ZAV7|Q5ZAV7_ORYSJ	gi|115440349	Os01g0783500 protein	1.066	1.434	1.03	1.71	2.189
tr|A2Y628|A2Y628_ORYSI	gi|125552829	Cysteine proteinase inhibitor	0.96	1.775	1.056	1.438	2.205
Osmotic stress responsive							
tr|A2XHR1|A2XHR1_ORYSI	gi|125544232	Sucrose synthase	0.82	1.102	0.545	0.366	1.005
tr|B8B835|B8B835_ORYSI	gi|115473055	NADH-dehydrogenase	0.992	1.182	0.879	1.282	1.538
tr|Q2RBD1|Q2RBD1_ORYSJ	gi|115483847	Non-specific lipid-transfer protein	0.988	1.244	0.894	1.274	2.009
tr|Q0IQK7|Q0IQK7_ORYSJ	gi|297612544	Non-specific lipid-transfer protein	1.226	2.979	0.78	1.023	2.235
tr|B8B936|B8B936_ORYSI	gi|218201512	Putative uncharacterized protein	0.871	1.281	0.93	0.976	1.51
tr|B8AII1|B8AII1_ORYSI	gi|218190145	Putative uncharacterized protein	0.904	1.55	0.954	1.167	1.478
tr|Q9SNL7|Q9SNL7_ORYSJ	gi|6006382	Putative SAM-protoporphyrin IX methyltransferase	0.935	0.974	0.965	1.05	1.216
sp|Q10LR9|DCUP2_ORYSJ	gi|115452897	Uroporphyrinogen decarboxylase 2	1.265	1.835	0.902	0.95	1.425
tr|A2X8B7|A2X8B7_ORYSI	gi|242062934	2-C-methyl-D-erythritol 2,4-cyclodiphosphate synthase	1.362	1.399	0.746	1.202	1.538
tr|Q2RBD1|Q2RBD1_ORYSJ	gi|115483847	Non-specific lipid-transfer protein	0.988	1.244	0.894	1.274	2.009
Ethylene responsive							
tr|B9G3V3|B9G3V3_ORYSJ	gi|222641669	Uncharacterized protein	1.837	2.313	1.837	1.825	1.982
sp|Q8W3D9|PORB_ORYSJ	gi|75248671	Protochlorophyllide reductase B	0.881	1.621	0.891	1.192	2.065
tr|Q0IQK7|Q0IQK7_ORYSJ	gi|297612544	Non-specific lipid-transfer protein (Fragment)	1.226	2.979	0.78	1.023	2.235
tr|Q2RBD1|Q2RBD1_ORYSJ	gi|115483847	Non-specific lipid-transfer protein	0.988	1.244	0.894	1.274	2.009

Table 3. *Cont.*

Protein ID	NCBI Accession	Protein Name	NPBA		LYP9		
			LS vs. CK	MS vs. CK	LS vs. CK	MS vs. CK	HS vs. CK
Metabolic responsive							
tr|Q0D572|Q0D572_ORYSJ	gi|297607511	Os07g0577300 protein	1.28	1.719	1.105	0.899	2.422
tr|A2YIJ5|A2YIJ5_ORYSI	gi|50509727	Os07g0168600 protein	0.779	0.93	0.952	1.118	1.317
tr|B9F240|B9F240_ORYSJ	gi|222622048	Uncharacterized protein	0.739	1.149	1.372	1.24	1.422
tr|B9F7T1|B9F7T1_ORYSJ	gi|222624734	Uncharacterized protein	1.389	1.083	1.317	0.854	0.954

2.6. Identification of Differential Expressive Proteins in LYP9 and NPBA Subjected to Different Salt Stress Levels

From the iTRAQ-based identified proteins in both rice genotypes, the proteins that showed a relative abundance of >1.2 fold or <0.8 fold in the salt stressed plants, as compared to the control, were considered to be differential expressive proteins (DEPs). In LYP9 rice, 1927 DEPs were identified under various salt levels. For instance, 93 (28 up-regulated, 65 down-regulated) DEPs were identified in the LS condition, 782 (368 up-regulated, 414 down-regulated) DEPs were identified in the MS condition, and 1052 (561 up-regulated, 491 down-regulated) DEPs were identified in the HS plants, as compared to the control (Figure 4A). On the other hand, 1154 DEPs were identified in the NPBA rice under the applied salt stress levels. Briefly, 432 (239 up-regulated, 193 down-regulated) DEPs were identified in the LS condition and 722 (385 up-regulated, 337 down-regulated) DEPs were identified in the MS plants, as compared with the control (Figure 4B). Identification of the DEPs in both rice genotypes indicated that, with an increase in the salt levels, the number of DEPs was also increased in both rice types. Further, under LS stress levels, the number of DEPs was significantly less in the salt tolerant LYP9 genotype than in the salt sensitive NBPA rice.

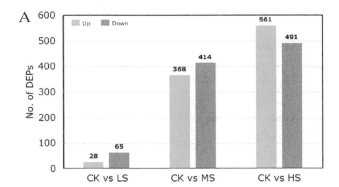 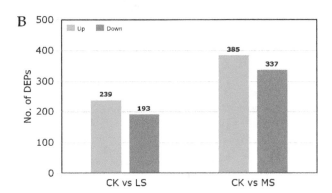

Figure 4. Identification of the differential expressive proteins (DEPs). (**A**) DEPs in LYP9 rice under various salt stress levels as compared with the control plants. (**B**) DEPs in NPBA under various salt stress levels as compared with the control plants. CK: control, LS: low salt, MS: medium salt, HS: high salt.

2.7. Gene Ontology (GO) and Kyoto Encyclopedia of Genes and Genomes (KEGG) Enrichment of the DEPs

To deduce the functionality and biological processes associated with the identified DEPs in the rice genotypes, GO analysis, Clusters of Orthologous Group (COG) annotations, and Kyoto Encyclopedia of Genes and Genomes (KEGG) enrichments were performed. The GO analysis revealed that the identified DEPs were associated with different molecular and biological processes (Figure 5A). Most of the identified DEPs in both rice genotypes were involved in cellular and metabolic processes (biological process). At the molecular level, most of the identified DEPs were involved in catalytic activity, binding, transporter and carrier activity, and structural molecule activity. Similarly, at the cellular component level, the identified DEPs were linked to the cell (membrane and cytoplasm) and organelles. In addition to that, COG analysis of the DEPs grouped them into 24 specific categories on the basis of their functional annotations (Figure 5B). Most of the DEPs were clustered in the "general functional prediction only" category, whereas the post-translational modifications, translation, energy production, carbohydrate metabolism, and amino acid metabolism clusters were found to be the other abundant ones. Altogether, these results suggest that, under salt stress in the rice, salt-responsive proteins might be involved in different metabolic and cellular processes and localize in different cell parts and organelles.

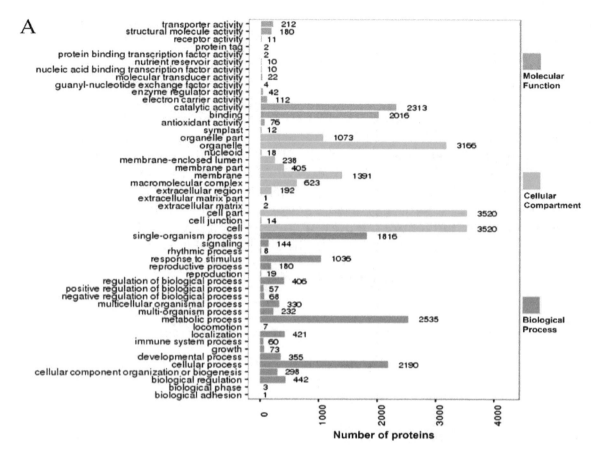

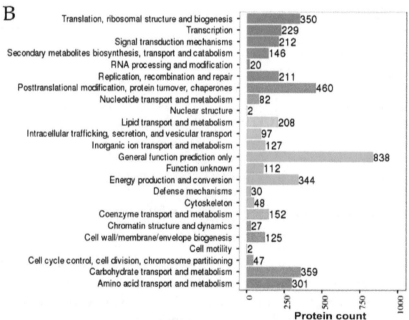

Figure 5. Gene ontology (GO) and Clusters of Orthologous Group (COG) analysis of the differentially responsive proteins in response to salt stress. (**A**) The distribution of number of differentially responsive proteins alongside their corresponding GO terms. Different colors represent different GO categories. (**B**) The distribution of number of differentially responsive proteins alongside their different functions as annotated by COG analysis.

In addition, the KEGG enrichment of the identified DEPs in both rice genotypes revealed their functionality as per the associated pathways. The KEGG pathways, including the metabolic pathway, oxidative phospohorylation, photosynthesis, lysine degradation, glyoxylate metabolism, carbon fixation, photosynthesis-antenna proteins, chlorophyll metabolism, pyruvate metabolism, and ribosomes were found to be the top 10 annotated pathways for the DEPs (Figure 6). From these, the metabolic pathways were found to be the primary enriched pathways in both the rice genotypes. Moreover, analysis of the detail of the KEGG enrichments and associated GO terms revealed that DEPs involved in the salt stress response, redox reactions, photosynthesis, and osmotic stress response were the most abundant in the rice genotypes (Figure 7). For instance, in LYP9, 41 salt-responsive proteins were found to be upregulated under various salt levels, whereas 26 upregulated DEPs were found in the NPBA rice. Similarly, 24 DEPs associated with carbohydrate metabolism were found to be upregulated in LYP9 rice, while 16 DEPs involved with carbohydrate metabolism were found to be upregulated in NPBA. The DEPs from both rice genotypes, with their corresponding fold changes as compared to the controls and their associated physiological pathways, are listed in Table 3. In addition, prediction of the subcellular localizations of the identified DEPs in both the rice genotypes revealed that most of the DEPs localize in the cytoplasm and chloroplasts (Figure 8).

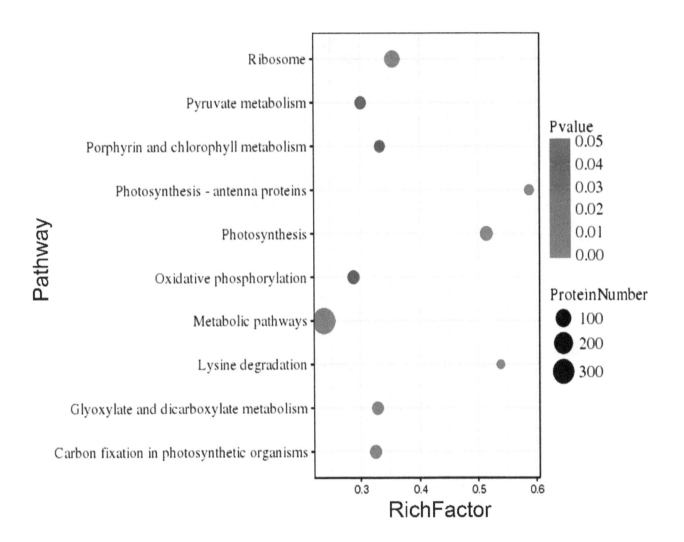

Figure 6. Top 10 pathway enrichments of the identified DEPs in LYP9 and NPBA by KEGG analysis. The corresponding pathways are listed on the Y-axis and the Rich factor values are mentioned along X-axis. Different sized dots represent the distribution of DEPs for a corresponding pathway, whereas, their color represents the *p* value.

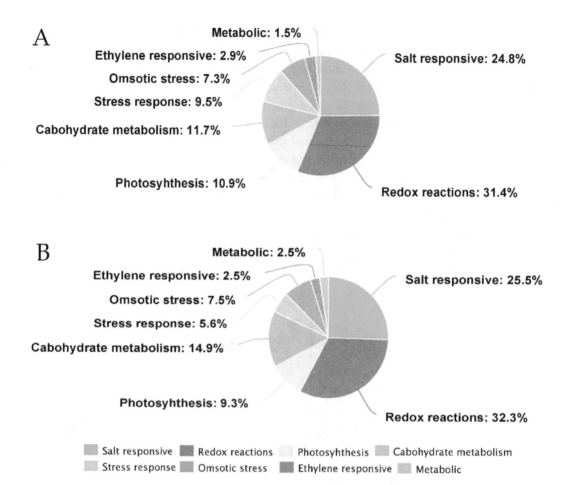

Figure 7. The major pathway annotations of the identified DEPs in LYP9 and NPBA rice. (**A**) Different pathways and their annotated DEP percentages in NPBA rice. (**B**) Different pathways and their annotated DEP percentages in LYP9.

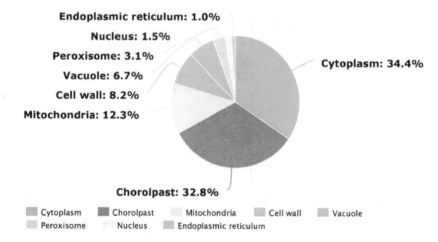

Figure 8. The predicted subcellular localization and compartmentation of the identified DEPs in LYP9 and NPBA.

3. Discussion

3.1. Biochemical Responses of Rice Plants to Salt Stress

Salt stress is a major concern in agriculture, affecting crop productivity across the world. Nutrient imbalance, due to the competition of Na^+ and Cl^- with other nutrients, including potassium (K^+), calcium (Ca^{2+}), and nitrate (NO^{3-}) ions, is a result of salt stress that compromises normal plant growth and development [8–12,32]. In addition, salt stress induces early leave-senescence and a decrease in photosynthesis area [33]. Moreover, osmotic imbalance, poor leaf growth, high CMI, and decreased root activity are associated with the typical salt stress responses in plants [31]. In the current study, the subjection of salt stress negatively affected rice growth in the early stages. All four levels of applied salt stress to both rice cultivars resulted in compromised growth parameters along with CMI. The degree of CMI was found to be higher in NPBA as compared with LYP9, suggesting LYP9 has a higher salt tolerance capacity than NPBA (Figure 1). Further, high rice root activity is usually associated with the interaction of the root with rhizosphere soil and the microbial environment [34], changes in physico-chemical status [35], and plant growth [36]. Further, by enhancing the root activity, plants cope better under an unfavorable environment [34] (Figure 3). In this study, the salt tolerance levels of LYP9 were found to be much higher than those of NPBA at high salt conditions (HS). LYP9 plants could survive by significantly increasing their root activities, whereas none of NPBA plants could survive at the same salt concentrations (Figure 2).

3.2. Proteomic Analysis in the Rice Genotypes Under Salt Stress

Both transcriptomic and proteomic dynamics occurring when subjected to salt stress have already been reported in several plants [37]. Further, the availability of substantial sequential information on rice has paved the way for the use of analytical proteomic studies, including iTRAQ analysis. In this study, iTRAQ-based protein identifications in LYP9 and NPBA cultivars revealed their proteome dynamics in response to salt stress. The comparative analysis of the total of identified proteins (5340) revealed that 93, 782, and 1052 proteins were differentially regulated in LYP9 as compared to the control (CK) under LS, MS, and HS salt stress conditions, respectively. On the other hand, in NPBA, 432 and 722 differentially expressed proteins were found as compared to CK under LS and MS salt stress conditions, respectively (Table 3). These results suggest that the numbers of identified proteins are in direct proportion to the increasing salt stress levels. In addition, the finding of increased numbers of differentially expressed proteins in between LS and MS in both cultivars, and in between LS and MS, and MS and HS in LYP9, further strengthens the proposed proportional relationship between differential protein expression and salt stress levels. Moreover, using the iTRAQ identified protein information, we compared the proteins expressed in LYP9 and NPBA, and thereby the biochemical pathways were identified, including salt stress-responsive protein synthesis, redox responses, photosynthesis, and other metabolic processes. Some of these pathways in response to salt stress have been confirmed in some of the previous studies [38,39]; therefore, the functions of the identified DEPs in this study are discussed further below.

The proteome dynamics and the DEPs in NPBA and LYP9 rice genotypes under different salt stress levels were determined by using iTRAQ analysis. Further, to detect and quantify the proteins in the rice genotypes, the high-resolution LC–MS/MS technique was employed. The identified proteins were quantified on automated software called IQuant [40]. Sequences of the identified DEPs were retrieved from the rice protein database based on the GI numbers, and a blastp algorithm was performed against the GO and KEGG databases. GO annotations of the DEPs were performed over three domains—cellular component, molecular function, and biological process—by using R software packages. Likewise, the COGs were delineated by using a PERL scripted pipeline. The pipeline of the iTRAQ-based protein identification and the subsequent bioinformatic characterizations are represented in Figure 9.

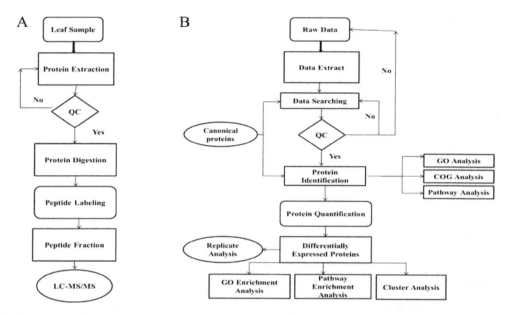

Figure 9. Schematic diagram of the experimental procedures and the complete pipeline for isobaric tags for relative and absolute quantitation (iTRAQ) bioinformatics quantification analysis. (**A**) Steps of the experiment of iTRAQ quantitative proteomics. (**B**) The bioinformatics analysis pipeline for the identified proteins from iTRAQ analysis. All the proteins (FDR < 0.01) proceeded with downstream analysis, including GO, COG, and Kyoto Encyclopedia of Genes and Genomes (KEGG).

3.2.1. Proteins Related to Salt Stress

The comparative proteomics study of both rice genotypes (LYP9 and NPBA) under salt stress revealed new insights into the salt resistance or sensitive mechanisms in rice. In both the rice genotypes, some of the major salt stress-responsive proteins exhibited differential up regulations as compared to the control, including malate dehydrogenase (gi | 115482534), glucanase (gi | 13249140), nascent polypeptide-associated complex (NAC) subunit (gi | 115450217), methyltransferase (gi | 115477769), and chloroplast inorganic pyrophosphatase (gi | 46805452) (Table 3). Plant malate dehydrogenase (MDH) (EC 1.1.1.37) is a member of the oxidoreductase group that catalyzes the inter-conversion of malate and oxaloacetate in a redox reaction [24]. Further, MDH has been shown to play a vital role in regulating the salt stress response in plants [41,42]. Likewise, glucanase and inorganic pyrophosphatases have been associated with salt resistance properties in plants [43,44]. NAC has been reported to be involved in the translocation of newly synthesized proteins from the ribosomes to the endoplasmic reticulum during various physiological conditions, by directly interacting with the signal recognition particles. Further, overexpression of SaβNAC from *Spartina alterniflora* has been reported to enhance the salt tolerance in *Arabidopsis* [45]. In addition, methylation is often utilized by plants under unfavorable conditions as a strategy for gene regulation, protein sorting, and repairs [46]. IbSIMT1, a methyltransferase gene, has been observed to be activated by salt stress, and confers salinity resistance in sweet potato [47]. On the contrary, DEPs associated with salt stress responses, including glutathione peroxidase (GP) (gi | 125540587), fructose-bisphosphate aldolase (FBA) (gi | 218196772), pyruvate dehydrogenase (gi | 125564321), and triosephosphate isomerase (TPI) (gi | 125528336) were found to be significantly upregulated in LYP9, but down regulated in NPBA. Recently, the rice GP gene (*OsGPX3*) has been reported to play a vital role in regulating the salt stress response [48]. Rice plants with silenced *OsGPX3* were found to be highly salt sensitive, confirming the positive role of GP in salinity tolerance. FBA is involved in plant glucose pathways, including glycolysis and gluconeogenesis, and also plays a role in the Calvin cycle [49]. However, the FBA gene has been reported to exhibit induced expressions under salt stress in plants, indicating its role in salt stress. The FBA genes in *Arabidopsis* and *Camellia oleifera* were found to be strongly upregulated under salt stress, conferring salinity tolerance [48,50]. Likewise, the transcription of TPI genes has been reported to become active in rice in response to salt stress [51,52]. The upregulated expression of these salt related proteins in the salt-tolerant genotype LYP9, and their down regulation in the salt-sensitive NPBA, suggests that these genotypes possess a different protein pool in response to salinity. Moreover, the difference in salt tolerance between these two rice genotypes might have resulted due to the differential expression of these key proteins. A functional validation study, such as the Western blot or protein interactions, will add further insights to this hypothesis.

3.2.2. Proteins Related to Redox Reactions

Salt stress in plants induces osmotic imbalances, disrupts ion-homeostasis, and triggers oxidative damage, including the generation of reactive oxygen species (ROS) [53,54]. A fitting response to these adversities caused by salinity stress includes physiological and developmental changes, reprograming of salt-induced gene or proteins, and activation of ROS scavenging pathways [55]. In the current study, the proteomic analysis of LYP9 and NBPA revealed that redox reactions and ROS signaling are involved in the salt stress response in rice. Major enzymes involved in ROS signaling and redox reactions, including peroxidases (POD) (gi I 125525683), superoxide dismutase (SOD) (gi I 125604340), and glutathione s-transferase (GST) (gi I 115459582), were found to be highly expressive in LYP9 and NPBA genotypes under the multiple salt stress levels we investigated (Table 3). Under salt stress, the cell membrane-bound peroxidases like NADPH oxidase and the diamine oxidases present in apoplast are activated, leading to generation of ROS [56,57]. In addition, SOD act as the first line of antioxidant defense in plants under multiple stress responses, and confer enhanced tolerance levels to oxidative stress [54]. Similarly, increased levels of GSTs in response to multiple stimuli have been reported in plants to mitigate oxidative stress [58]. Induced expressions and differential regulation of antioxidant enzymes, including PODs, SODs, and GSTs, have been reported by several studies in rice in response to salt stress [59,60]. Furthermore, comparative proteome analysis has confirmed the involvement of ROS and redox related protein in salt stress in plants, including alfalfa [61], searocket [62], maize [63], barley [64], and wheat [65]. Moreover, as many as 56 DEPs annotated with redox reaction functions were identified in both the rice genotypes under the various salt stress levels, suggesting oxidation and reduction reactions might be the key biochemical changes taking place in rice under salinity.

3.2.3. Proteins Related to Photosynthesis

Photosynthesis is a major physiological process accounting for sustainability and energy production in plants. However, salt stress has adverse effects on the plant photosynthesis process by causing a decrease in the leaf cellular CO_2 levels [7,66]. Additionally, salinity affects the Rubisco activity, retards chlorophyll synthesis, and destabilizes photosynthetic electron transport [66]. The findings from our study revealed that salt stress in rice affects the expression of the proteins involved in the photosynthesis process. These proteins, including the thylakoid lumenal protein (gi I 115477166), psbP domain-containing protein 6 (gi I 115440559), psbP-like protein 1 (gi I 38636895), ferredoxin-thioredoxin reductase (gi I 115447507), photosystem I 9K protein (gi I 218186547), photosystem II oxygen-evolving complex protein 2 (gi I 164375543), and protochlorophyllide reductase B (gi I 75248671), were found to be highly expressed under salt stress conditions (Table 3). Thylakoid luminal protein is required for the functioning of photosystem II (PspB), whereas ferredoxin reductase is a key enzyme that facilitates the conversion of ferredoxin to NADPH in the photosystem I (PSI) complex, and these are also affected by salt stress [67,68]. Moreover, the psbP proteins, thylakoid luminal proteins, and ferredoxin reductase have been reported to be differentially expressed under salt stress [68]. Likewise, differential expression of photosystem proteins was reported in tomatoes in response to salt stress [69]. Similarly, the differential protein expression of protochlorophyllide reductase between the salt stress-induced and control, and its effects on chlorophyll biosynthesis, has been reported in rice [70]. Usually, in salt sensitive plants, salinity causes the down-regulation of photosynthesis proteins, compromising plant sustainability [2,71]. However, the analysis of iTRAQ-based proteomics revealed that the proteins involved in photosynthesis were upregulated in both rice genotypes, which might have aided the rice types to withstand salinity pressures.

3.2.4. Proteins Related to Carbohydrate Metabolism

Apart from being the building blocks in plants, soluble carbohydrates act as osmolytes, and thereby participate in salt tolerance in plants [72]. Besides, the onset of salt stress affects the protein dynamics in plants, resulting in differential protein accumulations [73]. In this study, several carbohydrate metabolism related proteins, including xyloglucan endotransglycosylase/hydrolase protein (XTH) (gi | 115475445), β-glucosidase (gi | 115454825), and polygalacturonase (gi | 115479865), were found to be upregulated in both rice genotypes under various salt stress levels. XTH is known as a cell wall-modifying enzyme, however it also plays a role in salinity resistance responses in plants (Table 3). For instance, the constitutive and heterologous expression of CaXTH3 resulted in increased salt tolerance levels in *Arabidopsis* and tomato plants [74,75]. Similarly, β-glucosidase is a key enzyme in the cellulose hydrolysis process, and has been reported to be involved in the salt stress response. In barley, the activity of an extracellular β-glucosidase was reported to be highly induced in response to salt stress, and cause abscisic acid-glucose conjugate hydrolysis [76]. Further, the overexpression of *Thkel1*, a fungal gene that modulates β-glucosidase activity, improved the salt tolerance levels in transgenic *Arabidopsis* plants [77]. Polygalacturonase, another enzyme capable of hydrolyzing the α-1,4 glycosidic bonds, participates in the salt stress responses in plants. Characterization of the salt stress responses and the associated signal transduction pathways in *Arabidopsis* revealed the elevated transcript accumulation of a polygalacturonase gene (*At1g48100*) under salt stress [78]. However, several proteins related to carbohydrate metabolism, including xylanase inhibitor protein (XIP) (gi | 297605789, gi | 115467998) and MDH (gi | 116310134), were found to be downregulated in the NPBA rice, while being upregulated in the LYP9 rice. MDH is a key enzyme in stress responses and actively participate in the tricarboxylic acid (TCA) cycle [74]. In the current study, upregulated expression of MDH was found in LYP9, however down-regulation in NPBA suggests the inhibition of the TCA cycle in the salt sensitive NPBA, but not in the tolerant LYP9 genotype. Further, OsXIP was reported to be induced under various abiotic stresses, including salt stress, and to take part in the rice defense mechanisms against several biotic and abiotic stresses [79]. Moreover, the induced many-fold expression of the carbohydrate metabolism related proteins in LYP9, but their down regulation in NPBA, indicates that carbohydrate metabolism might be a major physiological process that is affected under salinity in rice, and can show the dynamic changes in protein expression depending on the salt tolerance capacity of a genotype.

3.2.5. Proteins Related to Osmotic Stress

Often, salt stress induces the reduction of cellular water potential, causing osmotic stress to the plant. Osmotic stress responses in plants can be very complex in higher plants, including rice [80]. In this study, 11 osmotic stress related proteins were differentially expressed in both rice genotypes under various salt levels, suggesting salt stress in rice leads to the onset of osmotic stress. For instance, a putative lipid transferase protein (gi | 297612544) identified as a DEP in both the rice genotypes was found to be upregulated under salt stress. The induced expression of *TSW12* and *SiLTP,* coding the lipid transferase proteins in tomato and foxtail millet plants, has been reported under salt stress [80,81]. Conversely, osmotic stress responsive proteins such as sucrose synthase (gi | 125544232) and NADH dehydrogenase (gi | 115473055) were found to exhibit an induced response in LYP9 rice under salt stress, but were not significantly induced in the NPBA rice. Sucrose synthase (Sus) is the major enzyme in sucrose metabolism, however it also plays a part in osmotic stress responses in plants. In *Arabidopsis*, up-regulation of Sus1 has been reported in response to osmotic stresses and water deficit conditions [82]. In addition, involvement of Sus in the osmotic stress response has been reported in Beta vulgaris [83]. On the other hand, NADH dehydrogenase facilitates electron transfer from NADH to the mitochondrial respiratory chain [84]. The up-regulation of NADH dehydrogenase under salt stress indicates an increase in the ATP pool in the LYP9 rice, subsequently aiding in sustainable plant growth and salinity tolerance. However, no induced expression of the same in NPBA suggests that, under salt stress, the ATP pool might decrease, resulting in declining plant growth (Table 3).

3.2.6. Proteins Related to Other Metabolic Processes

Salt stress alters the protein pool that contributes to many metabolic mechanisms, such as stress responses, energy metabolism, and phytohormone synthesis [23,85]. In this study, several DEPs have been identified in the rice genotypes under salt stress, with various physiological and metabolic functions. For instance, putative glucan endo-1,3-β-glucosidase 4 (gi | 297607511) was found to be up-regulated in both rice types under salt stress conditions. Similar findings were reported in cotton plants, where the subjected salt stress caused an increased accumulation of glucan endo-1,3-β-glucosidase [23]. Further, the strong induced response of a putative zinc finger protein (gi | 28971968) was found under salt stress in both rice genotypes. Induced expression of gene finger proteins has been associated with several stresses, including salt stress. Overexpression of a rice zinc-finger protein OsISAP1 in transgenic tobacco resulted in enhanced abiotic stress tolerance levels, including salinity, dehydration, and cold [86]. Recently, OsZFP213 was reported to interact with OsMPK3, conferring salinity tolerance in rice [87]. In addition, many other proteins with annotated functions or which are uncharacterized were found to be differentially regulated at various salt levels in the rice genotypes. Moreover, these results collectively suggest that salinity affects many physiological processes in rice, irrespective of their salt tolerance levels. Furthermore, the protein pool of a salt tolerant and a salt sensitive rice genotype might differ at a specific point of time, which could be the basic reason of their differential salt tolerance responses (Table 3).

4. Materials and Methods

4.1. Plant Material and Growth Conditions

A pot culture experiment was conducted in a greenhouse at China National Rice Research Institute ($39°4'49''$ N, $119°56'11''$ E), Zhejiang Province, China, during the rice growing season (May–November, 2017). Two rice cultivars (origin, China and Japan), Liangyoupiejiu (LYP9, Hybrid, *indica*) and Nipponbare (NPBA, *japonica*) were used as the planting materials. Thirty-day old seedlings were transplanted in pots (45×30 cm) with different salt stress levels and 23 kg air-dried soil. The experimental soil was loamy clay with an average bulk density of 1.12 g/cm, 4.7% organic matter, 0.0864 dS/m EC, and 5.95 pH. Each pot contained six rice seedlings with three replications.

Sodium chloride (NaCl) was used in each pot to develop artificial salinity in soil until the maximum tillering stage of the rice seeding was reached (about 45 days). The treatments were comprised of four NaCl levels: 0 (control, CK), 1.5 g NaCl/kg dry soil (low salt stress, LS), 4.5 g NaCl/kg dry soil (moderate salt stress, MS), and 7.5 g NaCl/kg dry soil (high salt stress, HS). After salinity development, the corresponding EC for these levels was 0.086 dS/m (CK), 1.089 dS/m (LS), 3.20 dS/m (MS), and 4.64 dS/m (HS).

Nitrogen was applied in the form of urea (N: 46%), phosphorous as superphosphate (P_2O_5: 12%), and potassium as potassium sulfate (K_2O: 54%). Urea was used at the rate of 4.02 g/pot in two splits: 50% was applied as the basal dose, and 50% was applied at the tillering stage. Potassium sulfate (3.08 g/pot) was applied in two equal splits, as a basal dose and at the tillering stage, while the whole amount of superphosphate (6.93 g/pot) was applied as a basal dose.

4.2. Soil and Plant Sampling

Rice flag leaves were collected at the maximum tillering stage and stored at -80 °C after being frozen in liquid nitrogen. Plants were collected for measurement of Na^+ and Cl^- contents in the roots and leaves at the maximum tillering stage. Soil samples were collected at the transplanting stage and at the maximum tillering stage to check the Na^+ and Cl^- contents in the soil. Five flag leaves with three replicates were collected to measure the cell membrane injury in rice leaves at the maximum tillering stage, while root samples were collected to measure the rice root activity. All these experiments were performed with three independent biological replicates.

4.3. Leaf Proteomics Analysis Pipeline

4.3.1. Protein Extraction

A total of 1–2 g of plant leaves with 10% PVPP were ground in liquid nitrogen and then sonicated on ice for 5 min in Lysis buffer 3 (8M Urea and 40 mM Tris-HCl containing 1 mM PMSF, 2 mM EDTA, 10 mM DTT, and pH 8.5) with 5 mL of samples. After centrifugation, 5 mL of 10% TCA/acetone with 10 mM DTT were added to the supernatant to precipitate the proteins. The precipitation step was repeated with acetone alone until the supernatant became colorless. The proteins were air dried and re-suspended in Lysis buffer 3. Ultra-sonication on ice for 5 min was used to improve protein dissolution with the help of Lysis buffer 3. After centrifugation, the supernatant was incubated at 56 °C for 1 h for reduction, and then alkylated by 55 mM iodoacetamide (IAM) in the dark at room temperature for 45 min. Acetone (5 mL) were used to precipitate the proteins and stored at –80 °C. The quality and quantity of the isolated proteins were estimated by performing Bradford assay and SDS-PAGE [88].

4.3.2. Digestion of Proteins and Peptide Labeling

About 100 µg of the protein solution with 8 M urea was diluted four times with 100 mM TEAB. For the digestion of the proteins, Trypsin Gold (Promega, Madison, WI, USA) was used at a ratio of trypsin: protein of 40:1, at 37 °C, and was put into the samples overnight. After the digestion with trypsin, Strata X C18 column (Phenomenex, Torrance, CA, USA) were used to desalt the peptides and vacuum-dry them according to the manufacturer's protocol. For peptide labeling, the peptides were dissolved in 30 µL 0.5 M TEAB. Then, the peptide labeling was performed by an iTRAQ reagent 8-plex kit. The labeled peptides with different reagents were combined and desalted with a Strata X C18 column (Phenomenex), and vacuum-dried.

4.3.3. Peptide Fractionation and HPLC

The peptide fractionations were performed by using a Shimadzu LC-20AB HPLC pump attached to a high pH RP column. About 2 mL of the reassembled peptides with buffer A (5% ACN, 95% H₂O, pH 9.8) was loaded on a 5 µm particulate column (Phenomenex). The flow rate was adjusted to 1 mL/min with a 5% buffer B (5% H₂O, 95% ACN, pH 9.8) gradient for 10 min, with 5–35% buffer B for 40 min, and with 35–95% buffer B for 1 min, to separate the peptides. An incubation of 3 min in 95% buffer B, and for 1 min in 5% buffer B, followed this, before the final equilibration with 5% buffer B. Each peptide fraction was collected at 1 min time intervals, and OD of the eluted fractions were measured at 214 nm. Twenty fractions were pooled together and vacuum dried. Post drying, the fractions were re-suspended in buffer A solution (2% CAN; 0.1% FA in water) individually and centrifuged. Then, the supernatant was collected and loaded onto a C18 trap column with a rate of 5 µL/min by using a LC-20AD nano-HPLC device (Shimadzu, Kyoto, Japan). Peptide elutions were performed afterwards and separated by using an analytical C18 column with an inner diameter of 75 µm. The gradients were run at 300 nL/min starting from 8 to 35% of buffer B (2% H₂O; 0.1% FA in ACN) for 35 min, with an increase up to 60% in 5 min, then were maintained at 80% buffer B for 5 min before returning to 5% in 6 s, with a final equilibration period of 10 min.

4.3.4. Mass Spectrometer Detection

The spectrometric data were acquired using a TripleTOF 5600 System (SCIEX, Framingham, MA, USA) fit to a Nano-Spray III source (SCIEX, Framingham, MA, USA) and a pulled quartz tip-type emitter (New Objectives, Woburn, MA, USA), which was controlled with the franchise software

Analyst v1.6 (AB-SCIEX, Concord, ON, Canada). The MS data procurements were undertaken as per the following conditions: the ion spray voltage was set to 2300 V, the curtain gas was set to 30, the nebulizer gas was set to 15, and the interface heater temperature was 150 °C. High sensitivity mode was used for the whole data acquisition process. The MS1 accumulation time was set to 250 ms, while 350–1500 Da was the allowed mass range. At least 30 product ion scans were collected based on the MS1 survey intensity, exceeding a threshold of 120 counts/s and a 2^+ to 5^+ charge-state. A value of 1/2 peak width was set for the dynamic exclusion. The collision energy was adjusted to all precursor ions for the collision-induced dissociation for the iTRAQ data acquisition, and the Q2 transmission window for 100 Da was at 100%. Three independent biological replicates were included for each sample in the experiment.

4.4. Cell Membrane Injury

Cell membrane injury (CMI) was determined by using flag leaves. Twenty pieces (1 cm diameter) were cut from these flag leaves, and were submerged into 20 mL distilled water (DI) contained in test tubes. The test tubes were kept at 10 °C in an incubator for 24 h. After 24 h, the samples were kept at 25 °C to warm the samples, and the electrical conductivity (C1) of the samples was measured. These samples were then autoclaved for 20 min at 120 °C and the electrical conductivity (C2) was determined again. Cellular injury was determined by using the following formula [89]:

$$Cell membrane injury = \frac{C1}{C2} \times 100 \qquad (1)$$

where, 'C' refers to EC 1 and 2. The experiment was performed with three independent biological replicates.

4.5. Rice Root Activity

Rice root activity was analyzed by the triphenyl tetrazolium chloride (TTC) method [90]. Briefly, rice root samples (0.5 g, root tips) were taken, and 5 mL of the phosphate buffer (pH 7) and 5 mL 0.4% TTC (Vitastain, $C_{19}H_{15}N_4Cl$) were added to keep the root activity alive. The samples were kept in an incubator in the dark at 37 °C for 3 h. After 3 h, the samples were taken out and 1 mL 1 mol/L H_2SO_4 was added to stop the reaction. The rice roots were then removed from the test tubes. These root samples were ground by adding a pinch of silica sand, and mixed with 8 mL ethyle acetate. The extract was transferred to test tubes and a 10 mL final volume was reached by adding ethylene acetate. These samples were analyzed using a spectrophotometer (UV-2600, UV-VIS Spectrophotometer Shimadzu) at 485 nm. The formula used for calculation of root activity is as follows:

$$Root Activty = \frac{C}{W/3} \qquad (2)$$

where C is the concentration of the samples calculated from a standard curve. W is the weight of the root samples. The experiment was performed with three independent biological replicates.

4.6. Na$^+$ Concentration in the Soil and Plants

Na$^+$ was extracted from the soil by ammonium acetate solution using Rihards (1954) method [91]. About 5 g ground (particle size ≤ 2mm) air dried soil was placed in 250 mL plastic bottles and 50 mL ammonium acetate (NH$_4$OAc, 1 mol/L) was added. These bottles were kept on a shaker for 30 min at 120 rpm. After that, the samples were filtered by using filter paper to obtain the soil solution.

Na$^+$ in the plants' parts was extracted by digestion with sulfuric acid (H$_2$SO$_4$) by following Rihards (1954) method [91] with the necessary modifications. About 0.3 g ground (particle size ≤ 2 mm) root and leaf were taken in 50 mL glass tubes and mixed with 5 mL H$_2$SO$_4$. These glass bottles were kept overnight. The samples were put into the fume hood and were incubated at 320 °C for 2 h. After 2 h, hydrogen peroxide solution (H_2O_2) was added drop by drop and the samples were mixed until a whitish or transparent color appeared. Then, the samples were cooled at room temperature before being filtered by using filter paper to get the plant part extracts.

The soil and plant extracts were used to measure the sodium ions (Na^+) by using a flame photometer. The standards used were 0, 2, 4, 6, 8, 10, 15, and 20 mL NaCl. The final soluble sodium (Na) in soil was measured by using the formula:

$$Na\left(\frac{\mu g}{g}\right) = \frac{A \times C}{W} \tag{3}$$

where A is the total volume of the extract (mL), C is the sodium concentration values given by the flame-photometer ($\mu g/mL$), and W is the weight of the air dried soil (g). The experiment was performed with three independent biological replicates.

4.7. Cl⁻ Concentration in the Soil and Plants

About 10 g air dried soil (particle size $\leq$ 2mm) was placed in 250 mL plastic bottles and mixed with 50 mL deionized water. These bottles were transferred onto a shaker and were shaken for 5 min at 180 rpm. The samples were then filtered by using filter paper to obtain the soil solution extract for Cl⁻.

Plant samples weighing approximately 0.1 g were placed in 50 mL glass tubes and mixed with 15 mL deionized water. The tubes were transferred into a hot water bath and kept for 1.5 h. The samples were then diluted with 25 mL deionized water after cooling at room temperature.

The soil and plant extracts were used to measure the chloride (Cl⁻) by using a chloride assay kit (QuantiChrom™ Chloride Assay Kit, 3191 Corporate Place Hayward, CA 94545, USA) following the manufacturer's instructions. The standards used were 0, 10, 20, 30, 40, 60, 80, and 100 mL. The final chloride concentration in the solution was measured by the formula:

$$Chloride = \frac{ODsample - ODblank}{Slop} \times n\left(\frac{mg}{dL}\right) \tag{4}$$

where ODsample is the OD 610 nm values of the samples, and ODblank is the OD 610 nm values of the blanks (water). The experiment was performed with three independent biological replicates.

4.8. Statistical Analysis

The statistical software package IBS SPSS Statistics 19.0 was used for the analyses of data. For evaluating the statistical significance of the biochemical parameters, a one-way ANOVA was employed with LSD at the level of $p = 0.05$. For the iTRAQ-based protein quantification, all identified DEPs were required to satisfy the t-test at $p \leq 0.05$, and with a fold change ratio of >1.2 or <0.8.

5. Conclusions

Using comparative iTRAQ-based protein quantification, the proteome dynamics of LYP9 and NPBA rice were explored in this study. The results from the study suggest that rice cell membrane integrity was inversely correlated and root activity was positively correlated with the concentration of salinity. Furthermore, the physiological processes, including carbohydrate metabolism, redox reactions, and photosynthesis, made significant contributions towards the salt tolerance in rice. The number of differentially expressed proteins—salt responsive proteins in particular—suggested that the protein pool in response to salt stress is different in a salt tolerant compared to a susceptible rice genotype. Finally, the *indica* rice LYP9 showed promising results under the subjected salt stress levels, and can be selected over the *japonica* NPBA for salt tolerance. Further works deciphering the functions of some particular proteins of interest will add new insights into their roles in salt tolerance in rice.

Author Contributions: S.H. (Sajid Hussain), J.Z., and Q.J. conceived and designed research. S.H. (Sajid Hussain), J.Z., and Q.J. conducted experiments. J.H., X.C., Z.B., and L.Z. contributed analytical tools. S.H. and C.Z. analyzed the data. S.H. (Sajid Hussain), S.N., and C.Z. wrote the manuscript. S.N., S.H. (Saddam Hussain), and A.R. revised the manuscript. Q.L., L.W., and Y.L. help in formal analysis. The manuscript has been read and approved by all authors.

Abbreviations

LYP9	Liangyoupeijiu
NPBA	Nipponbare
iTRAQ	Isobaric tags for relative and absolute quantitation
CMI	Cell membrane injury
RRA	Rice root activity
DEPs	Differentially expressed proteins
GO	Gene ontology
KEGG	Kyoto encyclopedia of genes and genomes
PSI	Photosystem I
LS	Low salt stress
MS	Moderate salt stress
HS	High salt stress
COG	Cluster of orthologous groups

References

1. Yamori, W.; Hikosaka, K.; Way, D.A. Temperature response of photosynthesis in C3, C4, and CAM plants: Temperature acclimation and temperature adaptation. *Photosynth. Res.* **2013**, *119*, 101–117. [CrossRef] [PubMed]

2. UNFPA. Linking Population, Poverty and Development. 2014. Available online: http://www.unfpa.org/pds/trends.htm (accessed on 28 January 2019).

3. Ray, D.K.; Mueller, N.D.; West, P.C.; Foley, J.A. Yield trends are insufficient to double global crop production by 2050. *PLoS ONE* **2013**, *8*, e66428. [CrossRef] [PubMed]

4. Nachimuthu, V.V.; Sabariappan, R.; Muthurajan, R.; Kumar, A. *Breeding Rice Varieties for Abiotic Stress Tolerance: Challenges and Opportunities*; Springer: Singapore, 2017.

5. Pushpam, R.; Rangasamy, S.R.S. In vivo response of rice cultivars to salt stress. *J. Ecol.* **2002**, *14*, 177–182.

6. Joseph, B.; Jini, D.; Sujatha, S. Biological and Physiological Perspectives of Specificity in Abiotic Salt Stress Response from Various Rice Plants. *Asian J. Agric. Sci.* **2010**, *2*, 99–105.

7. Hussain, S.; Chu, Z.; Zhigang, B.; Xiaochuang, C.; Lianfeng, Z.; Azhar, H.; Chunquan, Z.; Shah, F.; Allen, B.J.; Junhua, Z.; et al. Effects of 1-Methylcyclopropene on Rice Growth Characteristics and Superior and Inferior Spikelet Development Under Salt Stress. *J. Plant Growth Regul.* **2018**, *37*, 1368–1384. [CrossRef]

8. Islam, T.; Manna, M.; Reddy, M.K. Glutathione peroxidase of Pennisetum glaucum (PgGPx) is a functional Cd21 dependent peroxiredoxin that enhances tolerance against salinity and drought stress. *PLoS ONE* **2015**, *10*, e0143344. [CrossRef] [PubMed]

9. Islam, F.; Yasmeen, T.; Ali, S.; Ali, B.; Farooq, M.A.; Gill, R.A. Priming-induced antioxidative responses in two wheat cultivars under saline stress. *Acta Physiol. Plant.* **2015**, *37*, 153–161. [CrossRef]

10. Islam, F.; Yasmeen, T.; Arif, M.S.; Ali, S.; Ali, B.; Hameed, S.; Zhou, W. Plant growth promoting bacteria confer salt tolerance in Vigna radiata by up-regulating antioxidant defense and biological soil fertility. *Plant Growth Regul.* **2016**, *80*, 23–36. [CrossRef]

11. Islam, F.; Ali, B.; Wang, J.; Farooq, M.A.; Gill, R.A.; Ali, S.; Wang, D.; Zhou, W. Combined herbicide and saline stress differentially modulates hormonal regulation and antioxidant defense system in *Oryza sativa* cultivars. *Plant Physiol. Biochem.* **2016**, *107*, 82–95. [CrossRef] [PubMed]

12. Naeem, M.S.; Jin, Z.L.; Wan, G.L.; Liu, D.; Liu, H.B.; Yoneyama, K.; Zhou, W.J. 5-Aminolevulinic acid improves photosynthetic gas exchange capacity and ion uptake under salinity stress in oilseed rape (*Brassica napus* L.). *Plant Soil* **2010**, *332*, 405–415. [CrossRef]

13. Hussain, S.; Xiaochuang, C.; Chu, Z.; Lianfeng, Z.; Maqsood, A.K.; Sajid, F.; Junhua, Z.; Qianyu, J. Sodium chloride stress during early growth stages altered physiological and growth characteristics of rice. *Chil. J. Agric. Res.* **2018**, *78*, 183–197. [CrossRef]

14. Islam, F.; Ali, S.; Farooq, M.A.; Wang, J.; Gill, R.A.; Zhu, J.; Ali, B.; Zhou, W. Butachlor-induced alterations in ultrastructure, antioxidant, and stress-responsive gene regulations in rice cultivars. *Clean—Soil Air Water* **2017**, *45*, 1500851. [CrossRef]

15. Islam, F.; Farooq, M.A.; Gill, R.A.; Wang, J.; Yang, C.; Ali, B.; Wang, G.X.; Zhou, W. 2,4-D attenuates salinity-induced toxicity by mediating anatomical changes, antioxidant capacity and cation transporters in the roots of rice cultivars. *Sci. Rep.* **2017**, *7*, 10443. [CrossRef] [PubMed]

16. Cui, P.; Liu, H.; Islam, F.; Li, L.; Farooq, M.A.; Ruan, S.; Zhou, W. OsPEX11, a peroxisomal biogenesis factor 11, contributes to salt stress tolerance in *Oryza sativa*. *Front. Plant Sci.* **2016**, *7*, 1357. [CrossRef] [PubMed]

17. Shafi, A.; Chauhan, R.; Gill, T.; Swarnkar, M.K.; Sreenivasulu, Y.; Kumar, S.; Kumar, N.; Shankar, R.; Ahuja, P.S.; Singh, A.K. Expression of SOD and APX genes positively regulates secondary cell wall biosynthesis and promotes plant growth and yield in *Arabidopsis* under salt stress. *Plant Mol. Biol.* **2015**, *87*, 615–631. [CrossRef]

18. Biswas, M.S.; Mano, J.I. Lipid peroxide-derived short-chain carbonyls mediate hydrogen peroxide induced and salt-induced programmed cell death in plants. *Plant Physiol.* **2015**, *168*, 885–898. [CrossRef]

19. Ali, I.; Liu, B.; Farooq, M.A.; Islam, F.; Azizullah, A.; Yu, C.; Su, W.; Gan, Y. Toxicological effects of bisphenol A on growth and antioxidant defense system in Oryza sativa as revealed by ultrastructure analysis. *Ecotoxicol. Environ. Saf.* **2016**, *124*, 277–284. [CrossRef]

20. Ali, I.; Jan, M.; Wakeel, A.; Azizullah, A.; Liu, B.; Islam, F.; Ali, A.; Daud, M.K.; Liu, Y.; Gan, Y. Biochemical responses and ultrastructural changes in ethylene insensitive mutants of *Arabidopsis* thialiana subjected to bisphenol A exposure. *Ecotoxicol. Environ. Saf.* **2017**, *144*, 62–71. [CrossRef]

21. Li, H.; Chang, J.; Chen, H.; Wang, Z.; Gu, X.; Wei, C.; Zhang, Y.; Ma, J.; Yang, J.; Zhang, X. Exogenous Melatonin Confers Salt Stress Tolerance to Watermelon by Improving Photosynthesis and Redox Homeostasis. *Front. Plant Sci.* **2017**, *8*, 295. [CrossRef]

22. Khare, T.; Kumar, V.; Kishor, P.K. Na$^+$ and Cl$^-$ ions show additive effects under NaCl stress on induction of oxidative stress and the responsive antioxidative defense in rice. *Protoplasma* **2015**, *252*, 1149–1165. [CrossRef]

23. Li, W.; Zhao, F.; Fang, W.; Xie, D.; Hou, J.; Yang, X.; Zhao, Y.; Tang, Z.; Nie, L.; Lv, S. Identification of early salt stress responsive proteins in seedling roots of upland cotton (*Gossypium hirsutum* L.) employing iTRAQ-based proteomic technique. *Front. Plant Sci.* **2015**, *6*, 732. [CrossRef] [PubMed]

24. Wang, Z.Q.; Xu, X.Y.; Gong, Q.Q.; Xie, C.; Fan, W.; Yang, J.L.; Lin, Q.S.; Zheng, S.J. Root proteome of rice studied by iTRAQ provides integrated insight into aluminum stress tolerance mechanisms in plants. *J. Proteom.* **2014**, *98*, 189–205. [CrossRef] [PubMed]

25. Hu, X.; Li, N.; Wu, L.; Li, C.; Li, C.; Zhang, L.; Liu, T.; Wang, W. Quantitative iTRAQ-based proteomic analysis of phosphoproteins and ABA regulated phosphoproteins in maize leaves under osmotic stress. *Sci. Rep.* **2015**, *27*, 15626. [CrossRef] [PubMed]

26. Guo, G.; Ge, P.; Ma, C.; Li, X.; Lv, D.; Wang, S.; Ma, W.; Yan, Y. Comparative proteomic analysis of salt response proteins in seedling roots of two wheat varieties. *J. Proteom.* **2012**, *75*, 1867–1885. [CrossRef] [PubMed]

27. Gong, B.; Zhang, C.; Li, X.; Wen, D.; Wang, S.; Shi, Q.; Wang, X. Identification of NaCl and NaHCO3 stress-responsive proteins in tomato roots using iTRAQ-based analysis. *Biochem. Biophys. Res. Commun.* **2014**, *446*, 417–422. [CrossRef] [PubMed]

28. Kim, D.W.; Rakwal, R.; Agrawal, G.K.; Jung, Y.H.; Shibato, J.; Jwa, N.S.; Iwahashi, Y.; Iwahashi, H.; Kim, D.H.; Shim, I.S.; et al. A hydroponic rice seedling culture model system for investigating proteome of salt stress in rice leaf. *Electrophoresis* **2005**, *26*, 4521–4539. [CrossRef] [PubMed]

29. Dooki, A.D.; Mayer-Posner, F.J.; Askari, H.; Zaiee, A.; Salekdeh, G.H. Proteomic responses of rice young panicles to salinity. *Proteomics* **2006**, *6*, 6498–6507. [CrossRef]

30. Lee, D.G.; Kee, W.P.; Jae, Y.A.; Young, G.S.; Jung, K.H.; Hak, Y.K.; Dong, W.B.; Kyung, H.L.; Nam, J.K.; Byung-Hyun, L.; et al. Proteomics analysis of salt-induced leaf proteins in two rice germplasms with different salt sensitivity. *Can. J. Plant Sci.* **2011**, *91*, 337–349. [CrossRef]

31. Zhang, X.; Huang, G.; Bian, X.; Zhao, Q. Effects of root interaction and nitrogen fertilization on the chlorophyll content, root activity, photosynthetic characteristics of intercropped soybean and microbial quantity in the rhizosphere. *Plant Soil Environ.* **2013**, *59*, 80–88. [CrossRef]

32. Reddy, I.N.B.L.; Kim, S.M.; Kim, B.K.; Yoon, I.S.; Kwon, T.R. Identification of rice accessions associated with K1/Na1 ratio and salt tolerance based on physiological and molecular responses. *Rice Sci.* **2017**, *24*, 36–364. [CrossRef]

33. Amirjani, M.R. Effect of salinity stress on growth, sugar content, pigments and enzyme activity of rice. *Int. J. Bot.* **2011**, *7*, 73–81. [CrossRef]

34. Hinsinger, P.; Betencourt, E.; Bernard, L.; Brauman, A.; Plassard, C.; Shen, J.; Tang, X.; Zhang, F. P for two, sharing a scarce resource: Soil phosphorus acquisition in the rhizosphere of intercropped species. *Plant Physiol.* **2011**, *156*, 1078–1086. [CrossRef] [PubMed]

35. Song, Y.N.; Zhang, F.S.; Marschner, P.; Fan, F.L.; Gao, H.M.; Bao, X.G.; Sun, J.H.; Li, L. Effect of intercropping on crop yield and chemical and microbiological properties in rhizosphere of wheat (*Triticum aestivum* L.), maize (*Zea mays* L.), and faba bean (*Vicia faba* L.). *Biol. Fertil. Soils* **2007**, *43*, 565–574. [CrossRef]

36. Zhang, N.N.; Sun, Y.M.; Li, L.; Wang, E.T.; Chen, W.X.; Yuan, H.L. Effects of intercropping and Rhizobium inoculation on yield and rhizosphere bacterial community of faba bean (*Vicia faba* L.). *Biol. Fertil. Soils* **2010**, *46*, 625–639. [CrossRef]

37. Salekdeh, G.H.; Siopongco, J.; Wade, L.J.; Ghareyazie, B.; Bennett, J. A proteomic approach to analyzing drough and salt responsiveness in rice. *Field Crop Res.* **2002**, *76*, 199–219. [CrossRef]

38. Jiang, Q.; Xiaojuan, L.; Fengjuan, N.; Xianjun, S.; Zheng, H.; Hui, Z. iTRAQ-based quantitative proteomic analysis of wheat roots in response to salt stress. *Proteomics* **2017**, *17*, 1600265. [CrossRef]

39. Abbasi, F.M.; Komatsu, S. A proteomic approach to analyze salt-responsive proteins in rice leaf sheath. *Proteomics* **2004**, *4*, 2072–2081. [CrossRef]

40. Wen, B.; Zhou, R.; Feng, Q.; Wang, Q.; Wang, J.; Liu, S. IQuant: An automated pipeline for quantitative proteomics based upon isobaric tags. *Proteomics* **2014**, *14*, 2280–2285. [CrossRef]

41. Eprintsev, A.T.; Fedorina, O.S.; Bessmeltseva, Y.S. Response of the Malate Dehydrogenase System of Maize Mesophyll and Bundle Sheath to Salt Stress. *Russ. J. Plant Physiol.* **2011**, *58*, 448–453. [CrossRef]

42. Zhang, J.; Jia, W.; Yang, J.; Ismail, A.M. Role of ABA in integrating plant responses to drought and salt stresses. *Field Crops Res.* **2006**, *97*, 111–119. [CrossRef]

43. Su, Y.; Wang, Z.; Liu, F.; Li, Z.; Peng, Q.; Guo, J.; Xu, Q.Y. Isolation and Characterization of ScGluD2, a New Sugarcane beta-1,3-Glucanase D Family Gene Induced by Sporisorium scitamineum, ABA, H2O2, NaCl, and CdCl2 Stresses. *Front. Plant Sci.* **2016**, *7*, 1348. [CrossRef] [PubMed]

44. He, R.; Yu, G.; Han, X.; Han, J.; Li, W.; Wang, B.; Huang, S.; Cheng, X. ThPP1 gene, encodes an inorganic pyrophosphatase in Thellungiella halophila, enhanced the tolerance of the transgenic rice to alkali stress. *Plant Cell Rep.* **2017**, *36*, 1929. [CrossRef] [PubMed]

45. Karan, R.; Prasanta, K.S. Overexpression of a nascent polypeptide associated complex gene (SabNAC) of Spartina alterniflora improves tolerance to salinity and drought in transgenic *Arabidopsis*. *Bioch. Biophys. Res. Commun.* **2012**, *424*, 747–752. [CrossRef] [PubMed]

46. Singh, S.; Singh, C.; Tripathi, A.K. A SAM-dependent methyltransferase cotranscribed with arsenate reductase alters resistance to peptidyl transferase center-binding antibiotics in Azospirillum brasilense Sp7. *Appl. Microbiol. Biotechnol.* **2014**, *98*, 4625–4636. [CrossRef] [PubMed]

47. Paiva, A.L.S.; Passaia, G.; Lobo, A.K.M.; Jardim-Messeder, D.; Silveira, J.A.; Margis-Pinheiro, M. Mitochondrial glutathione peroxidase (OsGPX3) has a crucial role in rice protection against salt stress. *Environ. Exp. Bot.* **2018**, *158*, 12–21. [CrossRef]

48. Zeng, Y.; Tan, X.; Zhang, L.; Long, H.; Wang, B.; Li, Z.; Yuan, Z. A fructose-1,6-biphosphate aldolase gene from Camellia oleifera: Molecular characterization and impact on salt stress tolerance. *Mol. Breed.* **2015**, *35*, 1–17. [CrossRef]

49. Lu, W.; Tang, X.; Huo, Y.; Xu, R.; Qi, S.; Huang, J.; Zheng, C.; Wu, C.A. Identification and characterization of fructose 1,6-bisphosphate aldolase genes in *Arabidopsis* reveal a gene family with diverse responses to abiotic stresses. *Gene* **2012**, *503*, 65–74. [CrossRef]

50. Minhas, D.; Grover, A. Transcript levels of genes encoding various glycolytic and fermentation enzymes change in response to abiotic stresses. *Plant Sci.* **1991**, *146*, 41–51. [CrossRef]

51. Sharma, S.; Mustafiz, A.; Singla-Pareek, S.L.; Shankar, S.P.; Sopory, S.K. Characterization of stress and methylglyoxal inducible triose phosphate isomerase (OscTPI) from rice. *Plant Signal. Behav.* **2012**, *7*, 1337–1345. [CrossRef]

52. Zhu, J.K. *Plant Salt Stress*; John Wiley & Sons: Hoboken, NJ, USA, 2007.

53. Habib, S.H.; Kausar, H.; Saud, H.M. Plant growthpromoting Rhizobacteria enhance salinity stress tolerance in Okra through ROS-scavenging enzymes. *Biomed. Res. Int.* **2016**, *2016*, 6284547.

54. Hossain, M.S.; Dietz, K.J. Tuning of Redox Regulatory Mechanisms, Reactive Oxygen Species and Redox Homeostasis under Salinity Stress. *Front. Plant Sci.* **2016**, *7*, 548. [CrossRef] [PubMed]

55. Sharma, P.; Jha, A.B.; Dubey, R.S.; Pessarakl, M. Reactive oxygen species, oxidative damage, and antioxidative defense mechanism in plants under stressful conditions. *J. Bot.* **2012**, *2012*, 217037. [CrossRef]

56. Rejeb, K.B.; Benzarti, M.; Debez, A.; Bailly, C.; Savoure, A.; Abdelly, C. NADPH oxidase-dependent H2O2 production is required for saltinduced antioxidant defense in *Arabidopsis* thaliana. *J. Plant Physiol.* **2015**, *174*, 5–15. [CrossRef] [PubMed]

57. Chen, J.H.; Han-Wei, J.; En-Jung, H.; Hsing-Yu, C.; Ching-Te, C.; Hsu-Liang, H.; Tsan-Piao, L. Drought and Salt Stress Tolerance of an *Arabidopsis* Glutathione S-Transferase U17 Knockout Mutant Are Attributed to the Combined Effect of Glutathione and Abscisic Acid. *Plant Physiol.* **2012**, *158*, 340–351. [CrossRef] [PubMed]

58. Mishra, P.; Kumari, B.; Dubey, R.S. Differential responses of antioxidative defense system to prolonged salinity stress in salt-tolerant and salt-sensitive Indica rice (*Oryza sativa* L.) seedlings. *Protoplasma* **2013**, *250*, 3–19. [CrossRef] [PubMed]

59. Li, C.R.; Liang, D.D.; Li, J.; Duan, Y.B.; Li, H.; Yang, Y.C.; Qin, R.Y.; Li, L.I.; Wei, P.C.; Yang, J.B. Unravelling mitochondrial retrograde regulation in the abiotic stress induction of rice Alternative oxidase 1 gene. *Plant Cell Environ.* **2013**, *36*, 775–788. [CrossRef]

60. Gao, Y.; Cui, Y.; Long, R.; Sun, Y.; Zhang, T.; Yang, Q.; Kang, J. Salt-stress induced proteomic changes of two contrasting alfalfa cultivars during germination stage. *J. Sci. Food Agric.* **2019**, *99*, 1384–1396. [CrossRef]

61. Belghith, I.; Jennifer, S.; Tatjana, H.; Chedly, A.; Hans-Peter, B.; Ahmed, D. Comparative analysis of salt-induced changes in the root proteome of two accessions of the halophyte Cakile maritime. *Plant Physiol. Biochem.* **2018**, *130*, 20–29. [CrossRef]

62. Soares, A.L.C.; Christoph-Martin, G.; Sebastien, C.C. Genotype-Specific Growth and Proteomic Responses of Maize Toward Salt Stress. *Front Plant Sci.* **2018**, *9*, 661. [CrossRef]

63. Fatehi, F.; Abdolhadi, H.; Houshang, A.; Tahereh, B.; Paul, C.S. The proteome response of salt-resistant and salt-sensitive barley genotypes to long-term salinity stress. *Mol. Biol. Rep.* **2012**, *39*, 6387. [CrossRef]

64. Gao, L.; Yan, X.; Li, X.; Guo, G.; Hu, Y.; Ma, W.; Yan, Y. Proteome analysis of wheat leaf under salt stress by two-dimensional difference gel electrophoresis (2D-DIGE). *Phytochemistry* **2011**, *72*, 1180–1191. [CrossRef] [PubMed]

65. Li, B.; Tester, M.; Gilliham, M. Chloride on the move. *Trends Plant Sci.* **2017**, *22*, 236–248. [CrossRef] [PubMed]

66. Moolna, A.; Bowsher, C.G. The physiological importance of photosynthetic ferredoxin NADP+ oxidoreductase (FNR) isoforms in wheat. *J. Exp. Bot.* **2010**, *61*, 2669–2681. [CrossRef] [PubMed]

67. Ngara, R.; Roya, N.; Jonas, B.J.; Ole, N.J.; Bongani, N. Identification and profiling of salinity stress-responsive proteins in Sorghum bicolor seedlings. *J. Proteom.* **2012**, *75*, 4139–4150. [CrossRef]

68. Bai, J.; Yan, Q.; Jinghui, L.; Yuqing, W.; Rula, S.; Na, Z.; Ruizong, J. Proteomic response of oat leaves to long-term salinity stress. *Environ. Sci. Pollut. Res.* **2017**, *24*, 3387. [CrossRef] [PubMed]

69. Chen, S.; Natan, G.; Bruria, H. Proteomic analysis of salt-stressed tomato (*Solanum lycopersicum*) seedlings: Effect of genotype and exogenous application of glycinebetaine. *J. Exp. Bot.* **2009**, *60*, 2005–2019. [CrossRef] [PubMed]

70. Turan, S.; Baishnab, C.T. Salt-stress induced modulation of chlorophyll biosynthesis during de-etiolation of rice seedlings. *Physiol. Plantarum.* **2015**, *153*, 477–491. [CrossRef] [PubMed]

71. Wang, H.; Wang, H.; Shao, H.; Tang, X. Recent advances in utilizing transcription factors to improve plant abiotic stress tolerance by transgenic technology. *Front. Plant Sci.* **2017**, *7*, 4563. [CrossRef] [PubMed]

72. Boriboonkaset, T.; Theerawitaya, C.; Yamada, N.; Pichakum, A.; Supaibulwatana, K.; Cha-um, S.; Takabe, T.; Protoplasma, C.K. Regulation of some carbohydrate metabolism-related genes, starch and soluble sugar contents, photosynthetic activities and yield attributes of two contrasting rice genotypes subjected to salt stress. *Protoplasma* **2013**, *250*, 1157–1167. [CrossRef] [PubMed]

73. Wu, G.; Jin-Long, W.; Rui-Jun, F.; Shan-Jia, L.; Chun-Mei, W. iTRAQ-Based Comparative Proteomic Analysis Provides Insights into Molecular Mechanisms of Salt Tolerance in Sugar Beet (*Beta vulgaris* L.). *Int. J. Mol. Sci.* **2018**, *19*, 3866. [CrossRef]

74. Cho, S.K.; Jee, E.K.; Jong-A, P.; Tae, J.E. Woo Taek KimConstitutive expression of abiotic stress-inducible hot pepper CaXTH3, which encodes a xyloglucan endotransglucosylase/hydrolase homolog, improves drought and salt tolerance in transgenic *Arabidopsis* plants. *FEBS Lett.* **2006**, *580*, 3136–3144. [CrossRef] [PubMed]

75. Choi, J.Y.; Seo, Y.S.; Kim, S.J.; Kim, W.T.; Shin, J.S. Constitutive expression of CaXTH3, a hot pepper xyloglucan endotransglucosylase/hydrolase, enhanced tolerance to salt and drought stresses without phenotypic defects in tomato plants (*Solanum lycopersicum* cv. Dotaerang). *Plant Cell Rep.* **2011**, *30*, 867–877. [CrossRef] [PubMed]

76. Dietz, K.; Sauter, A.; Wichert, K.; Messdaghi, D.; Hartung, W. Extracellular β-glucosidase activity in barley involved in the hydrolysis of ABA glucose conjugate in leaves. *J. Exp. Bot.* **2000**, *51*, 937–944. [CrossRef] [PubMed]

77. Hermosa, R.; Leticia, B.; Emma, K.; Jesus, A.J.; Marta, M.B.; Vicent, A.; Aurelio, G.-C.; Enrique, M.; Carlos, N. The overexpression in *Arabidopsis* thaliana of a Trichoderma harzianum gene that modulates glucosidase activity, and enhances tolerance to salt and osmotic stresses. *J. Plant Physiol.* **2011**, *168*, 1295–1302. [CrossRef] [PubMed]

78. Liu, J.; Renu, S.; Ping, C.; Stephen, H.H. Salt stress responses in *Arabidopsis* utilize a signal transduction pathway related to endoplasmic reticulum stress signaling. *Plant J.* **2007**, *51*, 897–909. [CrossRef]

79. Takaaki, T.; Muneharu, E. Induction of a Novel XIP-Type Xylanase Inhibitor by External Ascorbic Acid Treatment and Differential Expression of XIP-Family Genes in Rice. *Plant Cell Phys.* **2007**, *48*, 700–714.

80. Upadhaya, H.; Sahoo, L.; Panda, S.K. Molecular Physiology of Osmotic Stress in Plants. In *Molecular Stress Physiology of Plants*; Rout, G., Das, A., Eds.; Springer: India, New Delhi, 2013.

81. Torres-Schumann, S.; Godoy, J.A.; Pintor-Toro, J.A. A probable lipid transfer protein gene is induced by NaCl in stems of tomato plants. *Plant Mol. Biol.* **1992**, *18*, 749–757. [CrossRef]

82. Pan, Y.; Li, J.; Jiao, L.; Li, C.; Zhu, D.; Yu, J. A Non-specific Setaria italica Lipid Transfer Protein Gene Plays a Critical Role under Abiotic Stress. *Front. Plant Sci.* **2016**, *7*, 1752. [CrossRef]

83. Dejardin, A.; Sokolov, L.N.; Kleczkowski, L.A. An *Arabidopsis* Stress-Responsive Sucrose Synthase Gene is UP-Regulated by Low Water Potential. In *Photosynthesis: Mechanisms and Effects*; Garab, G., Ed.; Springer: Dordrecht, The Netherlands, 1998.

84. Vastarelli, P.; Moschella, A.; Pacifico, D.; Mandolino, G. Water Stress in Beta vulgaris: Osmotic Adjustment Response and Gene Expression Analysis in ssp. Vulgaris and maritima. *Am. J. Plant Sci.* **2013**, *4*, 11–16. [CrossRef]

85. Sobhanian, H.; Razavizadeh, R.; Nanjo, Y.; Ehsanpour, A.A.; Jazii, F.R.; Motamed, N.; Komatsu, S. Proteome analysis of soybean leaves, hypocotyls and roots under salt stress. *Proteome Sci.* **2010**, *8*, 19. [CrossRef]

86. Sampath, K.I.; Ramgopal, R.S.; Vardhini, B.V. Role of Phytohormones during Salt Stress Tolerance in plants. *Curr. Trends Biotechnol. Pharm.* **2015**, *9*, 334–343.

87. Mukhopadhyay, A.; Shubha, V.; Akhilesh, K.T. Overexpression of a zinc-finger protein gene from rice confers tolerance to cold, dehydration, and salt stress in transgenic tobacco. *Proc. Natl. Acad. Sci. USA* **2004**, *101*, 6309–6314. [CrossRef] [PubMed]

88. Zhang, Z.; Huanhuan, L.; Ce, S.; Qibin, M.; Huaiyu, B.; Kang, C.; Yunyuan, X. A C2H2 zinc-finger protein OsZFP213 interacts with OsMAPK3 to enhance salt tolerance in rice. *J. Plant Phys.* **2018**, *229*, 100–110. [CrossRef] [PubMed]

89. Jamil, M.; Iqbal, W.; Bangash, A.; Rehman, S.; Imran, Q.M.; Rha, E.S. Constitutive Expression of OSC3H33, OSC3H50 AND OSC3H37 Genes in Rice under Salt Stress. *Pak. J. Bot.* **2010**, *42*, 4003–4009.

90. Wang, X.K.; Zhang, W.H.; Hao, Z.B.; Li, X.R.; Zhang, Y.Q.; Wang, S.M. *Principles and Techniques of Plant Physiological Biochemical Experiment*; Higher Education Press: Beijing, China, 2006; pp. 118–119. (In Chinese)

91. U.S. Salinity Laboratory Staff. *Diagnosis and Improvement of Saline and Alkali Soils*; Richards, L.A., Ed.; U.S. Goverment Publishing Office: Washington, DC, USA, 1954.

8

Rice Overexpressing *OsNUC1-S* Reveals Differential Gene Expression Leading to Yield Loss Reduction after Salt Stress at the Booting Stage

Chuthamas Boonchai [1], Thanikarn Udomchalothorn [1,2], Siriporn Sripinyowanich [3], Luca Comai [4], Teerapong Buaboocha [5,6] and Supachitra Chadchawan [1,6,*]

[1] Center of Excellence in Environment and Plant Physiology, Department of Botany, Faculty of Science, Chulalongkorn University, Bangkok 10330, Thailand; Gwang_desu@hotmail.com (C.B.); Thanikarn@sut.ac.th (T.U.)

[2] Surawiwat School, Suranaree University of Technology, Nakhon Ratchasima 30000, Thailand

[3] Faculty of Liberal Arts and Science, Kasetsart University, Kamphaeng Saen Campus, Nakhon Pathom 73140, Thailand; wanich_s@hotmail.com

[4] Department of Plant Biology and Genome Center, University of California Davis, Davis, CA 95616, USA; lcomai@ucdavis.edu

[5] Department of Biochemistry, Faculty of Science, Chulalongkorn University, Bangkok 10330, Thailand; Teerapong.B@chula.ac.th

[6] Omics Science Center, Faculty of Science, Chulalongkorn University, Bangkok 10330, Thailand

[*] Correspondence: Supachitra.C@chula.ac.th or s_chadchawan@hotmail.com

Abstract: Rice nucleolin (OsNUC1), consisting of two isoforms, OsNUC1-L and OsNUC1-S, is a multifunctional protein involved in salt-stress tolerance. Here, *OsNUC1-S*'s function was investigated using transgenic rice lines overexpressing *OsNUC1-S*. Under non-stress conditions, the transgenic lines showed a lower yield, but higher net photosynthesis rates, stomatal conductance, and transpiration rates than wild type only in the second leaves, while in the flag leaves, these parameters were similar among the lines. However, under salt-stress conditions at the booting stage, the higher yields in transgenic lines were detected. Moreover, the gas exchange parameters of the transgenic lines were higher in both flag and second leaves, suggesting a role for *OsNUC1-S* overexpression in photosynthesis adaptation under salt-stress conditions. Moreover, the overexpression lines could maintain light-saturation points under salt-stress conditions, while a decrease in the light-saturation point owing to salt stress was found in wild type. Based on a transcriptome comparison between wild type and a transgenic line, after 3 and 9 days of salt stress, the significantly differentially expressed genes were enriched in the metabolic process of nucleic acid and macromolecule, photosynthesis, water transport, and cellular homeostasis processes, leading to the better performance of photosynthetic processes under salt-stress conditions at the booting stage.

Keywords: RNA binding protein; nucleolin; salt stress; photosynthesis; light saturation point; booting stage; transcriptome

1. Introduction

Rice (*Oryza sativa* L.) is a staple food and a main source of energy for humans, especially in Asia. There are several biotic or abiotic stresses that limit rice growth and yield, such as soil salinity, drought, and soil nutrition [1]. Salinity is a severe abiotic stress worldwide that directly contributes to the economic outcome of agriculturists. It negatively affects plants at both physiological and cellular levels. The plant water absorption is disrupted, leading to reductions in plant growth and development [2]. Moreover, ion toxicity also causes changes in plant metabolism, including the photosynthesis processes and energy production [3]. It directly affects photosynthetic components, including chlorophyll *a*, chlorophyll *b*, and carotenoids, because salt stress increases enzyme activities involved in chlorophyll degradation, which leads to a decrease in chlorophyll levels [4]. Moreover, it can also induce reactive oxygen species (ROS) production, which can trigger protein and lipid damage [5]. The level of plant injury depends on species, developmental stage, age, and also the severity of the salinity.

Rice is strongly affected by salt stress at both the seedling and reproductive stages [6]. These can remarkably affect rice plants that are grown in the paddy field of rain-fed areas containing rock salt underneath, like the northeastern region of Thailand [7], because rice is normally germinated at the beginning of the rainy season, when the salinity is low due to a certain amount of rain. However, when the land is drying from the evaporation due to the lack of rain, the salinity can migrate from the rock salt underneath through the surface [7], causing salt stress at the certain developmental stage of rice. If this occurs at the reproductive stage, it becomes the major effect for plant yield reduction [8].

Nucleolin, a multifunctional protein, is localized in various cellular locations, including the nucleus. It consists of three domains, the N-terminus, containing several acidic stretches, the central region, containing an RNA-binding domain, and the C-terminus, containing a glycine/arginine-rich domain [9]. Because of differences in the numbers of each motif, different functions were found in different species. For example, the *NUC-L1* gene in *Arabidopsis thaliana* is related to its growth and development [10], whereas in pea, this gene is regulated by light [11].

In rice, two forms of Nucleolin1 were found, a longer (*OsNUC1-L*, GenBank Accession No. AK103446) and a shorter (*OsNUC1-S*, GenBank Accession No. AK063918) form. *OsNUC1-S* lacks an N-terminal region, but the cDNA sequences of the other regions are the same as the longer form. In 2013, Sripinyowanich and colleagues studied *OsNUC1-S* overexpression in *Arabidopsis* and found that this form promoted salt tolerance by increasing the number of lateral roots and enhancing root growth. In addition, the overexpression of this gene in rice leads to shoot fresh weight increasing when seedlings are grown under salt-stress conditions [12]. In the reproductive stage, the function of this gene is still unknown. Recently, the overexpression of the *OsNUC1-L* form resulted in an increase in salt tolerance in both rice and *Arabidopsis* by enhancing photosynthesis [13].

Here, we investigate the effects of *OsNUC1-S* overexpression on photosynthetic responses under normal and salt-stress conditions in transgenic rice at the reproductive stage and use a transcriptome analysis to explore the gene expression levels affected by *OsNUC1-S* overexpression. The experiments were performed in flag leaves and second leaves as both leaves have been reported to have a major role in generating carbohydrate resource from the photosynthesis process for seed production [14].

2. Results

2.1. Overexpression of OSNUC1-S Affects Rice Yield

The transgenic rice lines, TOSL1, TOSL2, and TOSL3, with *OsNUC1-S* overexpression were used in these experiments. *OsNUC1* expression was compared among wild type (WT) and the transgenic lines, when grown in the control and salt stress condition at reproductive stage. Significantly higher *OsNUC1* gene expression could be detected in transgenic lines when compared to WT, especially in the salt stress condition (Figure 1).

To investigate if the overexpression of *OsNUC1-S* affected rice productivity, the tiller number per plant, panicle number per plant, panicle length, fertility rate (%), and seed number per plant of the transgenic rice, TOSL1, TOSL2, TOSL3, and WT, were evaluated as shown in Table 1. In the normal grown condition (control), over-expression of *OsNUC1-S* caused the reduction in % fertility, leading to the reduction in seed number per plant. However, under salt stress, the transgenic lines tended to have the higher tiller number per plant, panicle number per plant, panicle length, % fertility, and seed number per plant, except the TOSL2 line that showed similar % fertility and seed number per plant to WT. The reduction percentage of seeds/plants in WT was 64%, while the transgenic lines had the seeds/plant reduction of 17%, 47%, and 27%. These data suggested that there should be some changes in metabolisms in the transgenic lines due to *OsNUC1-S* over-expression.

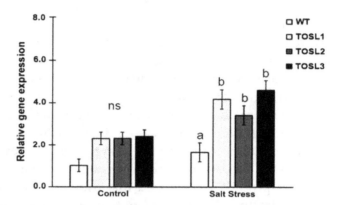

Figure 1. *OsNUC1* gene expression in wild type (WT) and *OsNUC1-S* over-expression lines, TOSL1, TOSL2, and TOSL3, in the control and salt stress condition. Analysis of variance was performed and means were compared with Tukey's range test analysis. The data were presented as the mean $\pm$ SE and a different letter above the bar showed the significant difference in means ($p < 0.05$). ns represents no statistically difference among means.

Table 1. Effects of salt stress at reproductive stage on tiller number per plant, panicle number per plant, panicle length, % fertility, and seed number per plant.

Reproductive Characters	Control *				Salt Stress *			
	WT	TOSL1	TOSL2	TOSL3	WT	TOSL1	TOSL2	TOSL3
Tiller number/pl	5.75 $\pm$ 0.55 ab	5.25 $\pm$ 0.55 abc	7.00 $\pm$ 0.55 a	5.50 $\pm$ 0.55 abc	3.00 $\pm$ 0.55 c	3.75 $\pm$ 0.55 bc	4.50 $\pm$ 0.55 abc	4.00 $\pm$ 0.55 bc
Panicle number/pl	5.25 $\pm$ 0.33 a	3.50 $\pm$ 0.33 b	3.50 $\pm$ 0.33 b	3.00 $\pm$ 0.33 bc	1.75 $\pm$ 0.33 c	2.50 $\pm$ 0.33 bc	2.75 $\pm$ 0.33 bc	2.00 $\pm$ 0.33 bc
Panicle length	10.15 $\pm$ 0.71	9.28 $\pm$ 0.71	8.87 $\pm$ 0.71	8.87 $\pm$ 0.71	7.38 $\pm$ 0.71	8.27 $\pm$ 0.71	9.00 $\pm$ 0.71	9.73 $\pm$ 0.71
% Fertility	65.36 $\pm$ 9.52 a	45.18 $\pm$ 9.52 ab	48.39 $\pm$ 9.52 ab	44.34 $\pm$ 9.52 ab	15.27 $\pm$ 9.52 b	47.03 $\pm$ 9.52 ab	14.17 $\pm$ 9.52 b	36.84 $\pm$ 9.52 ab
Seeds per plant	193.50 $\pm$ 11.15 a	108.75 $\pm$ 11.15 bc	130.00 $\pm$ 11.15 b	115.75 $\pm$ 11.15 bc	70.25 $\pm$ 11.15 c	89.75 $\pm$ 11.15 bc	68.50 $\pm$ 11.15 c	84.50 $\pm$ 11.15 bc

* The experiment was performed with random complete block design in four replicates, each of which consisted of two individuals. Analysis of variance was performed and means were compared with Tukey's range test analysis. The data were presented as the mean $\pm$ SE and a different letter above the bar showed the significant difference in means ($p < 0.05$).

2.2. Overexpression of OSNUC1-S Increased the Photosynthetic Rate, Stomatal Conductance, and Transpiration Rate under Salt-Stress Conditions

Salt stress can cause a decrease in productivity in rice, and the major organs generating the carbohydrate for grain filling are flag leaves and second leaves. Therefore, we investigated the photosynthetic activities in both of these leaves in WT and *OsNUC1-S* over-expressing lines under control and salt stress conditions. Under the control condition, only second leaves of transgenic rice overexpressing *OSNUC1-S* showed higher net photosynthetic rates (P_N) and stomatal conductance levels (g_s), while the flag leaves had similar levels, except TOSL2, which had a lower P_N than the other lines. These results may reflect positional effects of the transformation. After 9 days of salt stress, the overexpression of *OsNUC1-S* increased the P_N and g_s of all the transgenic lines (Figure 2). The P_N values in flag leaves of transgenic plants were approximately two-fold those of the WT (Figure 2A), while in second leaves, the P_N values of transgenic plants increased up to 2.5-fold those of WT (Figure 2B). Similar effects were also found for the g_s values of both flag and second leaves

(Figure 2C,D). There was no effect on the intercellular CO_2 concentration (Ci) (Figure 2E,F), but *OsNUC1-S*'s overexpression resulted in an increased transpiration rate of the second leaves under control conditions and in the flag leaves under salt-stress conditions (Figure 2G,H).

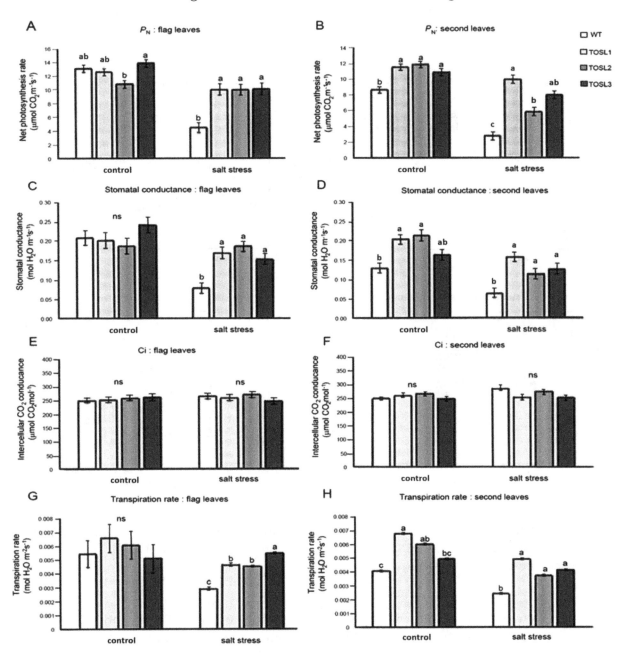

Figure 2. The net photosynthetic rate (P_N) (**A,B**), stomatal conductance (g_s) (**C,D**), intercellular CO_2 concentration (Ci) (**E,F**), and transpiration rate (E) (**G,H**) of flag leaves (**A,C,E,G**) and second leaves (**B,D,F,H**), when wild type (WT) and the *OsNUC1–S* overexpressing lines, TOSL1, TOSL2, and TOSL3, were grown under control or salt-stress conditions. The data were presented as the mean ± SE and a different letter above the bar showed the significant difference in means ($p < 0.05$) based on Tukey's range test analysis. ns represents no statistically difference among means.

2.3. Overexpression of OSNUC1-S Affected Both Light-Response Curves and CO2-Response Curves of Flag and Second Leaves under Salt-Stress Conditions

We investigated the light-response curves of these lines. The light-response curves of the flag leaves in all the lines were similar when the light intensity varied from 100 to 2000 $\mu mol \cdot m^{-2} \cdot s^{-1}$, except TOSL2, which had a lower P_N than other lines. The light-saturation points were approximately

1000 $\mu mol \cdot m^{-2} \cdot s^{-1}$ in all lines (Figure 3A). The salt stress caused decreases in the P_N values of all the lines, but the transgenic lines had significantly higher P_N values when the light intensity was greater than 200 $\mu mol \cdot m^{-2} \cdot s^{-1}$. The light-saturation point of the WT decreased to 600 $\mu mol \cdot m^{-2} \cdot s^{-1}$, while the transgenic lines had light-saturation points similar to those of plants grown under control conditions (1000 $\mu mol \cdot m^{-2} \cdot s^{-1}$) (Figure 3B).

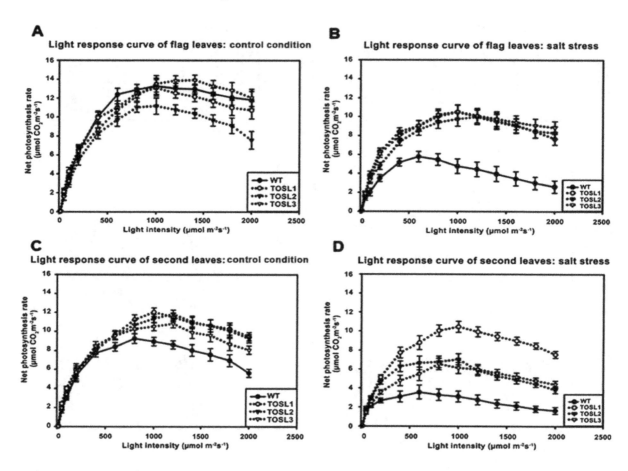

Figure 3. Light-response curves of flag leaves (**A,B**) and second leaves (**C,D**), when wild type and transgenic lines, TOSL1, TOSL2, and TOSL3, were grown under control (**A,C**) and salt-stress (**B,D**) conditions.

The second leaves of the WT had lower light-saturation points than those of the transgenic lines when grown under control conditions. The P_N values of the WT's second leaves started to decline when the light intensity was greater than 800 $\mu mol \cdot m^{-2} \cdot s^{-1}$; however, for the transgenic lines, the light-saturation point was 1000 $\mu mol \cdot m^{-2} \cdot s^{-1}$ (Figure 3C).

Salt stress caused a decrease in the light-saturation point in WT second leaves, but it did not affect the light-saturation points of the transgenic lines' second leaves. In WT second leaves, the light-saturation point declined to 600 $\mu mol \cdot m^{-2} \cdot s^{-1}$, and a more than two-fold reduction in the P_N was found. A similar reduction in the P_N was found in the transgenic lines, except TOSL1, which maintained a P_N in its second leaves that was similar to the P_N under control growth conditions. However, all the transgenic lines' second leaves had light-saturation points of 1000 $\mu mol \cdot m^{-2} \cdot s^{-1}$ (Figure 3D).

The CO_2-response curves of flag and second leaves of all the lines were also investigated as the CO_2 concentration increased from 200 to 1000 $\mu mol \cdot mol^{-1}$. Under control conditions, the flag leaves of all the lines showed similar CO_2-response curves. The CO_2-saturation point was ~800 $\mu mol \cdot mol^{-1}$

(Figure 4A). Under salt-stress conditions, significantly higher P_N values of the flag leaves were found in all the transgenic lines when the CO_2 concentration was greater than 200 µmol·mol^{-1}. However, salt stress had no effect on the CO_2-saturation points of flag leaves in any line (Figure 4B).

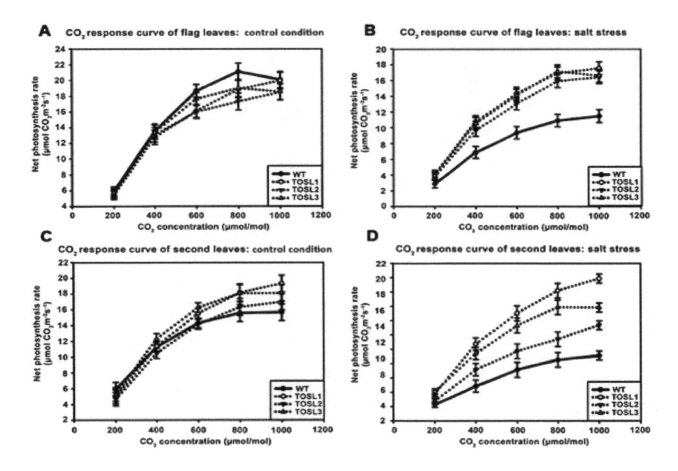

Figure 4. CO_2-response curves of flag leaves (**A,B**) and second leaves (**C,D**), when wild type and transgenic lines, TOSL1, TOSL2, and TOSL3, were grown under control (**A,C**) and salt-stress (**B,D**) conditions.

In the second leaves, the CO_2-response curves and CO_2-saturation points of all the lines were similar and consistent with the flag leaf's response to the lower P_N in plants grown under control conditions (Figure 4C). Salt stress caused decreases in the P_N values of all the lines, but it did not affect the CO_2 saturation point of WT second leaves. On the contrary, TOSL1 and TOSL2's second leaves had increased CO_2 saturation points to over 1000 mmol·mol^{-1}, while TOSL3 showed the same CO_2 saturation point as WT second leaves, when grown under salt-stress conditions (Figure 4D).

2.4. Salt Stress Affects Photosystem II (PSII) Photochemistry Efficiency and Photosynthetic Pigment Contents in Flag and Second Leaves

To investigate the effect of salt stress on the efficiency of PSII photochemistry, the F_v/F_m ratio was investigated. The salt-stress level did not significantly affect the PSII efficiency in the flag leaves of any lines (Figure 5A). However, in the second leaves, a significant reduction in the PSII efficiency (F_v/F_m) was found in WT after 9 days of salt treatment, while in all the transgenic lines, a reduced effect was found (Figure 5B). Thus, second leaves were more susceptible to salt stress than flag leaves, and *OsNUC1-S*'s overexpression contributed to the PSII photochemistry efficiency under salt-stress conditions.

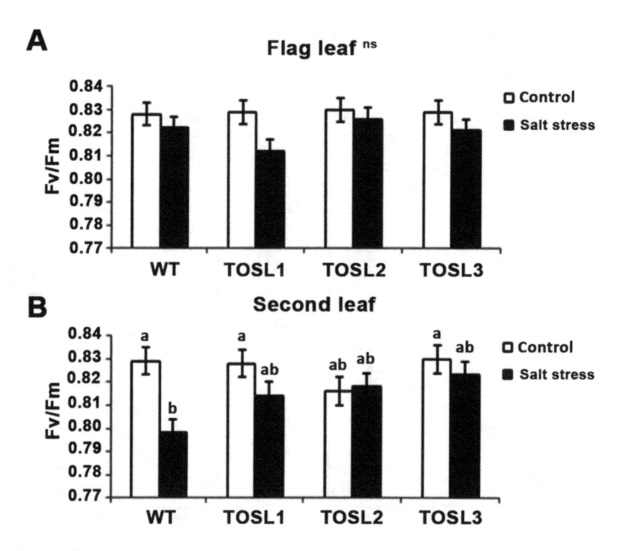

Figure 5. The F_v/F_m ratios of flag leaves (**A**) and second leaves (**B**) under control conditions and after being treated with 150 mM NaCl solution for 9 days. Three independent *OsNUC1-S* transgenic rice lines with difference transgene expression levels and WT plants were used. The data were presented as the mean ± SE and a different letter above the bar showed the significant difference in means ($p < 0.05$) based on Tukey's range test analysis. ns represents no statistically difference among means.

OsNUC1-S's overexpression tended to increase both the chlorophyll and carotenoid contents in flag leaves, but not in the second leaves, under optimal growth conditions (Figure 6A–F). Salt stress caused decreases in all of the pigments in the flag leaves, but significantly higher carotenoid contents were found in both flag and second leaves in the transgenic lines when compared with WT (Figure 6E,F).

2.5. OsNUC1-S's Overexpression Increased Carbohydrate Metabolism and Sugar Transport in Flag Leaves of Rice Grown Under Control Conditions

Because of the effects of *OsNUC1-S*'s overexpression on the photosynthetic characteristics, the transcriptome approach was used to investigate changes at the transcript level. TOSL3 was chosen as the representative for investigations of the flag leaf transcriptome. Based on a transcriptome comparison of DEGs (differentially expressed genes) between WT and TOSL3, when grown under control conditions, the DEGs were highly enriched in the cellular macromolecule metabolic processes, including the macromolecule biosynthetic process. Genes involved in transmembrane transport, regulation of cellular process, and the pigment metabolic process were also found.

Rice Overexpressing OsNUC1-S Reveals Differential Gene Expression Leading to Yield Loss...

141

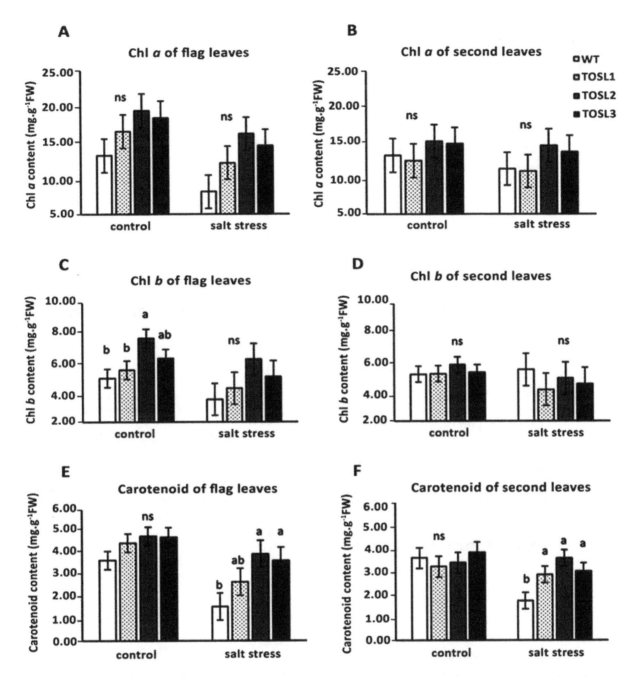

Figure 6. Photosynthetic pigments in flag leaves (**A,C,E**) and second leaves (**B,D,F**) 9 days after the salt-stress treatment at the booting stage. The data were presented as the mean ± SE and a different letter above the bar showed the significant difference in means ($p < 0.05$) based on Tukey's range test analysis. ns represents no statistically significant difference among means.

For the cellular component enrichment, a plasma membrane and chloroplast envelope were reported. This supported *OsNUC1-S*'s role in the enhancement of macromolecule production for grain filling in the flag leaves. Interestingly, the molecular function enrichment was found for only linoleate 13S-lipoxygenase activity. This enzyme is involved in plant growth and development [15], and also in the wounding response through jasmonic acid (JA) signaling [16,17] (Figure 7A).

Six days later, the DEGs resulting from *OsNUC1-S* expression changed. For the biological process, the genes functioning in carbohydrate transport were enriched, supporting the role of flag leaves in seed development. The consistent enrichment of molecular function in substrate-specific transmembrane transport was found (Figure 7B).

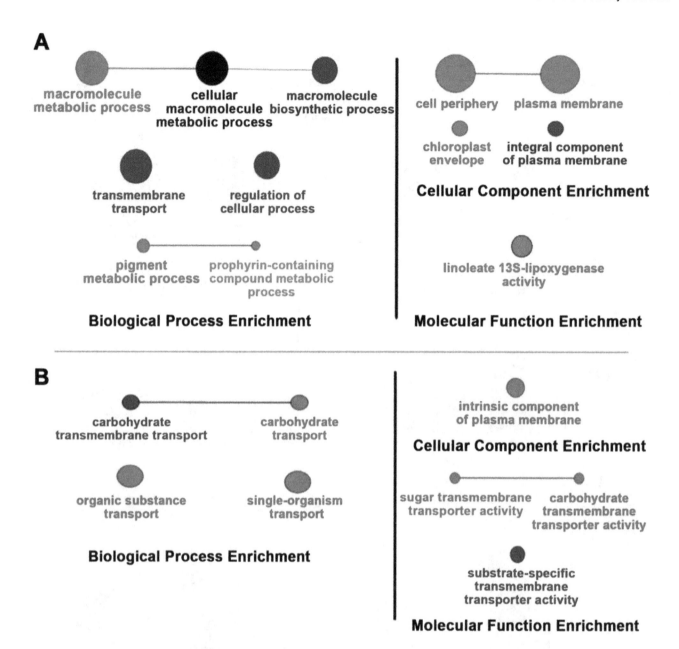

Figure 7. Gene enrichment analysis of the differentially expressed genes in the TOSL3 line, which overexpresses *OsNUC1-S*. The tissues for the transcriptome analysis were collected 3 days (**A**) and 9 days (**B**) after the first day of the booting stage. The darker colors represent the higher significance and the larger size of node represents the higher number of genes in the group.

2.6. The Overexpression of OsNUC1-S Increased the Expression of Genes Involved in Water Transport and Cellular Homeostasis, Nucleic Acid and Macromolecule Metabolic Processes, and Photosynthetic Processes

Under salt-stress conditions, the effects of *OsNUC1-S* overexpression were different from the effects found in normally grown plants. After 3 days of salt stress, the flag leaves of the transgenic lines were enriched with transcripts of genes involved in water transport and cellular homeostasis, nucleic and macromolecule metabolic processes, and photosynthetic processes. This was also consistent with the cellular enrichment in chloroplasts and nuclei. For the KEGG pathways, enrichment occurred in photosynthesis, carbon fixation, porphyrin and chlorophyll metabolism, and carotenoid biosynthesis, all of which involved activities in chloroplasts. Moreover, genes in glyoxylate and dicarboxylate metabolism, and terpenoid backbone biosynthesis, were also enriched (Figure 8).

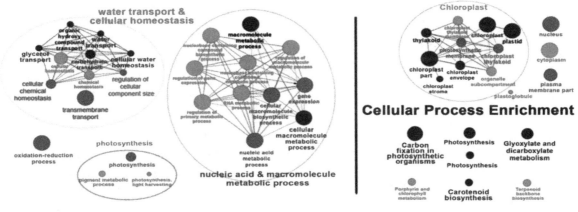

Figure 8. The enrichment for the biological process, cellular process, and Kyoto Encyclopedia of Genes and Genomes (KEGG) pathways of the differentially expressed genes in the flag leaves of transgenic rice overexpressing *OsNUC1-S* after 3 days of salt stress. The darker colors represent the higher significance and the larger size of node represents the higher number of genes in the group.

When the plants were under a prolonged stress for 9 days, similar processes, cellular compartments, and metabolic pathways were found to be affected, except that pigment and porphyrin-containing compound metabolic processes were detected, suggesting more pigment synthesis-related processes occurred, while the enrichment in light harvesting was not found. A new set of enriched genes found at this time point was in carboxylic acid metabolic processes and responses to oxidative stress. Chloroplasts were the main organelles that were enriched with the transcripts of genes affected by *OsNUC1-S* expression. The enrichment of the genes in the nucleus and plasma membrane also had a similar pattern to that found in flag leaves after 3 days of salt stress. The pathway enrichment after salt stress for 9 days was similar to that found after 3 days of stress, with more pathways in amino sugar and nucleotide sugar metabolism, pyruvate metabolism, ascorbate and aldarate metabolism, and pentose phosphate pathways (Figure 9). The list of the differentially expressed genes between wild type and transgenic rice with *OsNUC1-S* overexpression after salt stress treatment for 3 and 9 days is shown in Supplementary Table S1.

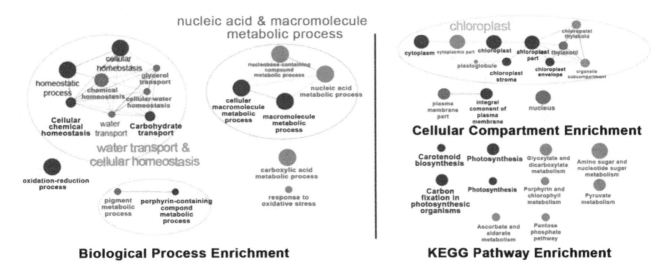

Figure 9. The enrichment for biological process, cellular process, and KEGG pathways of the differentially expressed genes in the flag leaves of transgenic rice overexpressing *OsNUC1-S* after 9 days of salt stress. The darker colors represent the higher significance and the larger size of the node represents the higher number of genes in the group.

To validate the RNA-Seq data, the up-regulated genes after 3 days of salt stress, *LOC_Os01g58470* (*CEST*) (Figure 10A), *LOC_Os03g39610* (*CAB*) (Figure 10B), and *LOC_Os04g33830* (*PSAO*) (Figure 10C), were chosen as representatives for quantitative RT-PCR (qRT-PCR). The expression in the transgenic line, TOSL3 was about two- to three-fold higher than WT under salt stress condition, based on RNA-Seq analysis (Supplementary Table S1). The comparable expression was also detected by qRT-PCR as shown in Figure 10A–C. Based on RNA-Seq analysis, *LOC_Os01g64960* (*PsbS1*) (Figure 10D) showed the highest fold change when compared to WT after salt stress for 9 days. The qRT-PCR of this gene expression revealed about five-fold higher expression than WT after 9 days (Figure 10D). The increase in the expression of these genes was correlated with the RNA-Seq data.

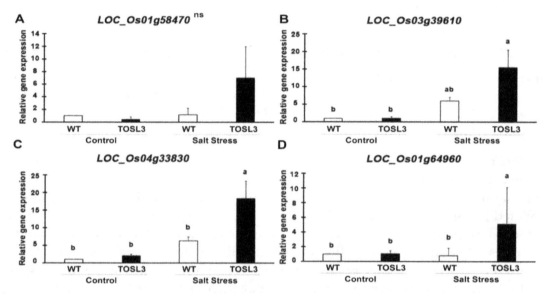

Figure 10. Gene expression analysis of *LOC_Os01g58470* (**A**), *LOC_Os03g39610* (**B**), *LOC_Os04g33830* (**C**), and *LOC_Os01g64960* (**D**) in WT and TOSL3 plants under control and salt stress conditions. Bar represents standard deviation of three biological replicates. The measurement was performed with three technical replicates. The data were presented as the mean ± SE and a different letter above the bar showed the significant difference in means ($p < 0.05$) based on Tukey's range test analysis. ns represents no statistically significant difference among means.

3. Discussion

OsNUC1-S, one of the mRNA splice forms, lacks an N-terminal region that contains several acidic stretches and a nuclear localization signal. However, based on the localization study [10], the protein localizes to both the cytoplasm and nucleus. The overexpression of *OsNUC1-S* did not change the photosynthetic activity levels in flag leaves when plants were grown under control conditions (Figure 2), but it resulted in reduced rice seed numbers and fertility rates (Table 1). The transcriptomic analysis found increased levels of carbohydrate metabolism and sugar transmembrane transport (Figure 7). This was in agreement with the role of flag leaves as the energy source for rice grain development [18]. Interestingly, TOSL3 was enriched in linoleate 13S-lipoxygenase activity (Figure 7). Lipoxygenase is the enzyme involved in JA biosynthesis and in responding to wounding and stress [17]. The exogenous application of JA induces flag leaf senescence by regulating chlorophyll degradation, membrane deterioration, and SAG (senescence associated gene) expression levels [19]. Moreover, the overexpression of the *TIFY* gene could enhance the rice grain yield, possibly owing to a reduction in JA sensitivity [20]. Therefore, the increase in lipoxygenase activity in the *OsNUC1-S* overexpression line could increase the JA level, causing decreases in the rice grain yield and fertility rate (Table 1). Both flag and second leaves contribute to the grain-filling process [14], but flag leaves provide more than 50% of the assimilates for grain filling [21]. In a comparison between WT and transgenic lines, a greater P_N was found only in the second leaves of the transgenic lines when grown under salt-stress conditions. This may be related to the maximization of the photosynthetic capacity of the flag leaves.

Under salt stress, a threefold reduction of P_N was detected in WT, while only a 10–15% reduction was found in the flag leaves of transgenic lines (Figure 2A). This suggested a role for *OsNUC1-S* in photosynthetic enhancement. The increase in photosynthetic activity was supported by the enrichment of genes involved in the photosynthetic processes, as well as in water transport and cellular homeostasis activities (Figure 8). The enrichment in nucleic acid and macromolecule metabolic processes supports a role for nucleolin that involves RNA modifications. Thus, OsNUC1-S may have a specific target for its activity, which results in greater changes in the expression levels of some genes compared with others when *OsNUC1-S* is overexpressed. The enrichment in carbon fixation, as well as porphyrin and chlorophyll metabolism, suggests that *OsNUC1-S* enhanced both the light reaction and the carbon fixation process. The increased light-saturation point (Figure 3) and carbon fixation activity (Figure 4) were consistent with the transcriptome data. Interestingly, when plants were impacted by salt, the function of *OsNUC1-S* changed. Both the transcriptome and physiological data indicated that this gene promoted salt tolerance through enhanced photosynthesis. Genes encoding proteins in the photosynthetic processes were strongly expressed in transgenic lines compared with WT during salt stress (Supplementary Table S1, Figure 10). Most of the up-regulated genes were located in PSI and PSII of the light reaction. Moreover, some genes in the Calvin cycle also increased, possibly causing the higher P_N values in transgenic plants. The enriched genes encoding chlorophyll A-B-binding protein (Supplementary Table S1), which functions in light harvesting, were also up-regulated in transgenic plants. This supported the light-response curve results in which a higher light-saturation point occurred in transgenic plants under salt stress.

The second leaves in WT were more susceptible to salt stress than flag leaves. The former had a significant decrease in F_v/F_m after 9 days of salt stress, while salt stress had no effect on flag leaves. Na^+ can be transferred from root to shoot, and in rice, *OsHKT1;5* excludes Na^+ from the phloem to prevent Na^+ transfer to the younger leaf blades [22]. Therefore, the second leaves were affected by salt stress to a larger extent than the flag leaves. This may be the mechanism that prevents Na^+ toxicity from reaching the flag leaf, which has the main role in carbon fixation for grain development.

4. Materials and Methods

4.1. Plant Material and Salt Treatment

The experiment was conducted in a planting house for transgenic plants in the Botany Department, Faculty of Science, Chulalongkorn University in Thailand. Three independent transgenic rice lines expressing *OsNUC1-S* driven by the Ubiquitin promoter have been produced by Sripinyowanich and colleagues [12] in a "Nipponbare" rice genetic background, and the homozygous T_3 generations of these transgenic lines were used in this experiment. "Nipponbare" rice seeds were obtained from the National Laboratory for Protein Engineering and Plant Genetic Engineering, Peking-Yale Joint Research Center for Plant Molecular Genetics and AgroBiotechnology, Peking University, People's Republic of China. Either wild type (WT) or transgenic seeds were germinated for 7 days. Then, seedlings were transferred to pots containing 5 kg of soil. When plants were at the booting stage, 150 mM NaCl solution was added to cause salt stress at 8–10 ds·m^{-1}. All photosynthetic parameters were measured after 9 days salt treatment using a Gas Analysis System (LI-COR, LI-6400, USA) with a 1200 μmol·m^{-2}·s^{-1} light intensity and 380 μmol mol^{-1} carbon dioxide (CO_2) concentration.

For chlorophyll fluorescence parameters, either the flag leaf or second leaf was used to determine maximum quantum yield of photosystem II (PSII) photochemistry (F_v/F_m) using Pocket PEA (Hansatech Instruments, Ltd., Norfolk, UK). In total, 30 min was used for dark adaptation, and then a 1-s saturation flash was used to measure the potential maximum photochemical efficiency of PSII.

4.2. RNA-seq and Data Analysis

For the transcriptomic analysis, flag leaves of WT and TOSL3, the transgenic line with *OsNUC1-S* overexpression, were collected 3 and 9 days after the first day of the booting stage. Total RNA was extracted using Invitrogen's Concert™ Plant RNA Reagent and then treated with DNaseI (NEB). A Dynabeads mRNA purification kit (Invitrogen, Carlsbad, CA, USA) and a KAPA Stranded mRNA-Seq Kit were used for mRNA isolation and cDNA libraries' preparation, respectively. Fragment sizes of ~300 bp were selected and connected with adaptors. Then, the fragments were enriched by PCR for 12 cycles. All libraries were sequenced using the Genome Analyzer (Illumina HiSeq4000, San Diego, CA, USA). Adaptors were subsequently removed from all short-sequence reads before grouping, following the protocol of Missirian et al. [23]. All the sequences were aligned and mapped to the rice genome, and then, the differentially expressed genes (DEGs) were identified using the DESeq program [24]. The genes showing the differential expression were selected to validate with quantitative RT-PCR. *LOC_Os01g58470*, *LOC_Os01g64960*, *LOC_Os03g39610*, and *LOC_Os04g33830* predicted to localize in plastids were selected. The primers for detection of the gene expression are shown in Table 2. *OsEF-1α* gene expression was used as an internal control. For q-RT-PCR, briefly, 1 µL of RNA extraction from either WT's or transgenic rice's flag leaves was used to synthesize the cDNA with an AccuPower® RT PreMix (BIONEER, Oakland, CA, USA). Gene expression of the target genes was detected by quantitative PCR using Luna® Universal qPCR Master MiX (BioLabs, San Diego, CA, USA). The thermal cycle was performed at 95 °C for 60 s, then 40 cycles of 95 °C for 15 s, 58.5 °C for 30 s, followed by 95 °C for 30 s, and the extension was done at 70 °C for 5 s. The relative expression of interesting genes was calculated using the Pfaffl method [25] with the formula:

$$\text{Ratio} = (E_{target})^{\,\Delta CP target(control-sample)} / (E_{ref})^{\,\Delta CPref(control-sample)}$$

Table 2. Primers for gene expression detection.

Gene	Primer	Sequence (5′–3′)	Tm (°C)
OsEF-1α	EF-1α-F	ATGGTTGTGGAGACCTTC	53.7
	EF-1α-R	ATGGTTGTGGAGACCTTC	58.2
LOC_Os01g58470	Os01g58470-F	AGGCATTGATCCTGAGACAG	54.3
	Os01g58470-R	AGAGCAGAATATCCCACTGC	54.4
LOC_Os01g64960	Os01g64960-F	GCATCGCCTTCTCCATCA	57.1
	Os01g64960-R	GAAGACGACGTTGAAGAGGA	57.3
LOC_Os03g39610	Os03g39610-F	GGAGGCGGTGTGGTTCAAGG	61.0
	Os03g39610-R	GCGGTAGCCCTCGACGAATC	60.3
LOC_Os04g3383	Os04g33830-F	CCGTTCTGGCTGTGGTT	55.4
	Os04g33830-R	CGTCCGTACAGTCAAGCTAA	54.3

4.3. Gene Ontology (GO) Term Analysis

Genes that had a *p*-value < 0.5 and $\log_2$fold change less than -1 or more than 1 were classified into GO terms using the ClueGO tool [26]. They were analyzed into three GO terms, cellular compartment, biological process, and molecular function, and subjected to a KEGG [27] pathway analysis.

4.4. Pigment Extraction and Quantification

Chlorophyll *a*, chlorophyll *b*, and total carotenoid contents were studied according to the method of Lichtenthaler [28]. Briefly, 50 mg fresh weight of either flag leaves or second leaves were extracted with 10 mL of 80% acetone and then incubated in the dark for 24 h at room temperature. Pigment extracts were measured at 470, 646.8, and 663.2 nm using a spectrophotometer (Agilent Technology, Santa Clara, CA, USA).

4.5. Yield Collection

After a 9 days salt-stress treatment, plants were recovered by the addition of water until soil salinity was below 2 ds·m^{-1}. Seeds were harvested when they were fully developed and then desiccated. The numbers of panicles per plant and seeds per panicle, as well as the fertility rates (%), were determined.

4.6. Statistical Analysis

A randomized complete block design with four replications was used for the experimental plots. Analysis of variance was performed to detect the differences among means and a Tukey's range test was used to detect significant differences between each mean at p-value < 0.05 using SPSS 21.0 statistical software. All the results are presented as mean ± standard error of the mean.

5. Conclusions

Based on the experimental results, a role of *OsNUC1-S* in the salt tolerance of rice during the reproductive stage was suggested to be due to the enhancement of photosynthetic processes in both flag and second leaves through the modification of gene expression levels in water transport, photosynthesis, cellular homeostasis, and carotenoid biosynthesis. These could help maintain the grain yield after a salt stress (Figure 11).

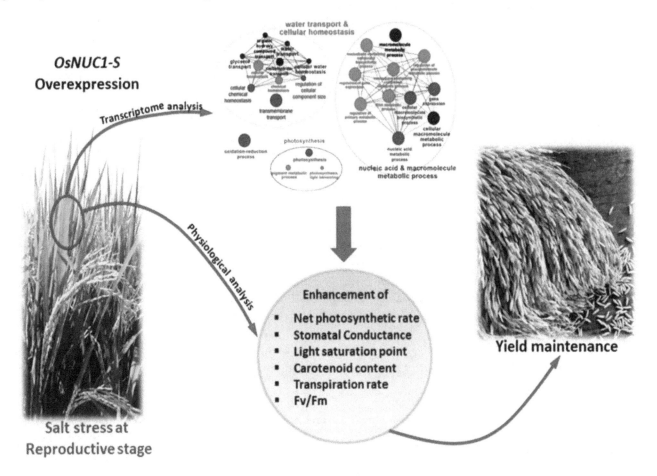

Figure 11. The scheme summarized the role of *OsNUC1-S* overexpression contributing to salt tolerance in rice.

Author Contributions: Conceptualization, S.C. and L.C.; methodology, S.C., T.B. and L.C.; software, L.C. and T.U.; validation, C.B. and S.S.; formal analysis, C.B. and S.C.; investigation, C.B.; resources, S.S.; data curation, S.C. and T.B.; writing—original draft preparation, C.B.; writing—review and editing, S.C., T.B. and L.C.; supervision, S.C., L.C. and T.B.; project administration, S.C.; funding acquisition, S.C.

Acknowledgments: We thank Lesley Benyon from Edanz Group (www.edanzediting.com/ac) for editing a draft of this manuscript.

Abbreviations

Ci	intercellular CO_2 concentration
F_v/F_m	intercellular CO_2 concentration
g_s	stomatal conductance
OsNUC1	rice *NUCLEOLIN1*
P_N	net photosynthesis rate
WT	wild type

References

1. Fageria, N. Yield physiology of rice. *J. Plant Nutr.* **2007**, *30*, 843–879. [CrossRef]
2. Hakim, M.; Juraimi, A.; Begum, M.; Hanafi, M.; Ismail, M.R.; Selamat, A. Effect of salt stress on germination and early seedling growth of rice (*Oryza sativa* L.). *Afr. J. Biotechnol.* **2010**, *9*, 1911–1918.
3. Bethke, P.C.; Drew, M.C. Stomatal and nonstomatal components to inhibition of photosynthesis in leaves of *Capsicum annuum* during progressive exposure to NaCl salinity. *Plant Physiol.* **1992**, *99*, 219–226. [CrossRef] [PubMed]
4. Reddy, M.; Vora, A. Changes in pigment composition, Hill reaction activity and saccharides metabolism in Bajra (*Pennisetum typhoides* S & H) leaves under NaCl salinity. *Photosynthetica* **1986**, *20*, 50–55.
5. Petrov, V.; Hille, J.; Mueller-Roeber, B.; Gechev, T.S. ROS-mediated abiotic stress-induced programmed cell death in plants. *Front. Plant Sci.* **2015**, *66*, 69. [CrossRef] [PubMed]
6. Lutts, S.; Kinet, J.; Bouharmont, J. Changes in plant response to NaCl during development of rice (*Oryza sativa* L.) varieties differing in salinity resistance. *J. Exp. Bot.* **1995**, *46*, 1843–1852. [CrossRef]
7. Touch, S.; Pipatpongsa, T.; Takeda, T.; Takemura, J. The relationships between electrical conductivity of soil and reflectance of canopy, grain, and leaf of rice in northeastern Thailand. *Int. J. Remote Sens.* **2015**, *36*, 1136–1166.
8. Zhang, J.; Lin, Y.J.; Zhu, L.F.; Yu, S.M.; Sanjoy, K.K.; Jin, Q.Y. Effects of 1-methylcyclopropene on function of flag leaf and development of superior and inferior spikelets in rice cultivars differing in panicle types. *Field Crops Res.* **2015**, *177*, 64–74. [CrossRef]
9. Tajrishi, M.M.; Tuteja, R.; Tuteja, N. Nucleolin: The most abundant multifunctional phosphoprotein of nucleolus. *Commun. Integr. Biol.* **2011**, *4*, 267–275. [CrossRef]
10. Petricka, J.J.; Nelson, T.M. Arabidopsis nucleolin affects plant development and patterning. *Plant Physiol.* **2007**, *144*, 173–186. [CrossRef]
11. Reichler, S.A.; Balk, J.; Brown, M.E.; Woodruff, K.; Clark, G.B.; Roux, S.J. Light differentially regulates cell division and the mRNA abundance of pea nucleolin during de-etiolation. *Plant Physiol.* **2001**, *125*, 339–350. [CrossRef] [PubMed]
12. Sripinyowanich, S.; Chamnanmanoontham, N.; Udomchalothorn, T.; Maneeprasopsuk, S.; Santawee, P.; Buaboocha, T.; Qu, L.-J.; Gu, H.; Chadchawan, S. Overexpression of a partial fragment of the salt-responsive gene *OsNUC1* enhances salt adaptation in transgenic *Arabidopsis thaliana* and rice (*Oryza sativa* L.) during salt stress. *Plant Sci.* **2013**, *213*, 67–78. [CrossRef] [PubMed]
13. Udomchalothorn, T.; Plaimas, K.; Sripinyowanich, S.; Boonchai, C.; Kojonna, T.; Chutimanukul, P.; Comai, L.; Buaboocha, T.; Chadchawan, S. *OsNucleolin1-L* expression in Arabidopsis enhances photosynthesis via transcriptome modification under salt stress conditions. *Plant Cell Physiol.* **2017**, *58*, 717–734. [CrossRef] [PubMed]

14. Lee, S.; Jeong, H.; Lee, S.; Lee, J.; Kim, S.-J.; Park, J.-W.; Woo, H.R.; Lim, P.O.; An, G.; Nam, H.G. Molecular bases for differential aging programs between flag and second leaves during grain-filling in rice. *Sci. Rep.* **2017**, *7*, 8792. [CrossRef]

15. Junghans, T.G.; De Almeida-Oliveira, M.G.; Moreira, M.A. Lipoxygenase activities during development of root and nodule of soybean. *Pesqui. Agropecu. Bras.* **2004**, *39*, 625–630. [CrossRef]

16. Lenglet, A.; Jaślan, D.; Toyota, M.; Mueller, M.; Müller, T.; Schönknecht, G.; Marten, I.; Gilroy, S.; Hedrich, R.; Farmer, E.E. Control of basal jasmonate signalling and defence through modulation of intracellular cation flux capacity. *New Phytol.* **2017**, *216*, 1161–1169. [CrossRef]

17. Brodhun, F.; Cristobal-Sarramian, A.; Zabel, S.; Newie, J.; Hamberg, M.; Feussner, I. An iron 13S-lipoxygenase with an α-linolenic acid specific hydroperoxidase activity from *Fusarium oxysporum*. *PLoS ONE* **2013**, *8*, e64919. [CrossRef]

18. Kholupenko, I.; Voronkova, N.; Burundukova, O.; Zhemchugova, V. Demand for assimilates determines the productivity of intensive and extensive rice crops in Primorskii krai. *Russ. J. Plant Physiol.* **2003**, *50*, 112–118. [CrossRef]

19. Liu, L.; Li, H.; Zeng, H.; Cai, Q.; Zhou, X.; Yin, C. Exogenous jasmonic acid and cytokinin antagonistically regulate rice flag leaf senescence by mediating chlorophyll degradation, membrane deterioration, and senescence-associated genes expression. *J. Plant Growth Regul.* **2016**, *35*, 366–376. [CrossRef]

20. Hakata, M.; Muramatsu, M.; Nakamura, H.; Hara, N.; Kishimoto, M.; Iida-Okada, K.; Kajikawa, M.; Imai-Toki, N.; Toki, S.; Nagamura, Y. Overexpression of *TIFY* genes promotes plant growth in rice through jasmonate signaling. *Biosci. Biotechnol. Biochem.* **2017**, *81*, 906–913. [CrossRef]

21. Li, Z.; Pinson, S.R.; Stansel, J.W.; Paterson, A.H. Genetic dissection of the source-sink relationship affecting fecundity and yield in rice shape (*Oryza sativa* L.). *Mol. Breed.* **1998**, *4*, 419–426. [CrossRef]

22. Kobayashi, N.I.; Yamaji, N.; Yamamoto, H.; Okubo, K.; Ueno, H.; Costa, A.; Tanoi, K.; Matsumura, H.; Fujii-Kashino, M.; Horiuchi, T.; et al. OsHKT1;5 mediates Na+ exclusion in the vasculature to protect leaf blades and reproductive tissues from salt toxicity in rice. *Plant J.* **2017**, *91*, 657–670. [CrossRef] [PubMed]

23. Missirian, V.; Henry, I.; Comai, L.; Filkov, V. POPE: Pipeline of parentally-biased expression. In Proceedings of the ISBRA: 2012 Bioinformatics Research and Applications, Dallas, TX, USA, 21–23 May 2012; pp. 177–188.

24. Anders, S.; Huber, W. Differential expression analysis for sequence count data. *Genome Biol.* **2010**, *11*, R106. [CrossRef] [PubMed]

25. Pfaffl, M.W. A new mathematical model for relative quantification in real-time RT-PCR. *Nucleic Acids Res.* **2001**, *29*, e45. [CrossRef] [PubMed]

26. Bindea, G.; Mlecnik, B.; Hackl, H.; Charoentong, P.; Tosolini, M.; Kirilovsky, A.; Fridman, W.-H.; Pagès, F.; Trajanoski, Z.; Galon, J. ClueGO: A Cytoscape plug-in to decipher functionally grouped gene ontology and pathway annotation networks. *Bioinformatics* **2009**, *25*, 1091–1093. [CrossRef] [PubMed]

27. Kanehisa, M.; Goto, S. KEGG: Kyoto encyclopedia of genes and genomes. *Nucleic Acids Res.* **2000**, *28*, 27–30. [CrossRef] [PubMed]

28. Lichtenthaler, H.K. Chlorophylls and carotenoids: Pigments of photosynthetic biomembranes. *Methods Enzymol.* **1987**, *148*, 350–380.

9

Melatonin: A Small Molecule but Important for Salt Stress Tolerance in Plants

Haoshuang Zhan [1,†], Xiaojun Nie [1,†], Ting Zhang [1], Shuang Li [1], Xiaoyu Wang [1], Xianghong Du [1], Wei Tong [1,*] and Weining Song [1,2,*]

[1] State Key Laboratory of Crop Stress Biology in Arid Areas, College of Agronomy and Yangling Branch of China Wheat Improvement Center, Northwest A&F University, Yangling 712100, China; zhanhaoshuang@nwsuaf.edu.cn (H.Z.); small@nwsuaf.edu.cn (X.N.); zhangting@nwsuaf.edu.cn (T.Z.); Lishuang@nwsuaf.edu.cn (S.L.); xiaoyuw@nwsuaf.edu.cn (X.W.); xianghongdu@nwsuaf.edu.cn (X.D.)

[2] ICARDA-NWSUAF Joint Research Center for Agriculture Research in Arid Areas, Yangling 712100, China

* Corresponding authors: tongw@nwsuaf.edu.cn (W.T.); sweining2002@nwsuaf.edu.cn or sweining2002@yahoo.com (W.S.)

† These authors contributed equally to this work.

Abstract: Salt stress is one of the most serious limiting factors in worldwide agricultural production, resulting in huge annual yield loss. Since 1995, melatonin (*N*-acetyl-5-methoxytryptamine)—an ancient multi-functional molecule in eukaryotes and prokaryotes—has been extensively validated as a regulator of plant growth and development, as well as various stress responses, especially its crucial role in plant salt tolerance. Salt stress and exogenous melatonin lead to an increase in endogenous melatonin levels, partly via the phyto-melatonin receptor CAND2/PMTR1. Melatonin plays important roles, as a free radical scavenger and antioxidant, in the improvement of antioxidant systems under salt stress. These functions improve photosynthesis, ion homeostasis, and activate a series of downstream signals, such as hormones, nitric oxide (NO) and polyamine metabolism. Melatonin also regulates gene expression responses to salt stress. In this study, we review recent literature and summarize the regulatory roles and signaling networks involving melatonin in response to salt stress in plants. We also discuss genes and gene families involved in the melatonin-mediated salt stress tolerance.

Keywords: antioxidant systems; ion homeostasis; melatonin; salt stress; signal pathway

1. Introduction

Salinity represents an environmental stress factor affecting plant growth and development, and a destructive threat to global agricultural production [1], which damages more than 400 million hectares of land—over 6% of the world's total land area. Of the irrigated farmland areas, currently 19.5% are salt-affected, with increasing numbers facing the threat of salinization (http://www.plantstress.com/Articles/index.asp). The effects of salt stress on plants mainly include osmotic stress, specific ion toxicity, nutritional imbalance, and reactive oxygen species [2]. Osmotic stress is a rapid process caused by salt concentrations around the roots, which is induced at the initial stage of salt stress [1–3]. Na^+ accumulation at a later stage causes nutrient imbalance, leading to specific ion toxicity [4]. Plants' exposure to salt stress induces overproduction of reactive oxygen species (ROS), which results in membrane injury [5,6].

Melatonin is a multi-regulatory molecule likely to be present in most plants and animals [7]. It was first identified in 1958, in the bovine pineal gland [8], and is a well-known animal hormone regulating various biological processes, such as the circadian rhythm [9,10], antioxidant activity [11], immunological enhancement [12], seasonal reproduction [13], emotional status, and physical conditions [14]. In 1995, melatonin was discovered in vascular plants [15,16], which initiated this field of study. Melatonin was found to have many physiological functions similar to indole-3-acetic acid (IAA), such as regulating plant photoperiod and protecting chlorophyll [17]. More importantly, it acts as a powerful antioxidant, thus protecting plants from various biotic/abiotic stresses [18,19].

In recent years, more functions of melatonin have been identified in higher plants, mainly its roles as a stress responses regulator. In this review, we systematically discuss the functional and potential regulatory mechanisms of melatonin in response to salt stress. We also focus on the putative genes involved in the melatonin-induced salt stress resistance. Furthermore, we summarized plant melatonin receptors, thus outlining the current situation and further directions for promoting the study of plant salt stress tolerance.

2. Function and Mechanism of Melatonin Effects on Plant Salt Tolerance

Extensive studies have revealed the crucial and indispensable roles that melatonin plays in increasing salt tolerance in diverse plant species (Table 1). These functions regulate antioxidant systems to protect plants from the salt stress-induced water deficits and physiological damages, improve photosynthetic efficiency and ion homeostasis, and behave as an activator mediating NO signaling and the polyamine metabolism pathway [7,17,33].

Table 1. The reported roles melatonin plays in response to salt and other stresses in plants.

Plant Species	Stress Condition	References
Actinidia deliciosa	Salt	[20]
Malus hupehensis	Salt	[21]
Arabidopsis thaliana	salt	[22]
Arabidopsis thaliana	Salt, drought and cold	[23]
Arabidopsis thaliana	Salt	[24]
Cynodon dactylon (L). Pers.	Salt, drought and cold	[25]
Chara australis	Salt	[26]
Chlamydomonas reinhardtii	Salt	[27]
Citrus aurantium L.	Salt	[28]
Cucumis sativus L.	Salt	[29]
Cucumis sativus L.	Salt	[17]
Cucumis sativus L.	Salt	[30]
Zea mays L.	Salt	[31]
Zea mays L.	Salt	[32]
Zea mays L.	Salt	[33]
Raphanus sativus L.	Salt	[34]
Raphanus sativus L.	Salt	[35]
Brassica napus L.	Salt	[36]
Brassica napus L.	Salt	[37]
Oryza sativa L.	Leaf senescence and salt	[38]
Oryza sativa L.	Salt	[39]
Glycine max	Salt and drought	[40]
Helianthus annuus	Salt	[41]
Helianthus annuus	Salt	[42]
Ipomoea batatas	Salt	[43]
Solanum lycopersicum	Salt	[44]
Vicia faba L.	Salt	[45]
Citrullus lanatus L.	Salt	[46]
Triticum aestivum L.	Salt	[47]

2.1. Melatonin Activates Antioxidant Systems in Response to Salt Stress

Salinity induces reactive oxygen species (ROS) production, including superoxide anion (O_2^-), hydrogen peroxide (H_2O_2), hydroxyl radical (OH^-), and singlet oxygen ($^1O_2$) [47]. Excess ROS usually leads to cell damage and oxidative stress [22]; it also acts as signaling molecules fundamentally involved in mediating salt tolerance [48]. Plants have developed two antioxidant systems to alleviate ROS-triggered damages: the enzymatic and non-enzymatic systems [49]. In response to salt stress, plants have evolved a complex antioxidant enzyme system, including superoxide dismutase (SOD), guaiacol peroxidase (POD), catalase (CAT), glutathione peroxidases (GPX), glutathione S-transferase (GST), dehydroascorbate reductase (DHAR), glutathione reductase (GR), and ascorbate peroxidase (APX) [17]. The non-enzymatic system, including ascorbic acid (AsA), α-tocopherols, glutathione (GSH), carotenoids, and phenolic compounds, is also essential for ROS elimination [50].

Exogenous melatonin treatment significantly reduced salinity-induced ROS. Following 12 days of salt stress, H_2O_2 concentration increased by 37.5%, while melatonin pre-treatment of cucumber maintained a low H_2O_2 concentration throughout the experiment [17]. Similar results were also observed in salt-stressed rapeseed seedlings, and the application of exogenous melatonin decreased H_2O_2 content by 11.2% [36]. Liang et al. [38] discovered inhibitory effects of melatonin resulting in an increased rate of H_2O_2 production in rice seedlings under salt stress, showing that melatonin works in a concentration-dependent manner. Melatonin scavenges ROS, mainly triggered by salt stress, via three pathways. Melatonin acts as a broad-spectrum antioxidant that interacts with ROS and directly scavenges it [51]. The primary function of melatonin is to act as a free radical scavenger and an antioxidant. Through the free radical scavenging cascade, a single melatonin molecule can scavenge up to 10 reactive oxygen species (ROS)/reactive nitrogen species (RNS), which differs from other conventional antioxidants [51]. Exogenous melatonin decreases H_2O_2 and O_2^- concentrations by activating antioxidant enzymes. This function has been confirmed in many plant species, such as rapeseed, radish, cucumber, rice, maize, bermudagrass, soybean, watermelon, kiwifruit, and *Malus hupehensis* [36]. In cucumber, the activity of major protective antioxidant enzymes—including SOD, CAT, POD, and APX—in melatonin pre-treated plants was significantly higher than control plants [17]. Under salt stress, exogenous melatonin application also significantly increased the activities of APX, CAT, SOD, POD, GR, and GPX in melatonin-treated seedlings compared to their non-treated counterparts [31,33]. Moreover, melatonin interacts with ROS by improving concentrations of antioxidants (AsA-GSH) [17]. In cucumber, AsA and GSH concentrations in melatonin pre-treated plants were 1.7- and 1.3-fold higher, respectively, compared to control plants [17]. Other studies have reported a marked melatonin-dependent induction of AsA and GSH in maize seedlings under salt stress [31]. These findings suggest that exogenous melatonin could activate enzymatic and non-enzymatic antioxidants to scavenge salt stress-induced ROS, thus improving salt stress tolerance in plants.

2.2. Melatonin Improves Plant Photosynthesis under Salt Stress

Photosynthesis, an important physio-chemical process responsible for energy production in higher plants, can be indirectly affected by salt stress [46,52]. For many plant species suffering salt stress, decline in productivity is often associated with lower photosynthesis levels [52]. There are two possible reasons for the salt-induced photosynthesis decline: stomatal closure and affected photosynthetic apparatus [52]. Salt stress can cause stomatal closure, and stomatal conductance (Gs) is one of the parameters for evaluating photosynthesis [52]. The parameters of chlorophyll fluorescence include maximum photochemical efficiency of PSII (Fv/Fm), photochemical quenching (qP), non-photochemical quenching [Y(NPQ)], and actual photochemical efficiency of PSII [Y(II)], etc. [46].

In addition to its broad-spectrum antioxidant effects, melatonin participates in the regulation of plant photosynthesis under salt stress. Pretreatment with various concentrations (50–500 μM) of melatonin clearly improved salt tolerance in watermelons, where the leaf net photosynthetic rate (Pn), Gs, chlorophyll content, Y(II) and qP were significantly decreased under salt stress. However, this decrease was alleviated by melatonin pretreatment. Melatonin can also protect watermelon photosynthesis by alleviating stomatal limitation [46]. Similar results were observed in salt-stressed cucumber seedlings, where the photosynthetic capacity of cucumber was significantly improved by exogenous melatonin at 50–150 μM concentrations. Photosynthesis improvement is manifested by increased P_N, maximum quantum efficiency of PSII, and total chlorophyll content [17]. In radish seedling, chlorophyll a, chlorophyll b and total chlorophyll contents increased upon melatonin treatment under salt stress, and the 100 μM dose was the best [34]. Melatonin also enhanced rice seedlings' salt tolerance by decreasing chlorophyll's degradation rate [38]. Even though the chlorophyll content in melatonin-treated maize seedlings did not change, an obvious increase in Pn was observed under salt stress [33]. Exogenous melatonin's protective roles in photosynthesis were also observed in soybean, apple, and tomato [21,40,44]. Overall, exogenous melatonin improves photosynthesis by effectively alleviating chlorophyll degradation and stomatal closure caused by salt stress, therefore enhancing salt stress tolerance.

2.3. Melatonin Promotes Ion Homeostasis under Salt Stress

Ion homeostasis refers to the ability of living organisms to maintain stable ion concentrations in a defined space [53]. Na^+, K^+, Ca^{2+}, and H^+ are major intracellular ions [53,54]. In salt-stressed plants, Na^+ can enter into plant cells, which at high concentrations is harmful to cytosolic enzymes [55]. Therefore, regulation of K^+ and Na^+ concentrations to maintain high of K^+ and low Na^+ cytosolic levels has a significant impact on salt-stressed plants [54,55]. Restriction of Na^+ influx, active Na^+ efflux, and compartmentalization of Na^+ into the vacuole are three major mechanisms of preventing Na^+ accumulation in the cytoplasm [56]. The *NHX1* gene encodes a vacuolar Na^+/H^+ exchanger, whose homologue in *Arabidopsis*, *AtNHX1*, was upregulated by salt stress resulting in excess transfer of Na^+ into vacuolar [57]. Salt Overly Sensitive1 (*SOS1*) encodes a transmembrane protein, identified as a plasma membrane Na^+/H^+ antiporter. SOS signaling is responsible for transporting Na^+ out of the cells [37,56]. The *Arabidopsis SOS1* gene possesses 12 transmembrane domains. Similar to *AtNHX1*, *AtSOS1* was also upregulated by salt stress [56]. Besides Na^+/H^+ antiporters, the involvement of K^+ channels has also been reported in plants' salt stress response. The *AKT1* gene encoding a Shaker type K^+ channel protein is responsible for absorbing K^+ from the soil and transporting it into the roots [58]. Under salt stress, *NHX1*, *SOS1* and *AKT1* upregulated gene expression leads to an increase of K^+ and decreased Na^+ in plant cells, thereby improving plants' salt stress tolerance.

Recently, studies have shown that the exogenous application of melatonin improves plants' ion homeostasis under salt stress. Melatonin significantly increased K^+ and decreased Na^+ contents in shoots of maize seedlings, leading to a significantly higher K^+/Na^+ ratio in shoots under melatonin-mediated salinity [33]. Improved ion homeostasis may be related to the upregulation of several genes, such as *NHX*, *SOS* and *AKT*. Under salt stress, *MdNHX1* and *MdAKT1* transcript levels were greatly upregulated by melatonin, which is consistent with the relatively high K^+ levels and K^+/Na^+ ratio in melatonin pretreated *Malus hupehensis* seedlings [21]. Similarly, *NHX1* and *SOS2* expression was higher in melatonin-treated rapeseed seedlings compared to non-treated plants, which correlated with the lower Na^+/K^+ ratio [37]. Ca^{2+} signaling plays critical roles in plant biotic and abiotic stress responses; however, no evidence regarding the involvement of Ca^{2+} signaling in melatonin-triggered salinity tolerance exists.

2.4. Melatonin Regulates Plant Hormones Metabolism

Plant hormones are important signals for plant growth and development [30]. Melatonin widely participates in the metabolism of most plant hormones, such as indole-3-acetic acid (IAA), abscisic acid (ABA), gibberellic acids (GA), cytokinins (CK), and ethylene [59].

The melatonin molecule shares chemical similarities with IAA, both using tryptophan as a substrate in their biosynthesis pathways [60]. It is reported that melatonin acts as a growth regulator and exhibits auxin-like activities [61]. Melatonin promotes vegetative growth and root development in many plant species, such as wheat, barley, rice, *Arabidopsis*, soybean, maize, tomato, etc. [59]. Under stress conditions, the growth-promoting effects of melatonin are higher compared to those in control plants [59]. Melatonin has been proposed to regulate lateral root formation through an

IAA-independent pathway in *Arabidopsis* [61]. In contrast, others suggest a certain relationship between melatonin and IAA; for example, a slight increase in endogenous IAA content was observed in *Brassica juncea* [59,62] when treated with exogenous melatonin. Furthermore, application of low concentrations of IAA increases endogenous melatonin levels. At the same time, high concentrations of melatonin inhibit PIN1,3,7 expression and decrease IAA levels in *Arabidopsis* roots, suggesting that melatonin may regulate root growth in *Arabidopsis,* completely or partially, through auxin synthesis and polar auxin transport [60].

Abscisis acid (ABA) and gibberellic acids (GA) are important plant hormones in stress responses. The dynamic balance of endogenous ABA and GA levels is crucial for seed germination [30,63]. Genes related to ABA synthesis—such as *ZEP* and *NCED1*—were upregulated during abiotic stresses, resulting in increased endogenous ABA levels [64]. GA acts as an ABA antagonist [65], and plays essential roles in plant stress tolerance [66]. Studies show that melatonin mediates ABA biosynthesis and metabolism regulation, thus decreasing ABA content under stress conditions. For example, in two drought-stressed *Malus* species, melatonin selectively downregulates *MdNCED3*, a key ABA biosynthesis gene, and upregulates *MdCYP707A1* and *MdCYP707A2*, ABA catabolic genes [67]. Similarly, in perennial ryegrass, exogenous melatonin downregulates ABA biosynthesis genes under heat stress, thereby decreasing ABA content [64]. However, melatonin treatment has no effect on water stress-induced ABA accumulation in maize [68]. Under salt stress, melatonin increased endogenous ABA content in *Elymus nutans*, which was significantly suppressed by fluridone. ABA and fluridone pretreatments had no effect on endogenous melatonin concentration, indicating that ABA might act as a downstream signal that participates in the melatonin-induced cold tolerance. Interestingly, melatonin can also activate the expression of cold-responsive genes to improve plant cold-stress tolerance in an ABA-independent manner. This suggests that both ABA-dependent and ABA-independent pathways might be involved in melatonin-induced cold tolerance [69]. These data suggest that, similar to the heat-related results, under drought and cold stresses, exogenous melatonin can also alleviate salt stress by regulating ABA biosynthesis and catabolism. Under salt stress, *CsNCED1* and *CsNCED2*—ABA synthesis-related genes—transcript levels were reduced in melatonin-pretreated seeds, and genes related to ABA catabolism were significantly increased, thus leading to a decreased ABA content. On the contrary, *GA20ox* and *GA3ox*—genes involved in GA synthesis—were significantly upregulated by melatonin, which is consistent with the increased GA content [30]. Overall, hormone biosynthesis- and catabolism-related research is helpful for understanding melatonin's mechanisms in response to salt stress.

2.5. Melatonin Mediates NO Signaling Pathway

Nitric oxide (NO) is an important messenger and ubiquitous signaling molecule, which participates in various plant physiological processes [70], and responds to abiotic and biotic stresses [41,42,71,72]. In animals, NO is synthesized by NO synthase (NOS) [72], and whether NOS-like proteins exist in plants remains controversial. NOS-like proteins were first identified in plants by Ninnemann and Maier [73]. Initially, *Arabidopsis* nitric oxide associated 1 (*NOA1*) was characterized as a NOS-like gene with NOS activity. However, further research indicated that these proteins function as a GTPases, involved in binding RNA/ribosomes [74]. There are at least seven different NO biosynthetic pathways found in plants, which can be classified as oxidative or reductive based on the operation [75]. Oxidative routes of NO biosynthesis use L-arginine, polyamine, or droxylamine as substrates [75]. S-nitrosylation refers to the process of covalently binding a NO group to its target proteins via cysteine (Cys) residues, and producing an S-nitrosothiol [76]. *S*-nitrosylation, with NO, is widely used to explain NO signaling in both animals and plants [77,78].

Studies have shown that melatonin, through its interaction with NO, plays important roles in plant stress responses. For examples, NO acts as a downstream signal for melatonin mitigated sodic alkaline stress in tomato seedlings [79]. In addition, exogenous melatonin significantly induces the accumulation of polyamine-mediated NO in the roots of *Arabidopsis* under Fe deficiency conditions, and increases the plants' tolerance to Fe deficiency [80]. Melatonin-induced NO production is also involved in the innate immune response of *Arabidopsis* against P. syringe pv. tomato (Pst) DC3000 infection [81]. In rapeseed seedlings, the possible roles of NO in melatonin-enhanced salt stress tolerance have been reported. Salt stress firstly induces the increase in melatonin and NO serves as the downstream signal. In addition, both melatonin and sodium nitroprusside (SNP) increased salinity-induced S-nitrosylation. Increased S-nitrosylation could be partially impaired by 2-phenyl-1-4,4,5,5-tetramethylimidazoline-1-oxyl-3-oxide (PTIO), an NO scavenger. Application of melatonin increased *NHX1* and *SOS2* transcript levels, which was blocked by NO removal. These data suggest that NO is involved in the maintenance of ion homeostasis in plant salt stress tolerance. NO is also involved in the improvement of the antioxidant systems triggered by melatonin [37]. However, the above research still lacks S-nitrosylation target protein identification. In addition, the interactions between NO and other substances, such as hormones, chlorophyll, polyamines, etc., in melatonin-enhanced salt stress tolerance requires further exploration.

2.6. Melatonin Regulates Polyamine Metabolism

Polyamines (PAs) are small aliphatic polycations that have been found in almost all living organisms. They play important roles in plant growth and development, and responses to various biotic and abiotic stimuli [82–84]. Spermidine (Spd), putrescine (Put), and spermine (Spm) are three main polyamines in plants [84]. Both the application of exogenous polyamines and modulating endogenous polyamine contents effectively enhance plant stress tolerance [83,84].

Studies have shown that melatonin plays a key role in polyamine-mediated signaling pathways under various abiotic stresses, such as alkaline stress, cold, oxidative, and iron deficiency tolerance [7]. For example, polyamines mediate the melatonin-induced alkaline stress tolerance of *Malus hupehensis*. Under alkaline stress, melatonin application significantly upregulated the expression of six polyamine synthesis-related genes, including *SAMDC1, -3, -4,* and *SPDS1, -3, -5, -6*. Moreover, melatonin-treated *Malus hupehensis* exhibited more polyamine accumulation compared to the untreated seedlings [85]. Exogenous melatonin also modulates polyamine and ABA metabolisms of cucumber seedlings during chilling stress. The melatonin-related cold tolerance improvement is consistent with the increased PA content [24]. PA modulation by melatonin under a salt stress response was also described by Ke et al. [7], where they show that melatonin treatment increases PAs content by accelerating the conversion of arginine and methionine to polyamines in wheat seedlings. At the same time, melatonin suppresses PAO (polyamine oxidase) and DAO (diamine oxidase) activities—two enzymes involved in polyamines metabolism—which decrease melatonin-induced polyamine degradation, thus improving salt stress tolerance [7]. This provides initial evidence that exogenous melatonin treatment enhances plant salt tolerance by regulating PAs, whether the proposed mechanisms are applicable to other plant species requires further investigation. In addition, polyamines are involved in the melatonin-induced NO production in the roots of Fe deficient *Arabidopsis*, and increase the plant tolerance to Fe deficiency [80]. Thus, the interaction between PAs and NO in melatonin-induced salt stress tolerance of plants requires further confirmation.

3. Melatonin Correlated Genes and Gene Families in Plants

To further investigate melatonin's mechanism in regulating salt tolerance in plants, melatonin biosynthesis- and metabolism-related genes, transcription factors, and other related genes and gene families were summarized.

3.1. Putative Genes Involved in Melatonin-Mediated Salt Stress Tolerance

In a wide range of plant species, the melatonin biosynthesis pathway begins with tryptophan, which is converted to tryptamine by tryptophan decarboxylase. Subsequently, tryptamine is converted to serotonin by tryptamine 5-hydroxylase (T5H). In some of the other plant species, the first two steps of the melatonin biosynthesis pathway are reversed. Tryptophan is first converted into 5-hydroxytrytophan by tryptophan 5-hydroxylase (TPH), and then to serotonin by aromatic-L-amino-acid decarboxylase (TDC/AADC) [86]. Although no TPH enzyme been cloned, the presence of ^{14}C-5-hydroxytryptophan and ^{14}C-serotonin have been detected when using ^{14}C-tryptophan as substrate in *Hypericum perforatum* [87]. In the following two steps, three distinct enzymes and two inversed routes were involved. Serotonin N-acetyltransferase (SNAT) catalyzes serotonin into N-acetylserotonin, and N-acetylserotonin was then converted into melatonin by N-acetylserotonin methyl-transferase (ASMT) or caffeic acid O-methyltransferase (COMT). As ASMT/COMT exhibits substrate affinity towards serotonin, and SNAT has substrate affinity toward 5-methoxytryptamine, serotonin could have been first methylated to 5-methoxytryptamine by ASMT/COMT and then to melatonin by SNAT. Different steps involved in the melatonin biosynthesis pathways may occur in different subcellular locations d. In total, six enzymes are involved in plant melatonin biosynthesis, which are related to four different routes. In an *Arabidopsis AtSNAT* mutant, endogenous melatonin content was lower than that in wild-type *Arabidopsis* seedlings. Moreover, the *AtSNAT* mutant was salt hypersensitive compared to wild-type [22]. The possible functions of apple *MzASMT9* were investigated in *Arabidopsis*. Under salt stress, *MzASMT9* transcript levels were upregulated, and melatonin levels were also increased by the ectopic expression of *MzASMT9*, thus leading to an enhanced salt tolerance in transgenic *Arabidopsis* lines [88]. Although there is no direct evidence about the possible roles of TDC, T5H, and COMT in plant salt tolerance, overexpression and suppression of these genes obviously affected endogenous plant melatonin levels [89–92].

The catabolism of phyto-melatonin has also been reported in recent years. Unlike the biosynthesis of melatonin, the metabolism of phyto-melatonin is either through an enzymatic or non-enzymatic pathway [41]. The major melatonin metabolites in plants are N^1-acetyl-N^2-formyl-5-methoxykynuramine (AFMK) and melatonin hydroxylated derivatives, such as 2-hydroxymelatonin and cyclic-3-hydroxymelatonin (3-OHM) [41,93,94].

In rice, melatonin is catabolized into 2-hydroxymelatonin by melatonin 2-hydroxylase (M2H), which belongs to the 2-oxoglutarate-dependent dioxygenase (2-ODD) superfamily [95]. The first *M2H* gene was cloned from rice in 2015 [96].

Except for genes involved in the biosynthesis and catabolism of phyto-melatonin, transcription factors also play critical roles in the melatonin-mediated salt stress response. Under abiotic stress (salt, drought, and cold), exogenous melatonin significantly improves endogenous melatonin levels and upregulates the expression of C-repeat binding factors (CBFs)/Drought response element Binding 1 factors (DREB1s), thus leading to an increase in transcript levels of multiple stress-responsive genes, including *COR15A*, *RD22*, and *KIN1* [23]. RNA sequencing was performed in cucumber roots with or without melatonin treatment under salt stress. The results show that many transcription factors including WRKY, MYB, NAC, and the ethylene-responsive transcription factor were differentially expressed in melatonin-treated plants compared to control plants under NaCl-induced stress [97].

The effects of melatonin on the expression of genes involved in ROS scavenging under NaCl stress were investigated. The application of 1 mM melatonin induced the expression of *CsCu-ZnSOD*, *CsFe-ZnSOD*, *CsPOD*, and *CsCAT* in cucumber under salt stress [30]. Similar results were also observed in rapeseed, and studies showed that antioxidant defense-related genes such as *APX*, *Cu/ZnSOD* and *MnSOD* were involved in melatonin-induced salt stress tolerance [37]. In tomato seedlings under salt stress, melatonin significantly improved *TRXf* gene expression, which participates in the redox regulation of many physiological processes [44]. Genes responsible for maintaining ion homeostasis were also involved in melatonin-enhanced salt stress. *MdNHX1* and *MdAKT1*, two ion-channel genes, were upregulated by exogenous melatonin in *Malus hupehensis* under salinity [21]. *NHX1* and *SOS2 expression* was also modulated by melatonin in salt-stressed rapeseed. Several studies have shown that melatonin alleviates salinity stress by regulating hormone biosynthesis and metabolism gene expression. Melatonin induced the expression of GA biosynthesis genes (*GA20ox* and *GA3ox*). Meanwhile, the ABA catabolism genes, *CsCYP707A1* and *CsCYP707A2*, were obviously upregulated, whereas the ABA biosynthesis gene *CsNECD2* was downregulated by melatonin in salt-stressed cucumber seedlings [30].

3.2. Comparative and Phylogenetic Analysis of TDC, T5H, SNAT, and ASMT Gene Families in Plants

TDC, T5H, SNAT, and *ASMT* correlate with melatonin biosynthesis in most plant species [86]. Recently, a genome-wide expression, classification, phylogenetic, and expression profiles of the tryptophan decarboxylase (TDC) gene family was conducted in *Solanum lycopersicum* [98]. A total of five *TDC* genes were obtained from the tomato genome. Among the five candidate genes, *SlTDC3* was expressed in all the tested tissues, whereas *SlTDC1* and *SlTDC2* were specifically expressed in the fruit and leaves of the tomato plant, respectively. *SlTDC4* and *SlTDC5* are not expressed in tomato. The study of *TDC* genes in rice is relatively clearer compared to other plants. Rice has at least three *TDC* genes [89]. *OsTDC1* (AK31) and *OsTDC2* (AK53) were first identified by Kang et al. [99]. Heterologous expression of *OsTDC1* and *OsTDC2* in *Escherichia coli* showed that both genes exhibited TDC activity [99]. The expression profiles of *OsTDC1, OsTDC2,* and *OsTDC3* have also been investigated in rice. *OsTDC1* and *OsTDC2* have similar expression profiles, with low expression in seedling shoots, and relatively high levels in leafs, stems, roots and flowers. In comparison, *OsTDC3* expression was very low in almost all tested organs, except the roots [89]. These results indicated that different *TDC* genes might play different roles during plant growth and development. Overexpression of *OsTDC1, OsTDC2,* and especially *OsTDC3* leads to improved melatonin levels in transgenic rice [89]. The phylogenetic relationships and gene structures of TDCs from algae to higher plants showed that they are found throughout the high plant kingdom with a small family size. The evolution of *TDC* genes in plants was mainly through gene expansion and intron loss events. This is the first research of its kind on the TDC gene family; however, the expression profiles of TDCs were not investigated under the salt stress condition [98–100]. The ASMT gene family was also analyzed in *Solanum lycopersicum* [101]. There are 14 candidate *ASMT* genes involved in tomato, three of which may be pseudogenes. The expression patterns of *SlASMTs* suggested that four *SlASMTs* were involved in tomato plant response to biotic stresses [101].

TDC, T5H, SNAT, and genes have been identified and functionally analyzed in many plants, especially in rice [89,91,102–104]. A systematic analysis of the tomato TDC gene family has been conducted, and the phylogenetic relationships between *TDC* genes in plants have also been analyzed. In addition to the *ASMT* gene families in tomato, genome-wide analysis of *SNAT, ASMT,* and *T5H* families has not been reported. Based on the methods described by Pang et al. [98] and Liu et al. [101], we searched *TDC* genes in wheat genome, as well as *SNAT* and *ASMT* genes in 10 plant species from algae to higher plants. We further validated these *TDC* and *ASMT* genes using the previously reported main residues [105–107]. Only BLASTP (identity >70%, coverage >70%) was conducted for *T5H* genes identification, using rice *T5H* genes as the query. A total of eight *T5H,* 37 *SNAT,* and 140 *ASMT* candidate genes were obtained in 10 plant species (Supplementary Table S1). Furthermore, there are 33 candidate *TDC* genes in wheat. Phylogenetic relationships of *SNAT* and *ASMT* are shown in Figures 1 and 2. Based on the phylogenetic tree topology, the SNAT gene family could be divided into four groups (Group I to IV). *SNAT* members in Group I are highly conserved across all species. Similar numbers of *SNAT* genes were found in different species, and no obvious gene expansion was observed. *OsSNAT2* of rice belongs to Group I, whose function is already revealed [108]. The *ASMT* gene family phylogenetic tree is similar to that of the *TDC* gene family [100]. One member from *Volvox carteri* clustered into a separate branch, indicating that *ASMT* genes originated before the divergence of green algae and land plant species. The average gene number of *ASMT* in algae, pteridophyta, gymnosperms, and angiosperms is 1, 4, 25, and 18.3, respectively, suggesting that gene expansion occurred during the evolution from algae to higher plants.

Furthermore, we specially investigated the expression profiles of *TDC, T5H, SNAT,* and *ASMT* genes in wheat under salt stress. RNA-sequencing data were downloaded from the NCBI Sequence Read Archive (SRA) ddabase (https://www.ncbi.nlm.nih.gov/sra/). FPKM (fragments per kilobase of transcript per million fragments mapped) values for all candidate genes in wheat were calculated using Hisat2 and Stringtie, and the heat maps were generated using the geom_tile method in ggplot2 [109]. As shown in Figure 3, there are four *TDC* genes, two *T5H* genes, one *SNAT* gene, and 10 *ASMT* genes specifically expressed under salt stress, and lots of genes are upregulated under salt conditions, indicating that these genes could be involved in the salt stress tolerance of wheat.

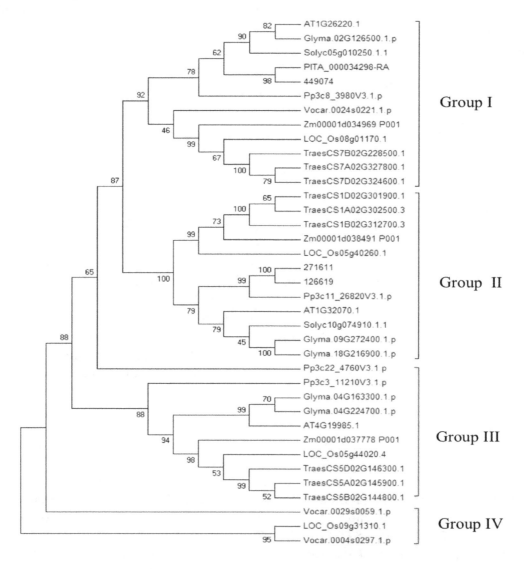

Figure 1. Phylogenetic relationship of the serotonin N-acetyltransferase (*SNAT*) genes from 10 plant species. The candidate *SNAT* genes involved in the phylogenetic tree include the dicots (*Arabidopsis.thaliana* (AT): AT1G26220.1, AT1G32070.1, and AT4G19985.1; *Solanum lycopersicum* (Solyc): Solyc05g010250.1.1, and Solyc10g074910.1.1; *Glyma max* (Glyma): Glyma.02G126500.1.p, Glyma.04G163300.1.p, Glyma.04G224700.1.p, Glyma.09G272400.1.p, and Glyma.18G216900.1.p), monocot (*Zea mays* (Zm): Zm00001d037778_P001, Zm00001d034969_P001, and Zm00001d038491_P001; *Oryza sativa* (LOC_Os): LOC_Os05g40260.1, LOC_Os05g44020.4, LOC_Os08g01170.1, and LOC_Os09g31310.1; *Triticum aestivum* (Traes): TraesCS5D02G146300.1, TraesCS7B02G228500.1, TraesCS7A02G327800.1, TraesCS7D02G324600.1, TraesCS5A02G145900.1, TraesCS1D02G301900.1, TraesCS5B02G144800.1, TraesCS1B02G312700.3, and TraesCS1A02G302500.3), Gymnospermae (*Pinus taeda* (PITA): PITA_000034298-RA), Pteridophyta (*Selaginella moellendorffii*: 271611, 449074, and 126619), Bryophyta (*Physcomitrella patens* (Pp): Pp3c22_4760V3.1.p, Pp3c3_11210V3.1.p, Pp3c11_26820V3.1.p, and Pp3c8_3980V3.1.p), and algae (*Volvox carteri* (Vocar): Vocar.0029s0059.1.p, Vocar.0004s0297.1.p, and Vocar.0024s0221.1.p)

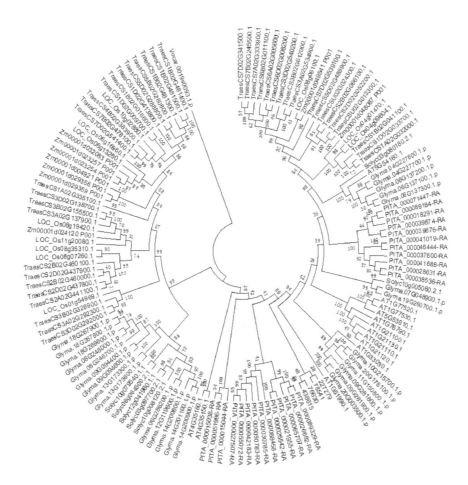

Figure 2. Phylogenetic relationship of the N-acetylserotonin methyl-transferase (*ASMT*) genes from 10 plant species. The 10 plant species include *A.thaliana, S.lycopersicum, G.max, Z.mays, O.sativa, T.aestivum, P.taeda, S.moellendorffii, P.patens,* and algae.

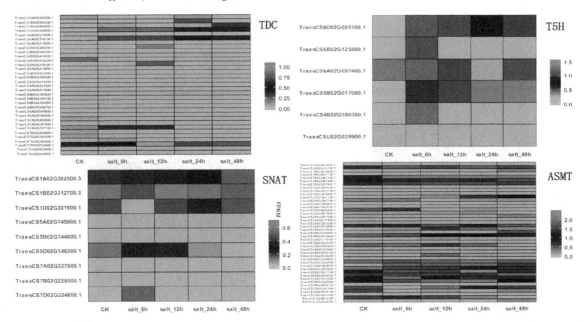

Figure 3. Expression profiles of *TDC, T5H, SNAT,* and *ASMT* genes in wheat under salt stress conditions. The red or green colors represent the higher or lower relative abundance of each transcript in each sample, respectively.

4. Phyto-Melatonin Receptor

It is clear that exogenous melatonin plays a considerable role during plant growth and development, and is associated with plant stress responses—including salt stress. However, the method by which plants perceive exogenous melatonin and convert it into downstream signals remains unknown. The phyto-melatonin receptor holds promise for better understanding melatonin's biological function and mechanism. Animal melatonin receptors were discovered earlier than the phyto-melatonin receptor. The first melatonin receptor (Mel1c) was cloned from frogs (*Xenopus laevis*) in 1994 [110]. Melatonin receptors belong to the G protein-coupled receptor (GPCR) superfamily, which possess seven transmembrane helices [111]. To date, a total of three melatonin receptor subtyoes have been reported in mammals; MT1 (Mel1a), MT2 (Mel1b), and MT3 (ML2) [112,113]. MT1 and MT2 are G protein-coupled receptors, which exhibit high-affinity for melatonin [112,114], while MT3 exhibits low affinity for melatonin and it belongs to the quinone reductases family [115].

AtCAND2/PMTR1, the first phyto-melatonin receptor, was recently discovered in *Arabidopsis*. When melatonin is perceived by CAND2/PMTR1, it triggers the dissociation of Gα form Gγβ, which activates the downstream H_2O_2 and Ca^{2+} signaling transduction cascade, leading to the phenotype of stomatal closure. Several studies have identified CAND2 as the first phyto-melatonin receptor. *AtCAND2* is a membrane protein with seven transmembrane helices. Interaction with the unique G proteinαsubunit (GPA1) of *Arabidopsis* proved that CAND2 is a G protein-coupled receptor. ^{125}I-melatonin can bind to CAND2 in a specific and saturated manner. *Arabidopsis AtCand2* mutant exhibits no changes in the stomatal aperture when treated with melatonin, while 10 μmol/L melatonin induced stomatal closure in the wild-type counterparts [114]. These data indicate that further research on CAND2/PMTR1-mediated signaling in salt stress is required. Moreover, the discovery of CAND2/PMTR1 provides a new method for finding other melatonin receptors in plants.

5. Conclusions and Future Perspectives

Melatonin, as an antioxidant and signaling molecule, modulates a wide range of physiological functions in bacteria, fungi, invertebrates, vertebrates, algae, and plants. It has been extensively studied in humans and other animals, while plant studies have lagged behind. In light of its importance and significance, more and more attention has focused on the biosynthesis and bio-function of melatonin in plants. It has become a research hotspot in the plant biology kingdom, with increasing research being conducted in recent years [116,117]. To promote related research in plant salt tolerance, we summarized the regulatory roles and mechanisms of melatonin in plants during salt stress resistance by reviewing recently published literature, and we finally propose a model (Figure 4).

First, salt stress or the application of exogenous melatonin improves endogenous melatonin levels in plants, which modulates the expression of genes involved in melatonin biosynthesis and metabolisms or assimilates exogenous melatonin directly [116]. Increased levels of endogenous melatonin occur mainly by upregulation of melatonin biosynthesis-related genes or absorption of exogenous melatonin by plants; both mechanisms require further investigation. Increased endogenous levels enhanced plant salt stress tolerance via several different pathways. The improvement of antioxidant capacity, ion homeostasis, photosynthetic capacity and the regulation of ROS, NO, hormone, and polyamine metabolism by melatonin in salt-stressed plants was discussed. Previous studies have shown that Ca^{2+} signaling plays important roles in salt stress tolerance [118]; however, little evidence of Ca^{2+} signaling was observed in the melatonin-induced salt stress tolerance. Therefore, whether melatonin enhances plants salinity resistance through Ca^{2+} signaling requires further investigation.

Genetic modification and RNA-sequencing analysis are effective tools in the identification of the putative target genes involved in melatonin-enhanced salt stress tolerance. TPH, a putative gene involved in serotonin biosynthesis, has not been cloned in plants yet. However, *TDC* and *T5H*, two genes involved in serotonin biosynthesis, have been identified in many plants, but have not been cloned in *Arabidopsis*. We suspect that other biosynthesis pathways of melatonin may also exist in plants.

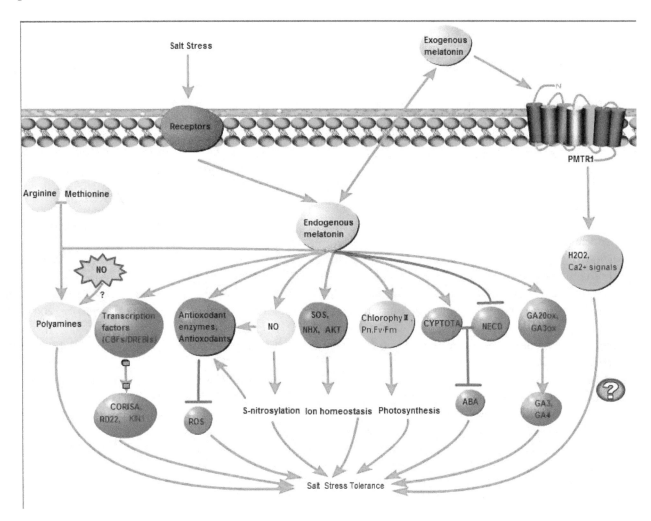

Figure 4. Melatonin-mediated salt stress response in plants. Abbreviation: NO, nitric oxide; ROS, reactive oxygen species; Pn, net photosynthetic rate; ABA, abscisic acid; GA, gibberellin acid. ⊥: represents inhibition; and →: represents promotion.

Plant melatonin receptors have been the bottleneck in the study of phyto-melatonin in the past few decades. With the first phytomelatonin receptor discovered recently in *Arabidopsis*, the involvement of PMTR1-mediated phytomelatonin signaling in salt stress response requires updated exploration. In addition, three melatonin receptors MT1, MT2, and MT3, have been identified in mammals, the identification of new phytomelatonin receptors is another exciting field to explore. Further studies in this field might deepen our understanding of the biological functions and molecular mechanisms governing melatonin's regulatory role during salt stress tolerance and beyond.

Author Contributions: Conceptualization, X.N.; Formal Analysis, H.Z., T.Z., and S.L.; Resources, W.T. and W.S.; Data Curation, H.Z., X.W. and X.D.; Writing—Original Draft Preparation, H.Z.; Writing—Review and Editing, X.N. and W.S.; Supervision, W.S. and W.T.; Funding Acquisition, W.S. and X.N.

Acknowledgments: We are grateful to Hong Yue for her help with the phylogeny analysis, and Kewei Feng for his help on executing the figures.

References

1.   Munns, R.; Tester, M. Mechanisms of salinity tolerance. *Annu. Rev. Plant Biol.* **2008**, *59*, 651–681. [CrossRef]
2.   Abbasi, H.; Jamil, M.; Haq, A.; Ali, S.; Ahmad, R.; Malik, Z.; Parveen. Salt stress manifestation on plants, mechanism of salt tolerance and potassium role in alleviating it: A review. *Zemdirbyste-Agriculture* **2016**, *103*, 229–238. [CrossRef]
3.   Rahnama, A.; James, R.; Poustini, K.; Munns, R. Stomatal conductance as a screen for osmotic stress tolerance in durum wheat growing in saline soil. *Funct. Plant Biol.* **2010**, *37*, 255–263. [CrossRef]
4.   Ashraf, M.; Wu, L. Breeding for salinity tolerance in plants. *Crit. Rev. Plant Sci.* **1994**, *13*, 17–42. [CrossRef]
5.   Shalata, A.; Mittova, V.; Volokita, M.; Guy, M.; Tal, M. Response of the cultivated tomato and its wild salt-tolerant relative Lycopersicon pennellii to salt-dependent oxidative stress: The root antioxidative system. *Physiol. Plant* **2001**, *112*, 487–494. [CrossRef] [PubMed]
6.   Hasanuzzaman, M.; Oku, H.; Nahar, K.; Bhuyan, M.H.M.B.; Mahmud, J.A.; Baluska, F.; Fujita, M. Nitric oxide-induced salt stress tolerance in plants: ROS metabolism, signaling, and molecular interactions. *Plant Biotechnol. Rep.* **2018**, *12*, 77–92. [CrossRef]
7.   Ke, Q.; Ye, J.; Wang, B.; Ren, J.; Yin, L.; Deng, X.; Wang, S. Melatonin mitigates salt stress in wheat seedlings by modulating polyamine metabolism. *Front. Plant Sci.* **2018**, *9*, 914. [CrossRef]
8.   Lerner, A.B.; Case, J.D.; Takahashi, Y.; Lee, T.H.; Mori, W. Isolation of melatonin, the pineal gland factor that lightens melanocyteS1. *J. Am. Chem. Soc.* **1958**, *80*, 2587. [CrossRef]
9.   Brainard, G.C.; Hanifin, J.P.; Greeson, J.M.; Byrne, B.; Glickman, G.; Gerner, E.; Rollag, M.D. Action spectrum for melatonin regulation in humans: evidence for a novel circadian photoreceptor. *J. Neurosci.* **2001**, *21*, 6405–6412. [CrossRef] [PubMed]
10. Mishima, K. Melatonin as a regulator of human sleep and circadian systems. *Nihon Rinsho* **2012**, *70*, 1139–1144, [Article in Japanese].
11. Rodriguez, C.; Mayo, J.C.; Sainz, R.M.; Antolín, I.; Herrera, F.; Martín, V.; Reiter, R.J. Regulation of antioxidant enzymes: a significant role for melatonin. *J. Pineal Res.* **2004**, *36*, 1–9. [CrossRef] [PubMed]
12. Calvo, J.R.; González-Yanes, C.; Maldonado, M.D. The role of melatonin in the cells of the innate immunity: a review. *J. Pineal Res.* **2013**, *55*, 103–120. [CrossRef] [PubMed]
13. Barrett, P.; Bolborea, M. Molecular pathways involved in seasonal body weight and reproductive responses governed by melatonin. *J. Pineal Res.* **2012**, *52*, 376–388. [CrossRef] [PubMed]
14. Dollins, A.B.; Zhdanova, I.V.; Wurtman, R.J.; Lynch, H.J.; Deng, M.H. Effect of inducing nocturnal serum melatonin concentrations in daytime on sleep, mood, body temperature, and performance. *Proc. Natl. Acad. Sci. USA* **1994**, *91*, 1824–1828. [CrossRef]
15. Hattori, A.; Migitaka, H.; Iigo, M.; Itoh, M.; Yamamoto, K.; Ohtani-Kaneko, R.; Hara, M.; Suzuki, T.; Reiter, R.J. Identification of melatonin in plants and its effects on plasma melatonin levels and binding to melatonin receptors in vertebrates. *Biochem. Mol. Biol. Int.* **1995**, *35*, 627–634. [PubMed]
16. Dubbels, R.; Reiter, R.J.; Klenke, E.; Goebel, A.; Schnakenberg, E.; Ehlers, C.; Schiwara, H.W.; Schloot, W. Melatonin in edible plants identified by radioimmunoassay and by high performance liquid chromatography-mass spectrometry. *J. Pineal Res.* **1995**, *18*, 28–31. [CrossRef] [PubMed]
17. Wang, L.Y.; Liu, J.L.; Wang, W.X.; Sun, Y. Exogenous melatonin improves growth and photosynthetic capacity of cucumber under salinity-induced stress. *Photosynthetica* **2016**, *54*, 19–27. [CrossRef]
18. Tan, D.-X.; Hardeland, R.; Manchester, L.C.; Korkmaz, A.; Ma, S.; Rosales-Corral, S.; Reiter, R.J. Functional roles of melatonin in plants, and perspectives in nutritional and agricultural science. *J. Exp. Bot.* **2012**, *63*, 577–597. [CrossRef] [PubMed]
19. Yu, Y.; Lv, Y.; Shi, Y.; Li, T.; Chen, Y.; Zhao, D.; Zhao, Z. The Role of Phyto-Melatonin and Related Metabolites in Response to Stress. *Molecules* **2018**, *23*, 1887. [CrossRef] [PubMed]
20. Xia, H.; Ni, Z.; Pan, D. Effects of exogenous melatonin on antioxidant capacity in Actinidia seedlings under salt stress. *IOP Conf. Ser. Earth Environ. Sci.* **2017**, *94*, 012024. [CrossRef]
21. Li, C.; Wang, P.; Wei, Z.; Liang, D.; Liu, C.; Yin, L.; Jia, D.; Fu, M.; Ma, F. The mitigation effects of exogenous melatonin on salinity-induced stress in Malus hupehensis. *J. Pineal Res.* **2012**, *53*, 298–306. [CrossRef] [PubMed]

22. Chen, Z.; Xie, Y.; Gu, Q.; Zhao, G.; Zhang, Y.; Cui, W.; Xu, S.; Wang, R.; Shen, W. The AtrbohF-dependent regulation of ROS signaling is required for melatonin-induced salinity tolerance in Arabidopsis. *Free Radic. Bio. Med.* **2017**, *108*, 465–477. [CrossRef] [PubMed]

23. Shi, H.; Qian, Y.; Tan, D.-X.; Reiter, R.J.; He, C. Melatonin induces the transcripts of CBF/DREB1s and their involvement in both abiotic and biotic stresses in Arabidopsis. *J. Pineal Res.* **2015**, *59*, 334–342. [CrossRef] [PubMed]

24. Zheng, X.; Tan, D.X.; Allan, A.C.; Zuo, B.; Zhao, Y.; Reiter, R.J.; Wang, L.; Wang, Z.; Guo, Y.; Zhou, J.; et al. Chloroplastic biosynthesis of melatonin and its involvement in protection of plants from salt stress. *Scientific reports* **2017**, *7*, 41236. [CrossRef] [PubMed]

25. Shi, H.; Jiang, C.; Ye, T.; Tan, D.-x.; Reiter, R.J.; Zhang, H.; Liu, R.; Chan, Z. Comparative physiological, metabolomic, and transcriptomic analyses reveal mechanisms of improved abiotic stress resistance in bermudagrass (*Cynodon dactylon* (L). Pers.) by exogenous melatonin. *J. Exp. Bot.* **2015**, *66*, 681–694. [CrossRef] [PubMed]

26. Beilby, M.J.; Al Khazaaly, S.; Bisson, M.A. Salinity-induced noise in membrane potential of Characeae chara australis: effect of exogenous melatonin. *J. Membrane Biol.* **2015**, *248*, 93–102. [CrossRef] [PubMed]

27. Zhang, Y.; Gao, W.; Lv, Y.; Bai, Q.; Wang, Y. Exogenous melatonin confers salt stress tolerance to Chlamydomonas reinhardtii (Volvocales, Chlorophyceae) by improving redox homeostasis. *Phycologia* **2018**, *57*, 680–691. [CrossRef]

28. Kostopoulou, Z.; Therios, I.; Roumeliotis, E.; Kanellis, A.K.; Molassiotis, A. Melatonin combined with ascorbic acid provides salt adaptation in *Citrus aurantium* L. seedlings. *Plant Physiol. Bioch.* **2015**, *86*, 155–165. [CrossRef]

29. Zhang, N.; Zhang, H.-J.; Sun, Q.-Q.; Cao, Y.-Y.; Li, X.; Zhao, B.; Wu, P.; Guo, Y.-D. Proteomic analysis reveals a role of melatonin in promoting cucumber seed germination under high salinity by regulating energy production. *Sci. Rep.* **2017**, *7*, 503. [CrossRef]

30. Zhang, H.-J.; Zhang, N.; Yang, R.-C.; Wang, L.; Sun, Q.-Q.; Li, D.-B.; Cao, Y.-Y.; Weeda, S.; Zhao, B.; Ren, S.; et al. Melatonin promotes seed germination under high salinity by regulating antioxidant systems, ABA and GA4 interaction in cucumber (*Cucumis sativus* L.). *J. Pineal Res.* **2014**, *57*, 269–279. [CrossRef]

31. Chen, Y.-E.; Mao, J.-J.; Sun, L.-Q.; Huang, B.; Ding, C.-B.; Gu, Y.; Liao, J.-Q.; Hu, C.; Zhang, Z.-W.; Yuan, S.; et al. Exogenous melatonin enhances salt stress tolerance in maize seedlings by improving antioxidant and photosynthetic capacity. *Physiol. Plant* **2018**, *164*, 349–363. [CrossRef] [PubMed]

32. Jiang, X.; Li, H.; Song, X. Seed priming with melatonin effects on seed germination and seedling growth in maize under salinity stress. *Pak. J. Bot.* **2016**, *48*, 1345–1352.

33. Jiang, C.; Cui, Q.; Feng, K.; Xu, D.; Li, C.; Zheng, Q. Melatonin improves antioxidant capacity and ion homeostasis and enhances salt tolerance in maize seedlings. *Acta Physiol. Plant.* **2016**, *38*, 82. [CrossRef]

34. Jiang, Y.; Liang, D.; Liao, M.A.; Lin, L. Effects of melatonin on the growth of radish Seedlings under salt stress. In Proceedings of the 3rd international conference on renewable energy and environmental technology (ICERE 2017), Hanoi, Vietnam, 25–27 February 2017.

35. Yao, H.; Wang, X.; Liao, M.A.; Lin, L. Effects of melatonin treated radish on the growth of following stubble lettuce under salt stress. In Proceedings of the 3rd international conference on renewable energy and environmental technology (ICERE 2017), Hanoi, Vietnam, 25–27 February 2017.

36. Zeng, L.; Cai, J.S.; Li, J.J.; Lu, G.Y.; Li, C.S.; Fu, G.P.; Zhang, X.K.; Ma, H.Q.; Liu, Q.Y.; Zou, X.L.; et al. Exogenous application of a low concentration of melatonin enhances salt tolerance in rapeseed (Brassica napus L.) seedlings. *J. Integ. Agr* **2018**, *17*, 328–335. [CrossRef]

37. Zhao, G.; Zhao, Y.; Yu, X.; Kiprotich, F.; Han, H.; Guan, R.; Wang, R.; Shen, W. Nitric oxide is required for melatonin-enhanced tolerance against salinity stress in rapeseed (*Brassica napus* L.) seedlings. *Int. J. MolSci.* **2018**, *19*, 1912. [CrossRef]

38. Liang, C.; Zheng, G.; Li, W.; Wang, Y.; Hu, B.; Wang, H.; Wu, H.; Qian, Y.; Zhu, X.-G.; Tan, D.-X.; et al. Melatonin delays leaf senescence and enhances salt stress tolerance in rice. *J. Pineal Res.* **2015**, *59*, 91–101. [CrossRef] [PubMed]

39. Li, X.; Yu, B.; Cui, Y.; Yin, Y. Melatonin application confers enhanced salt tolerance by regulating Na+ and Cl− accumulation in rice. *Plant Growth Regul.* **2017**, *83*, 441–454. [CrossRef]

40. Wei, W.; Li, Q.-T.; Chu, Y.-N.; Reiter, R.J.; Yu, X.-M.; Zhu, D.-H.; Zhang, W.-K.; Ma, B.; Lin, Q.; Zhang, J.-S.; et al. Melatonin enhances plant growth and abiotic stress tolerance in soybean plants. *J. Exp Bot.* **2015**, *66*, 695–707. [CrossRef] [PubMed]

41. Arora, D.; Bhatla, S.C. Melatonin and nitric oxide regulate sunflower seedling growth under salt stress accompanying differential expression of Cu/Zn SOD and Mn SOD. *Free Radical Biol. Med.* **2017**, *106*, 315–328. [CrossRef]

42. Kaur, H.; Bhatla, S.C. Melatonin and nitric oxide modulate glutathione content and glutathione reductase activity in sunflower seedling cotyledons accompanying salt stress. *Nitric Oxide* **2016**, *59*, 42–53. [CrossRef]

43. Yu, Y.; Wang, A.; Li, X.; Kou, M.; Wang, W.; Chen, X.; Xu, T.; Zhu, M.; Ma, D.; Li, Z.; et al. Melatonin-stimulated triacylglycerol breakdown and energy turnover under salinity stress contributes to the maintenance of plasma membrane H^+-ATPase activity and K^+/Na^+ homeostasis in sweet potato. *Front. Plant Sci.* **2018**, *9*, 256. [CrossRef] [PubMed]

44. Zhou, X.; Zhao, H.; Cao, K.; Hu, L.; Du, T.; Baluška, F.; Zou, Z. Beneficial roles of melatonin on redox regulation of photosynthetic electron transport and synthesis of D1 protein in tomato seedlings under salt stress. *Front. Plant Sci.* **2016**, *7*, 1823. [CrossRef] [PubMed]

45. Dawood, M.G.; El-Awadi, M.E. Alleviation of salinity stress on *Vicia faba* L. plants via seed priming with melatonin. *Acta Biológica Colombiana* **2015**, *20*, 223–235. [CrossRef]

46. Li, H.; Chang, J.; Chen, H.; Wang, Z.; Gu, X.; Wei, C.; Zhang, Y.; Ma, J.; Yang, J.; Zhang, X. Exogenous Melatonin Confers Salt Stress Tolerance to Watermelon by Improving Photosynthesis and Redox Homeostasis. *Front. Plant Sci.* **2017**, *8*, 295. [CrossRef] [PubMed]

47. El-Mashad, A.A.A.; Mohamed, H.I. Brassinolide alleviates salt stress and increases antioxidant activity of cowpea plants (*Vigna sinensis*). *Protoplasma* **2012**, *249*, 625–635. [CrossRef] [PubMed]

48. Zhang, M.; Smith, J.A.C.; Harberd, N.P.; Jiang, C. The regulatory roles of ethylene and reactive oxygen species (ROS) in plant salt stress responses. *Plant Mol. Biol.* **2016**, *91*, 651–659. [CrossRef] [PubMed]

49. Ahmad, P.; Abdul Jaleel, C.; A Salem, M.; Nabi, G.; Sharma, S. Roles of Enzymatic and non-enzymatic antioxidants in plants during abiotic stress. *Crit Rev Biotechnol.* **2010**, *30*, 161–175. [CrossRef]

50. Tan, D.X.; Manchester, L.C.; Terron, M.P.; Flores, L.J.; Reiter, R.J. One molecule, many derivatives: A never-ending interaction of melatonin with reactive oxygen and nitrogen species? *J. Pineal Res.* **2007**, *42*, 28–42. [CrossRef]

51. Campos, L.M.O.; Hsie, S.B.; Granja, A.J.A.; Correia, M.R.; Almeida-Cortez, J.; Pompelli, M.F. Photosynthesis and antioxidant activity in *Jatropha curcas* L. under salt stress. *Braz. J. Plant Physiol.* **2012**, *24*, 55–67. [CrossRef]

52. Meloni, D.A.; Oliva, M.A.; Martinez, C.A.; Cambraia, J. Photosynthesis and activity of superoxide dismutase, peroxidase and glutathione reductase in cotton under salt stress. *Environ. Exp. Bot.* **2003**, *49*, 69–76. [CrossRef]

53. Amtmann, A.; Leigh, R. Ion Homeostasis. In *Abiotic Stress Adaptation in Plants: Physiological, Molecular and Genomic Foundation*; Pareek, A., Sopory, S.K., Bohnert, H.J., Eds.; Springer: Dordrecht, The Netherlands, 2010.

54. Zhu, J.K. Regulation of ion homeostasis under salt stress. *Curr. Opin. Plant Biol.* **2003**, *6*, 441–445. [CrossRef]

55. Fukuda, A.; Nakamura, A.; Hara, N.; Toki, S.; Tanaka, Y. Molecular and functional analyses of rice NHX-type Na^+/H^+ antiporter genes. *Planta* **2011**, *233*, 175–188. [CrossRef] [PubMed]

56. Padan, E.; Venturi, M.; Gerchman, Y.; Dover, N. Na^+/H^+ antiporters. *BBA- Bioenergetics* **2001**, *1505*, 144–157. [CrossRef]

57. Shi, H.; Zhu, J.-K. Regulation of expression of the vacuolar Na^+/H^+ antiporter gene AtNHX1 by salt stress and abscisic acid. *Plant Mol. Biol.* **2002**, *50*, 543–550. [CrossRef] [PubMed]

58. Garriga, M.; Raddatz, N.; Véry, A.-A.; Sentenac, H.; Rubio-Meléndez, M.E.; González, W.; Dreyer, I. Cloning and functional characterization of HKT1 and AKT1 genes of *Fragaria* spp.—Relationship to plant response to salt stress. *J. Plant. Physiol.* **2017**, *210*, 9–17. [CrossRef] [PubMed]

59. Arnao, M.B.; Hernández-Ruiz, J. Melatonin and its relationship to plant hormones. *Ann. Bot* **2018**, *121*, 195–207. [CrossRef]

60. Wang, Q.; An, B.; Wei, Y.; Reiter, R.J.; Shi, H.; Luo, H.; He, C. Melatonin regulates root meristem by repressing auxin synthesis and polar auxin transport in Arabidopsis. *Front. Plant Sci.* **2016**, *7*, 1882. [CrossRef]

61. Pelagio-Flores, R.; Muñoz-Parra, E.; Ortiz-Castro, R.; López-Bucio, J. Melatonin regulates Arabidopsis root system architecture likely acting independently of auxin signaling. *J. Pineal Res.* **2012**, *53*, 279–288. [CrossRef]

62. Chen, Q.; Qi, W.-b.; Reiter, R.J.; Wei, W.; Wang, B.-m. Exogenously applied melatonin stimulates root growth and raises endogenous indoleacetic acid in roots of etiolated seedlings of Brassica juncea. *J. Plant Physiol.* **2009**, *166*, 324–328. [CrossRef]

63. Footitt, S.; Douterelo-Soler, I.; Clay, H.; Finch-Savage, W.E. Dormancy cycling in Arabidopsis seeds is controlled by seasonally distinct hormone-signaling pathways. *Proc. Natl. Acad. Sci. USA* **2011**, *108*, 20236–20241. [CrossRef]

64. Zhang, J.; Shi, Y.; Zhang, X.; Du, H.; Xu, B.; Huang, B. Melatonin suppression of heat-induced leaf senescence involves changes in abscisic acid and cytokinin biosynthesis and signaling pathways in perennial ryegrass (*Lolium perenne* L.). *Environ. Exp. Bot.* **2017**, *138*, 36–45. [CrossRef]

65. Yang, R.; Yang, T.; Zhang, H.; Qi, Y.; Xing, Y.; Zhang, N.; Li, R.; Weeda, S.; Ren, S.; Ouyang, B.; et al. Hormone profiling and transcription analysis reveal a major role of ABA in tomato salt tolerance. *Plant Physiol. Bioch.* **2014**, *77*, 23–34. [CrossRef] [PubMed]

66. Maggio, A.; Barbieri, G.; Raimondi, G.; De Pascale, S. Contrasting Effects of GA3 Treatments on Tomato Plants Exposed to Increasing Salinity. *J. Plant Growth Regul.* **2010**, *29*, 63–72. [CrossRef]

67. Li, C.; Tan, D.-X.; Liang, D.; Chang, C.; Jia, D.; Ma, F. Melatonin mediates the regulation of ABA metabolism, free-radical scavenging, and stomatal behaviour in two Malus species under drought stress. *J. Exp. Bot.* **2015**, *66*, 669–680. [CrossRef] [PubMed]

68. Jia, W.; Zhang, J. Water stress-induced abscisis acid accumulation in relation to reducing agents and sulfhydryl modifiers in maize plant. *Plant Cell Environ.* **2000**, *12*, 1389–1395. [CrossRef]

69. Fu, J.; Wu, Y.; Miao, Y.; Xu, Y.; Zhao, E.; Wang, J.; Sun, H.; Liu, Q.; Xue, Y.; Xu, Y.; et al. Improved cold tolerance in Elymus nutans by exogenous application of melatonin may involve ABA-dependent and ABA-independent pathways. *Scientific Reports* **2017**, *7*, 39865. [CrossRef] [PubMed]

70. Aydogan, S.; Yerer, M.B.; Goktas, A. Melatonin and nitric oxide. *J. Endocrinol. Invest.* **2006**, *29*, 281–287. [CrossRef] [PubMed]

71. Zhao, M.G.; Tian, Q.Y.; Zhang, W.H. Nitric oxide synthase-dependent nitric oxide production is associated with salt tolerance in Arabidopsis. *Plant Physiol.* **2007**, *144*, 206–217. [CrossRef]

72. Lozano-Juste, J.; León, J. Enhanced abscisic acid-mediated responses in *nia1nia2noa1-2* triple mutant impaired in NIA/NR- and *AtNOA1*-dependent nitric oxide biosynthesis in Arabidopsis. *Plant Physiol.* **2010**, *152*, 891–903. [CrossRef]

73. Ninnemann, H.; Maier, J. Indications for the occurrence of nitric oxide synthases in fungi and plants and the involvement in photoconidiation of *Neurospora crassa*. *Photochem. Photobiol.* **1996**, *64*, 393–398. [CrossRef]

74. Corpas, F.J.; Palma, J.M.; Del Río, L.A.; Barroso, J.B. Evidence supporting the existence of L-arginine-dependent nitric oxide synthase activity in plants. *New Phytologist* **2009**, *184*, 9–14. [CrossRef] [PubMed]

75. Gupta, K.J.; Fernie, A.R.; Kaiser, W.M.; van Dongen, J.T. On the origins of nitric oxide. *Trends Plant Sci.* **2011**, *16*, 160–168. [CrossRef] [PubMed]

76. Astier, J.; Rasul, S.; Koen, E.; Manzoor, H.; Besson-Bard, A.; Lamotte, O.; Jeandroz, S.; Durner, J.; Lindermayr, C.; Wendehenne, D. S-nitrosylation: An emerging post-translational protein modification in plants. *Plant Sci.* **2011**, *181*, 527–533. [CrossRef] [PubMed]

77. Gupta, K.J. Protein S-nitrosylation in plants: photorespiratory metabolism and NO signaling. *Sci Signal.* **2011**, *4*, jc1. [CrossRef] [PubMed]

78. Jaffrey, S.R.; Erdjument-Bromage, H.; Ferris, C.D.; Tempst, P.; Snyder, S.H. Protein S-nitrosylation: A physiological signal for neuronal nitric oxide. *Nat. Cell Biol.* **2001**, *3*, 193. [CrossRef]

79. Liu, N.; Gong, B.; Jin, Z.; Wang, X.; Wei, M.; Yang, F.; Li, Y.; Shi, Q. Sodic alkaline stress mitigation by exogenous melatonin in tomato needs nitric oxide as a downstream signal. *J. Plant Physiol.* **2015**, *186-187*, 68–77. [CrossRef] [PubMed]

80. Zhou, C.; Liu, Z.; Zhu, L.; Ma, Z.; Wang, J.; Zhu, J. Exogenous melatonin improves plant iron deficiency tolerance via increased accumulation of polyamine-mediated nitric oxide. *Int. J. Mol. Sci.* **2016**, *17*, 1777. [CrossRef]

81. Shi, H.; Chen, Y.; Tan, D.-X.; Reiter, R.J.; Chan, Z.; He, C. Melatonin induces nitric oxide and the potential mechanisms relate to innate immunity against bacterial pathogen infection in Arabidopsis. *J. Pineal Res.* **2015**, *59*, 102–108. [CrossRef]

82. Masson, P.H.; Takahashi, T.; Angelini, R. Editorial: Molecular mechanisms underlying polyamine functions in plants. *Front. Plant Sci.* **2017**, *8*, 14. [CrossRef]

83. Gill, S.S.; Tuteja, N. Polyamines and abiotic stress tolerance in plants. *Plant Signal. Behav.* **2010**, *5*, 26–33.

84. Sánchez-Rodríguez, E.; Romero, L.; Ruiz, J.M. Accumulation of free polyamines enhances the antioxidant response in fruits of grafted tomato plants under water stress. *J. Plant Physiol.* **2016**, *190*, 72–78. [CrossRef] [PubMed]

85. Gong, X.; Shi, S.; Dou, F.; Song, Y.; Ma, F. Exogenous melatonin alleviates alkaline stress in *Malus hupehensis* Rehd. by regulating the biosynthesis of polyamines. *Molecules* **2017**, *22*, 1542. [CrossRef] [PubMed]

86. Zhao, H.; Zhang, K.; Zhou, X.; Xi, L.; Wang, Y.; Xu, H.; Pan, T.; Zou, Z. Melatonin alleviates chilling stress in cucumber seedlings by upregulation of CsZat12 and modulation of polyamine and abscisic acid metabolism. *Sci. Rep.* **2017**, *7*, 4998. [CrossRef]

87. Back, K.; Tan, D.-X.; Reiter, R.J. Melatonin biosynthesis in plants: Multiple pathways catalyze tryptophan to melatonin in the cytoplasm or chloroplasts. *J. Pineal Res.* **2016**, *61*, 426–437. [CrossRef] [PubMed]

88. Murch, S.J.; KrishnaRaj, S.; Saxena, P.K. Tryptophan is a precursor for melatonin and serotonin biosynthesis in in vitro regenerated St. John's wort (*Hypericum perforatum* L. cv. Anthos) plants. *Plant Cell Rep.* **2000**, *19*, 698–704. [CrossRef]

89. Byeon, Y.; Park, S.; Lee, H.Y.; Kim, Y.-S.; Back, K. Elevated production of melatonin in transgenic rice seeds expressing rice tryptophan decarboxylase. *J. Pineal Res.* **2014**, *56*, 275–282. [CrossRef]

90. Zhao, D.; Wang, R.; Liu, D.; Wu, Y.; Sun, J.; Tao, J. Melatonin and expression of tryptophan decarboxylase gene (TDC) in Herbaceous peony (*Paeonia lactiflora* Pall.) flowers. *Molecules* **2018**, *23*, 1164. [CrossRef]

91. Park, S.; Byeon, Y.; Back, K. Transcriptional suppression of tryptamine 5-hydroxylase, a terminal serotonin biosynthetic gene, induces melatonin biosynthesis in rice (*Oryza sativa* L.). *J. Pineal Res.* **2013**, *55*, 131–137. [CrossRef]

92. Byeon, Y.; Choi, G.-H.; Lee, H.Y.; Back, K. Melatonin biosynthesis requires *N*-acetylserotonin methyltransferase activity of caffeic acid *O*-methyltransferase in rice. *J. Exp. Bot.* **2015**, *66*, 6917–6925. [CrossRef]

93. Hardeland, R. Taxon- and site-specific melatonin catabolism. *Molecules* **2017**, *22*, 2015. [CrossRef]

94. Kanwar, M.K.; Yu, J.; Zhou, J. Phytomelatonin: Recent advances and future prospects. *J. Pineal Res.* **2018**, *65*, e12526. [CrossRef] [PubMed]

95. Wei, Y.; Zeng, H.; Hu, W.; Chen, L.; He, C.; Shi, H. Comparative transcriptional profiling of melatonin synthesis and catabolic genes indicates the possible role of melatonin in developmental and stress responses in rice. *Front. Plant Sci.* **2016**, *7*, 676. [CrossRef] [PubMed]

96. Byeon, Y.; Back, K. Molecular cloning of melatonin 2-hydroxylase responsible for 2-hydroxymelatonin production in rice (Oryza sativa). *J. Pineal Res.* **2015**, *58*, 343–351. [CrossRef] [PubMed]

97. Zhang, N.; Zhang, H.J.; Zhao, B.; Sun, Q.Q.; Cao, Y.Y.; Li, R.; Wu, X.X.; Weeda, S.; Li, L.; Ren, S.; et al. The RNA-seq approach to discriminate gene expression profiles in response to melatonin on cucumber lateral root formation. *J. Pineal Res.* **2014**, *56*, 39–50. [CrossRef] [PubMed]

98. Pang, X.; Wei, Y.; Cheng, Y.; Pan, L.; Ye, Q.; Wang, R.; Ruan, M.; Zhou, G.; Yao, Z.; Li, Z.; et al. The Tryptophan Decarboxylase in *Solanum lycopersicum*. *Molecules* **2018**, *23*, 998. [CrossRef] [PubMed]

99. Kang, S.; Kang, K.; Lee, K.; Back, K. Characterization of rice tryptophan decarboxylases and their direct involvement in serotonin biosynthesis in transgenic rice. *Planta* **2007**, *227*, 263–272. [CrossRef] [PubMed]

100. Fan, J.B.; Xie, Y.; Zhang, Z.C.; Chen, L. Melatonin: A Multifunctional Factor in Plants. *Int. J. Mol. Sci.* **2018**, *19*, 1528. [CrossRef]

101. Liu, W.; Zhao, D.; Zheng, C.; Chen, C.; Peng, X.; Cheng, Y.; Wan, H. Genomic analysis of the ASMT gene family in *Solanum lycopersicum*. *Molecules* **2017**, *22*, 1984. [CrossRef]

102. Kang, K.; Lee, K.; Park, S.; Byeon, Y.; Back, K. Molecular cloning of rice serotonin N-acetyltransferase, the penultimate gene in plant melatonin biosynthesis. *J. Pineal Res.* **2013**, *55*, 7–13. [CrossRef]

103. Byeon, Y.; Lee, H.Y.; Lee, K.; Park, S.; Back, K. Cellular localization and kinetics of the rice melatonin biosynthetic enzymes SNAT and ASMT. *J. Pineal Res.* **2014**, *56*, 107–114. [CrossRef]

104. Kang, S.; Kang, K.; Lee, K.; Back, K. Characterization of tryptamine 5-hydroxylase and serotonin synthesis in rice plants. *Plant Cell Rep.* **2007**, *26*, 2009–2015. [CrossRef]

105. Torrens-Spence, M.P.; Liu, P.; Ding, H.; Harich, K.; Gillaspy, G.; Li, J. Biochemical evaluation of the decarboxylation and decarboxylation-deamination activities of plant aromatic amino acid decarboxylases. *J. Biol. Chem.* **2013**, *288*, 2376–2387. [CrossRef] [PubMed]

106. Torrens-Spence, M.P.; Lazear, M.; von Guggenberg, R.; Ding, H.; Li, J. Investigation of a substrate-specifying residue within Papaver somniferum and Catharanthus roseus aromatic amino acid decarboxylases. *Phytochemistry* **2014**, *106*, 37–43. [CrossRef]

107. Kang, K.; Kong, K.; Park, S.; Natsagdorj, U.; Kim, Y.S.; Back, K. Molecular cloning of a plant N-acetylserotonin methyltransferase and its expression characteristics in rice. *J. Pineal Res.* **2011**, *50*, 304–309. [CrossRef] [PubMed]

108. Byeon, Y.; Lee, H.Y.; Back, K. Cloning and characterization of the serotonin N-acetyltransferase-2 gene (SNAT2) in rice (*Oryza sativa*). *J. Pineal Res.* **2016**, *61*, 198–207. [CrossRef] [PubMed]

109. Maag, J.L.V. gganatogram: An R package for modular visualisation of anatograms and tissues based on ggplot2. *F1000Research* **2018**, *7*, 1576. [CrossRef] [PubMed]

110. Ebisawa, T.; Karne, S.; Lerner, M.R.; Reppert, S.M. Expression cloning of a high-affinity melatonin receptor from Xenopus dermal melanophores. *Proc. Natl. Acad. Sci. USA* **1994**, *91*, 6133–6137. [CrossRef]

111. Ng, K.Y.; Leong, M.K.; Liang, H.; Paxinos, G. Melatonin receptors: distribution in mammalian brain and their respective putative functions. *Brain Struct. Funct.* **2017**, *222*, 2921–2939. [CrossRef]

112. Witt-Enderby, P.A.; Bennett, J.; Jarzynka, M.J.; Firestine, S.; Melan, M.A. Melatonin receptors and their regulation: biochemical and structural mechanisms. *Life Sci.* **2003**, *72*, 2183–2198. [CrossRef]

113. Dubocovich, M.L.; Delagrange, P.; Krause, D.N.; Sugden, D.; Cardinali, D.P.; Olcese, J. International union of basic and clinical pharmacology. LXXV. Nomenclature, classification, and pharmacology of G protein-coupled melatonin receptors. *Pharmacol. Rev.* **2010**, *62*, 343–380. [CrossRef]

114. Wei, J.; Li, D.-X.; Zhang, J.-R.; Shan, C.; Rengel, Z.; Song, Z.-B.; Chen, Q. Phytomelatonin receptor PMTR1-mediated signaling regulates stomatal closure in Arabidopsis thaliana. *J. Pineal Res.* **2018**, *65*, e12500. [CrossRef] [PubMed]

115. Nosjean, O.; Ferro, M.; Cogé, F.; Beauverger, P.; Henlin, J.-M.; Lefoulon, F.; Fauchère, J.-L.; Delagrange, P.; Canet, E.; Boutin, J.A. Identification of the Melatonin-binding SiteMT 3 as the Quinone Reductase 2. *J. Biol. Chem.* **2000**, *275*, 31311–31317. [CrossRef] [PubMed]

116. Zhang, N.; Sun, Q.; Zhang, H.; Cao, Y.; Weeda, S.; Ren, S.; Guo, Y.-D. Roles of melatonin in abiotic stress resistance in plants. *J. Exp. Bot.* **2015**, *66*, 647–656. [CrossRef] [PubMed]

117. Tan, D.X.; Manchester, L.C.; Liu, X.; Rosales-Corral, S.A.; Acuna-Castroviejo, D.; Reiter, R.J. Mitochondria and chloroplasts as the original sites of melatonin synthesis: a hypothesis related to melatonin's primary function and evolution in eukaryotes. *J. Pineal Res.* **2013**, *54*, 127–138. [CrossRef] [PubMed]

118. Park, S.-Y.; B Seo, S.; J Lee, S.; G Na, J.; Kim, Y.J. Mutation in PMR1, a Ca^{2+}-ATPase in Golgi, confers salt tolerance in Saccharomyces cerevisiae by inducing expression of PMR2, an Na^{+}-ATPase in plasma membrane. *J. Biol. Chem.* **2001**, *276*, 28694–28699. [CrossRef] [PubMed]

10

Analysis of bZIP Transcription Factor Family and their Expressions under Salt Stress in *Chlamydomonas reinhardtii*

Chunli Ji, Xue Mao, Jingyun Hao, Xiaodan Wang, Jinai Xue, Hongli Cui and Runzhi Li *

Institute of Molecular Agriculture and Bioenergy, Shanxi Agricultural University, Taigu 030801, China; jichunnli@sxau.edu.cn (C.J.); maoxue@sxau.edu.cn (X.M.); haojingyun@stu.sxau.edu.cn (J.H.); wangxiaodan@sxau.edu.cn (X.W.); xuejinai@sxau.edu.cn (J.X.); cuihongli@sxau.edu.cn (H.C.)
* Correspondence: lirunzhi@sxau.edu.cn

Abstract: The basic leucine-region zipper (bZIP) transcription factors (TFs) act as crucial regulators in various biological processes and stress responses in plants. Currently, bZIP family members and their functions remain elusive in the green unicellular algae *Chlamydomonas reinhardtii*, an important model organism for molecular investigation with genetic engineering aimed at increasing lipid yields for better biodiesel production. In this study, a total of 17 *C. reinhardtii* bZIP (CrebZIP) TFs containing typical bZIP structure were identified by a genome-wide analysis. Analysis of the CrebZIP protein physicochemical properties, phylogenetic tree, conserved domain, and secondary structure were conducted. *CrebZIP* gene structures and their chromosomal assignment were also analyzed. Physiological and photosynthetic characteristics of *C. reinhardtii* under salt stress were exhibited as lower cell growth and weaker photosynthesis, but increased lipid accumulation. Meanwhile, the expression profiles of six *CrebZIP* genes were induced to change significantly during salt stress, indicating that certain CrebZIPs may play important roles in mediating photosynthesis and lipid accumulation of microalgae in response to stresses. The present work provided a valuable foundation for functional dissection of CrebZIPs, benefiting the development of better strategies to engineer the regulatory network in microalgae for enhancing biofuel and biomass production.

Keywords: *Chlamydomonas reinhardtii*; bZIP transcription factors; salt stress; transcriptional regulation; photosynthesis; lipid accumulation

1. Introduction

Microalgae are considered to be one of the most promising feedstocks for renewable biofuel production. However, the shortage of inexpensive algal biomass currently hampers microalgae-based biofuel industrialization [1]. Microalgae accumulate high level of lipids, mainly in the form of triacylglycerol (TAG), when subjected to nutrient deprivation and other stresses [2–5]. In parallel, these adverse conditions also limit algal biomass accumulation. Consequently, genetic engineering to achieve an optimized balance between oil accumulation and biomass growth may represent an effective strategy for the improvement of microalgae biofuel yield. Therefore, it is necessary to comprehensively analyze the underlying molecular mechanisms that mediate stress-induced accumulation of oil in microalgae, particularly to identify the key transcription factors (TFs). The unicellular algae *Chlamydomonas reinhardtii* is the de facto model organism for research in microalgae. Various types of omics data for *C. reinhardtii* are available, including its full genome [6], the proteomics and metabolomics analysis, and the phenotype transition during N starvation [7–13]. These achievements provide the basis for further investigation into oil metabolism and regulation in *C. reinhardtii*, which would shed light on the development of rational strategies for sustainable production of microalgae biofuel.

Transcription factor (TF) encoding genes are considered as contributing to the diversity and evolution in plants. Identification of the transcriptional factors and their cognate transcriptional factor binding-sites is essential in manipulating the regulatory network for desired traits of the target molecules [14]. Moreover, the control of transcription initiation rates by transcription factors is an important means to modulate gene expression, and then regulate the organism growth and development [15]. The basic region-leucine zipper (bZIP) family is one of the most conserved and wildly distributed TFs present in multiple eukaryotes. To date, they have been extensively investigated in many plants including *Arabidopsis*, rice, tomato, maize, sorghum, carrot, and so forth [16–22]. The bZIP TFs have been found to mediate various biological processes, such as cell elongation [23], organ and tissue differentiation [24–26], energy metabolism [27], embryogenesis and seed maturation [28], and so forth. The bZIP TFs also participate in plant responses to biotic and abiotic stresses, including pathogen defense [29,30], hormone and sugar signaling [31,32], light response [33,34], salt and drought tolerance [20,35], and so forth. Typically, bZIP TFs contain a conserved 40–80 amino acid (aa) domain which has two structure motifs: A DNA-binding basic region and a leucine zipper dimerization domain [15]. The basic region composing of around 20 amino acid residues with an invariant N-X7-R/K-X9 motif is highly conserved, and the main function of this region is for nuclear localization and DNA binding. The leucine zipper containing a heptad repeat of leucine is less conserved, with the property for recognition and dimerization [21]. The diversified leucine zipper region is located exactly 9 aa downstream from the C-terminal of the basic region. Although bZIP family members were intensively reported to mediate diverse stress responses in higher plants, little attention has been paid to studying bZIP TFs and their downstream target genes on a genome-wide scale in microalgae.

A total of 147 putative TFs of 29 different protein families have been identified in *C. reinhardtii*, including 1 WRKY, 4 bHLH, 5 C2H2, 11 MYB, 2 MADS, 7 bZIP TFs, and so forth [36]. However, functions remain unclear for the majority of these TFs. The bZIP family is also one of the four largest TF families in oleaginous microalgae *Nannochloropsis* [14], showing that some bZIPs were putatively related with the transcriptional regulation of TAG biosynthesis pathways in *Nannochloropsis*. Therefore, the present study focused on the genome-wide identification of bZIP TFs in *C. reinhardtii* and their functional analysis, with an objective to elucidate the mechanism underlying the regulation of fatty acid and oil accumulation, and photosynthesis in microalgae, particularly under stresses.

In this study, the bZIP sequences of *C. reinhardtii* were intensively identified using a proteomic database, and a total of 17 CrebZIP TFs were obtained after removing the redundancy. Bioinformatics tools were employed to perform a detailed analysis of their genetic structure, chromosome distribution, classification, protein domain, and motifs, as well as evolutionary relationship. Furthermore, the physiological and photosynthetic characteristics of *C. reinhardtii* under salt stress were measured, including biomass concentration, lipid and pigment contents, as well as chlorophyll fluorescence variation. Finally, to infer the potential functions of these CrebZIPs, the expression profiles of *CrebZIP* genes under salt stress were quantitatively examined using quantitative real time (qRT)-PCR. Thus, these integrated data would provide new insights into comprehensive understanding of the stress-adaptive mechanisms and oil accumulation mediated by bZIP TFs in *C. reinhardtii* and other microalgae.

2. Results and Discussion

2.1. Identification of C. reinhardtii bZIP Family Members

To perform genome-wide identification of bZIP proteins in *C. reinhardtii*, BLAST and the Hidden Markov Model (HMM) profiles of the bZIP domain were used to screen the *C. reinhardtii* genome and proteome database, with bZIP sequences from *Arabidopsis* as the query. A total of 17 *CrebZIP* genes in *C. reinhardtii* were identified and denominated as *CrebZIP1–CrebZIP17* based on their locations in the chromosome (Table 1).

The number of CrebZIPs obtained here was not consistent with previous reports. Corrêa et al. and Riano-Pachon et al. identified 7 putative CrebZIP TF coding sequences [15,36]. However, study on the evolution of bZIP family TFs among different plants by Que et al. detected 19 bZIP TFs in *C. reinhardtii*. They summarized that the number of bZIP TFs in algae (less than 20) and land plants (greater than 25) differed remarkably, and bZIP TFs might thus have expanded many times during plant evolution [22]. Such difference in bZIP numbers in *C. reinhardtii* might have resulted from different versions of the *C. reinhardtii* genome and protein database, and criteria used in those reports.

Conserved Domain Database (CDD) and Simple Modular Architecture Research Tool (SMART) analysis indicated that the 17 CrebZIP proteins all had typical bZIP conserved domains. Table 1 summarizes their physicochemical properties, including the protein length which ranged from 334 (CrebZIP13) to 2018 (CrebZIP1) amino acids, the corresponding molecular weight which varied from 3,4514.92 to 198,080.64 Da, and theoretical isoelectric point (pI) which varied from 4.96 (CrebZIP13) to 9.55 (CrebZIP12). The great difference in these properties may reflect their functional diversity in *C. reinhardtii*. The minus hydrophility of all CrebZIP proteins and their higher instability index (>40) showed that they were hydrophilic and unstable.

To get the protein structure information of these CrebZIP members, the secondary structure of the proteins was predicted by the PBIL LYON-GERLAND database. The secondary structure information is listed in Table 2, including α-helix, extended strand, and random coil. Of them, random coil accounted for a higher percentage (45.23–72.75%), while extended strand had the lowest proportion (0.42–8.73%). No β-bridge was detected in CrebZIPs.

Table 1. Physicochemical properties of *CrebZIP* gene coding proteins.

Gene Number	NCBI Accession Number	Phytozome Identifier	Chromosome Localization (bp)	Protein Length (aa)	Molecular Weight (Da)	Theoretical pI	Hydrophility	Instability Index
CrebZIP1	PNW88934.1	Cre01.g051174	Chr.1: 7078516–7088297	2018	198,080.64	6.00	−0.422	68.82
CrebZIP2	PNW83651.1	Cre05.g238250	Chr.5: 2933060–2936249	524	53,624.16	5.47	−0.045	58.98
CrebZIP3	PNW80451.1	Cre07.g318050	Chr.7: 782941–789359	802	82,929.15	6.16	−0.406	59.38
CrebZIP4	PNW80535.1	Cre07.g321550	Chr.7: 1255642–1260306	393	41,429.96	5.48	−0.510	44.72
CrebZIP5	PNW81157.1	Cre07.g344668	Chr.7: 4675427–4680202	750	72,176.41	6.18	−0.046	65.59
CrebZIP6	PNW79382.1	Cre09.g413050	Chr.9: 7331940–7342521	1053	106,414.13	6.42	−0.401	54.24
CrebZIP7	PNW77489.1	Cre10.g438850	Chr.10: 2769177–2772294	485	49,874.20	6.64	−0.678	58.82
CrebZIP8	PNW77864.1	Cre10.g454850	Chr.10: 4910181–4919296	1363	128,662.31	8.75	−0.249	64.61
CrebZIP9	PNW74780.1	Cre12.g510200	Chr.12: 2010086–2013670	353	35,933.29	6.36	−0.398	58.58
CrebZIP10	PNW74984.1	Cre12.g501600	Chr.12: 2939321–2944866	902	89,920.31	6.25	−0.274	56.84
CrebZIP11	PNW75863.1	Cre12.g557300	Chr.12: 7330581–7337724	1526	154,908.24	9.40	−0.719	65.07
CrebZIP12	PNW73681.1	Cre13.g568350	Chr.13: 981891–989596	1150	114,710.75	9.55	−0.563	68.90
CrebZIP13	XP_001693067.1	Cre13.g590350	Chr.13: 3885273–3888238	334	34,514.92	4.96	−0.172	58.83
CrebZIP14	PNW71231.1	Cre16.g692250	Chr.16: 591049–600037	1525	153,084.80	5.30	−0.558	72.50
CrebZIP15	PNW71414.1	Cre16.g653300	Chr.16: 1524635–1531081	867	90,393.04	6.11	−0.374	55.17
CrebZIP16	PNW72253.1	Cre16.g675700	Chr.16: 6123854–6131215	1172	115,285.37	6.48	−0.512	46.46
CrebZIP17	PNW71098.1	Cre17.g746547	Chr.17: 7009739–7013345	767	75,438.72	6.01	−0.379	55.72

Table 2. Secondary structure of CrebZIP proteins.

Gene Number	Alpha Helix (%)	Extended Strand (%)	Beta Bridge (%)	Random Coil (%)
CrebZIP1	24.58	2.68	0	72.75
CrebZIP2	53.24	1.53	0	45.23
CrebZIP3	36.41	8.73	0	54.86
CrebZIP4	45.29	1.53	0	53.18
CrebZIP5	41.60	0.80	0	57.60
CrebZIP6	43.49	2.94	0	53.56
CrebZIP7	45.80	0.42	0	53.78
CrebZIP8	30.01	3.67	0	66.32
CrebZIP9	36.54	8.22	0	55.24
CrebZIP10	38.03	3.77	0	58.20
CrebZIP11	26.34	4.59	0	69.07
CrebZIP12	39.57	3.22	0	57.22
CrebZIP13	50.30	3.59	0	46.11
CrebZIP14	39.08	1.70	0	59.21
CrebZIP15	46.25	1.85	0	51.90
CrebZIP16	33.19	1.88	0	64.93
CrebZIP17	27.38	3.00	0	69.62

2.2. Phylogenetic and Motif Analysis of CrebZIP Proteins

To explore the evolution and classification of CrebZIP TFs, we performed a phylogenetic analysis (Figure 1) of 17 CrebZIP and 11 AtbZIP protein sequences. AtbZIPs were selected according to the classification of *Arabidopsis* bZIP proteins. Ten groups of AtbZIP proteins named Group A, B, C, D, E, F, G, H, I, and S were defined according to the sequence similarity of the basic region and other conserved motifs [17]. Those AtbZIP proteins that did not fit into any group mentioned above were classified as Group U (unknown). One AtbZIP protein was selected from each group respectively, including AtbZIP12 (Group A), AtbZIP17 (Group B), AtbZIP9 (Group C), AtbZIP20 (Group D), AtbZIP34 (Group E), AtbZIP19 (Group F), AtbZIP16 (Group G), AtbZIP56 (Group H), AtbZIP18 (Group I), AtbZIP1 (Group S), and AtbZIP60 (Group U). As shown in Figure 1, most bZIP members from the same species tended to cluster together. Only three CrebZIPs (CrebZIP2, 7, and 15) were grouped together with three AtbZIPs (AtbZIP16, 17 and 20), respectively, forming three subfamilies. Analysis on sequence identity and the similarity of bZIP proteins between *C. reinhardtii* and *Arabidopsis* grouped in the same subfamilies, showed low levels of amino acid conservation between the two species. CrebZIP2 and AtbZIP16 had 10.8% identity and 14.4% similarity, while CrebZIP7 and AtbZIP17 exhibited 10.5% identity and 16.9% similarity. The third pair of CrebZIP15 and AtbZIP20 only had 6.6% identity and 11.6% similarity. The remaining 14 CrebZIPs were grouped into 8 subfamilies, with each containing two CrebZIP members except for CrebZIP1 and CrebZIP5, which were classified as two single-member subfamilies. It is possible that the 14 CrebZIPs may have independent ancestral origins different from the AtbZIPs, in consideration of the fact that *C. reinhardtii* is a lower plant, while *Arabidopsis* is a higher plant. To extend the bZIP analysis to larger lineages of green plants, bZIP members from two bryophytes including *Physcomitrella patens* and *Marchantia polymorpha* were also added into the phylogenetic analysis, as bryophytes are considered as one of the earliest diverging distant land-plant lineages. Table S1 shows 43 *PpbZIP* genes from *P. patens* and 14 *MpbZIP* genes from *M. polymorpha* that were identified. SMART analysis indicated that these bZIP proteins all had the typical bZIP conserved domains. The phylogenetic analysis of bZIP proteins from *C. reinhardtii*, *Arabidopsis*, *P. patens*, and *M. polymophra* indicated that most CrebZIP proteins were also not highly homologous with *P. patens* and *M. polymorpha* bZIPs (Figure S1). Two CrebZIPs (CrebZIP 2 and 6) and two MpbZIPs (MpbZIP4 and 12) clustered together, respectively. However, levels of identity and similarity were very low between the CrebZIPs and MpbZIPs grouped in the same subfamilies. The identity and similarity between CrebZIP2 and MpbZIP4 were 14.5% and 19.8%, respectively, while CrebZIP6 and MpbZIP12 only shared 8.6% identity and 13.4% similarity.

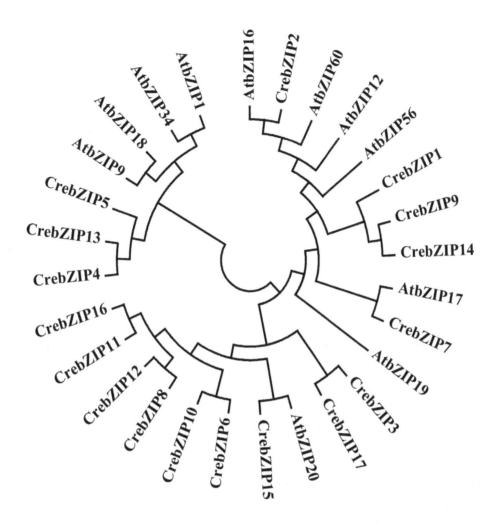

Figure 1. Phylogenetic analysis of *C. reinhardtii* and *Arabidopsis* basic leucine-region zipper (bZIP) proteins. ClustalW was employed to align the protein sequences of 17 CrebZIPs and 11 AtbZIPs representing subgroups A to I, S, and U in *Arabidopsis*. The phylogenetic tree was constructed using the neighbor-joining (NJ) method with MEGA7.0 software. The evolutionary distances were computed using the Poisson correction method with the number of amino acid substitutions per site as the unit. All positions containing gaps and missing data were excluded.

In view of the orthologous bZIP proteins playing a similar role, CrebZIP2 may function like AtbZIP16 (Group G), which mainly linked to light-regulated signal transduction and seed maturation. CrebZIP7 was aligned with AtbZIP17 (Group B), however no functional information was available for members of this group. CrebZIP15 was clustered together with AtbZIP20 from Group D. Members of this bZIP group mainly participated in defense against pathogens, and development [17].

To obtain insight into the divergence and function of CrebZIP TFs, the conserved motifs in the CrebZIPs were analyzed by MEME software. As depicted in Figure 2, all CrebZIP proteins contained the typical bZIP structure domain (motif 1). In addition, a glutamine (Q) enrichment region (motif 2) was detected in 10 CrebZIP proteins including CrebZIP1, 3, 6, 8, 10, 11, 12, 14, 15, and 16. In general, the basic region of bZIP protein has an invariant N-x7-R/K-x9 conserved motif residue rich in lysine (K) and arginine (R). The leucine zipper linked to the C-terminus of the basic region contains two sequential heptad repeat peptides, where a leucine (L) is located at the seventh of each peptide. In some cases, the leucine residue is replaced by isoleucine, valine, phenylalanine, or methionine. For CrebZIP proteins, the primary structure of the conserved domain was detected as N-x7-R/K-x9-L-x6-L-x6-L (motif 1), which is consistent with *Arabidopsis* bZIPs [17].

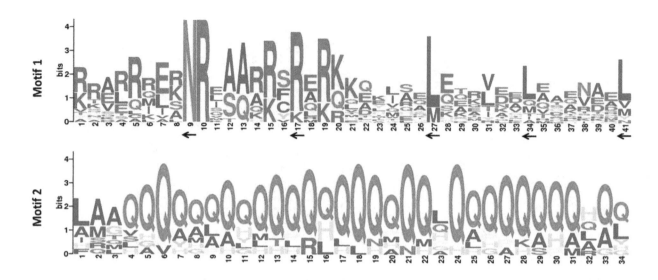

Figure 2. Analysis of the conserved domain in CrebZIP proteins. The conserved motifs in the CrebZIP proteins were identified with MEME software. Conserved sites in the key conserved domain (motif 1) are indicated by "←".

Previous studies showed that there were other special structural domains (e.g., proline-rich, glutamine-rich, and acidic domains) in plant bZIP proteins. These domains may have transcriptional activation function in regulating downstream target gene expressions [37]. For example, two glutamine-rich (~30% Gln) domains adjacent to the C terminal of a bZIP protein encoded by *PERIANTHIA* (*PAN*) in *Arabidopsis* were supposed to act as transcriptional activation domains [38]. A glutamine-rich region in the C-terminal halves of wheat bZIP family members HBP-1b (c38) and HBP-1b (c1), was reported to activate transcription of nuclear genes [39]. STGA1 (soybean TGA1), a member of the TGA (TGACG motif binding factor) subfamily of soybean bZIP TFs, contained a C-terminal glutamine-rich region as a putative transcription activation domain [40]. The conserved motifs shared by *Arabidopsis*, wheat, soybean, and other plants, suggested a similar function for the bZIP proteins. In this study, most *C. reinhardtii* bZIP protein sequences also contained a glutamine-rich region (motif 2), indicating that like higher plants, unicellular microalgae may retain structural domains of important functions during evolution, although the roles of these additional conserved motifs found in CrebZIP proteins are not yet clear.

2.3. Analysis of CrebZIP Gene Structure and Their Chromosomal Assignment

To further understand the evolutionary relationships among *CrebZIP* genes, GSDS (Gene Structure Display Server) was used to analyze their intron-exon structures. As shown in Figure 3, the number of exons varied from 3 to 16, demonstrating a great divergence among the 17 *CrebZIP* genes. The exon-intron structures of the genes were also highly different even in the same subfamily despite six exons were conserved in the subfamily composed of *CrebZIP8* and *CrebZIP12*. For instance, *CrebZIP6* and *CrebZIP10* grouped as a subfamily, with *CrebZIP6* having 16 exons and *CreZIP10* consisting of 5 exons. For the same subfamily of *CrebZIP9* and *CrebZIP14*, the former had 4 exons while the later contained 10 exons. Such variance in intron-exon structures was also found in *bZIP* genes of rice (*Oryza sativa*), soybean (*Glycine max*), and strawberry (*Fragaria ananassa*) plants. Among the *OsbZIP* genes having introns, the number of introns in open reading frames (ORF) varied from 1–12, 1–18, and 1–20 in rice [41], soybean [42], and strawberry plants [43], respectively. In agreement with these previous findings, this diversity in exon-intron organization indicated that both exon loss and gain occurred during the evolution of the *C. reinhardtii bZIP* gene family.

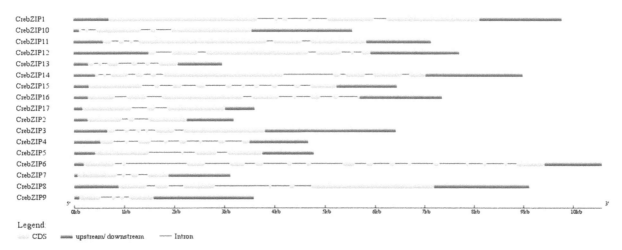

Figure 3. The Exon-intron organization of *CrebZIP* genes. The exons and introns are represented by yellow boxes and horizontal black lines, respectively. Untranslated regions are shown with blue boxes. The length of *CrebZIP* genes are indicated by the horizontal axis (kb).

Chromosome assignment of CrebZIP genes depicted by MapInspect software displayed 17 *CrebZIP* genes unevenly distributed on nine chromosomes of *C. reinhardtii*. Three *CrebZIP* genes were located on chromosomes 7, 12, and 16. Chromosomes 10 and 13 both had two *CrebZIP* genes. Only one *CrebZIP* gene was found on chromosomes 1, 5, 9, and 17 (Figure 4). In addition, several *CrebZIP* genes including *CrebZIP1, 2, 6, 13,* and *17*, were distributed near the ends of chromosomes.

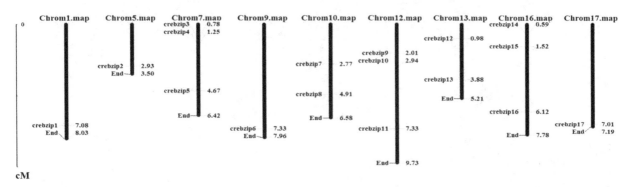

Figure 4. Distribution of *CrebZIP* gene family members on the *C. reinhardtii* chromosomes. The chromosome number is indicated at the top of each chromosome. Values next to the *bZIP* genes indicate the location on the chromosome.

2.4. Characterization of Cell Growth and Lipid Accumulation in C. reinhardtii Under Salt Stress

The effects of salt stress on *C. reinhardtii* growth and lipid accumulation are shown in Figure 5. Salt stress (150 mM NaCl treatment) significantly affected the cell growth, with OD values increasing slowly from 0.127 ± 0.002 to 0.276 ± 0.044 during 48 h cultivation, while the cell growth curve showed a rapid increase in the control with OD values from 0.127 ± 0.002 to 1.242 ± 0.052 (Figure 5a). In contrast, total lipid content in *C. reinhardtii* cells under salt stress significantly increased from 0.284 ± 0.029 to 0.437 ± 0.012 (Figure 5b), whereas lipid content in the control only increased at a small scale from 0.284 ± 0.029 to 0.348 ± 0.022, after 48 h cultivation. Consistent with these findings, Kato et al. observed that salinity stimulated lipid accumulation, but negatively affected biomass production in microalgae [44].

Oil accumulation in *C. reinhardtii* cells under salt stress was also examined by fluorescent microscopy using Nile Red staining (Figure 6). Notably, salt stress resulted in more oil droplets in *C. reinhardtii* cells, which was consistent with the total lipid content (Figure 5b) measured for the stressed cells. Similarly, other reports also showed that salinity stress induced enhancement of total lipid content and neutral lipid fractions within microalgae cells by affecting the fatty acid metabolism [45–47]. It is possible that in response to the decrease of cell membrane osmotic pressure and fluidity caused by salt stress, microalgae could employ an adaptive strategy to accumulate neutral lipid TAG, so as to maintain membrane integrity.

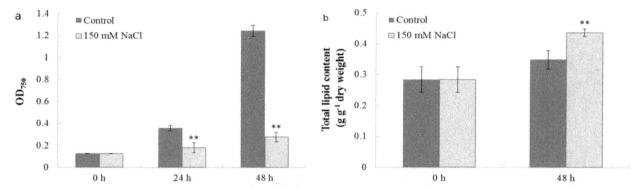

Figure 5. The effects of salt stress on (**a**) cell growth and (**b**) total lipid content of *C. reinhardtii*. Cell samples harvested after 0, 24, and 48 h cultivation from both the salt treatment and the control were used for measurements of cell growth and lipid content. Each value is the mean ± SD of three biological replicates. Asterisks indicate statistical significance (* $0.01 < p < 0.05$, ** $p < 0.01$) between control and salt-treated cells according to the Tukey's test.

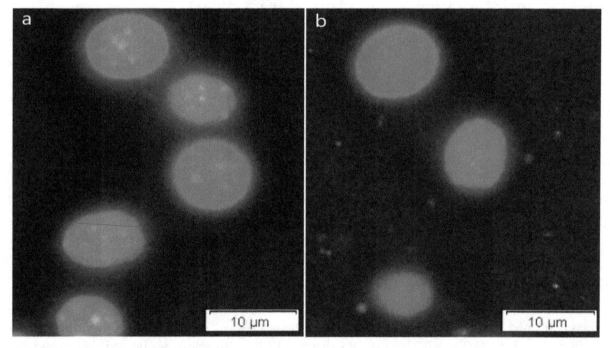

Figure 6. Nile Red staining lipid drops in *C. reinhardtii* under (**a**) salt stress and (**b**) the control treatments. The cell and lipid droplet colors were visualized with red and yellow, respectively, under fluorescent light.

2.5. Photosynthetic Properties of C. reinhardtii Cells under Salt Stress

To investigate the effects of salt stress on *C. reinhardtii* photosynthesis, the contents of three pigments, Chlorophyll a (Chla), Chlorophyll b (Chlb), and carotenoids (Car), were measured. As shown in Figure 7, salt stress also led to significant changes in pigment content of *C. reinhardtii*. Under salt stress, Chla and Chlb contents in *C. reinhardtii* both decreased from 6.147 ± 0.409 and 2.680 ± 0.161 to 4.903 ± 0.258 and 2.427 ± 0.0.055 mg 10^{-10} cells, respectively, during 48 h cultivation, showing significant difference from the control where Chla and Chlb contents increased during cultivation.

Chlorophyll is the main light-harvesting molecule for photosynthetic organisms. The reduction of chlorophyll content indicated weakened photosynthesis in *C. reinhardtii* under salt stress, possibly being the result of decreased synthesis or enhanced degradation of chlorophylls caused by the stress. It is known that salinity caused limitations in photosynthetic electron transport and then photosynthetic rate [48]. Moreover, salinity may stimulate chlorophyllase activity and accelerate chlorophyll degradation [49]. Unlike the chlorophyll case, the carotenoid content in *C. reinhardtii* grown under salinity increased from 2.240 ± 0.295 to 4.263 ± 0.180 mg 10^{-10} cells after 48 h cultivation, in comparison with the control where carotenoid content exhibited no obvious change during the cultivation. Higher carotenoid content under salt stress in *C. reinhardtii* was possibly due to positive adaptation to the stress, which was also observed in microalgae *Botryococcus braunii* by Rao et al. [50]. It was proved that carotenoids could protect the photosynthetic apparatus from photo-oxidative damage [51], providing the protection mechanism for microalgae cells against adverse stresses.

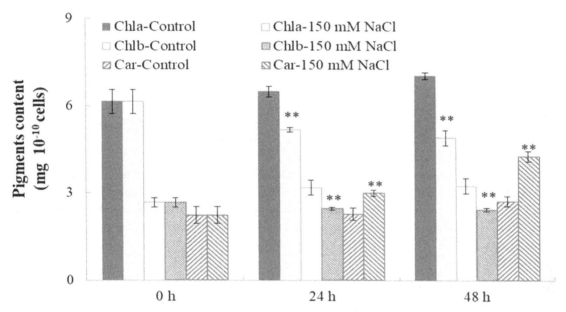

Figure 7. The effect of salt stress on pigment contents of *C. reinhardtii*. Cell samples harvested after 0, 24, and 48 h cultivation from both the salt treatment and the control were used to measure the contents of chlorophyll a, Chlorophyll b, and carotenoid. Each value is the mean $\pm$ SD of three biological replicates. Asterisks indicate statistical significance (* $0.01 < p < 0.05$, ** $p < 0.01$) between control and salt-treated cells according to the Tukey's test.

The photosynthetic performance in algae and plants was widely monitored by measuring chlorophyll fluorescence. For example, the F_v/F_m parameter was used for estimating the maximum quantum yield of PSII photochemistry. Non-photochemical quenching (NPQ) was employed to examine changes in the apparent rate constant for excitation decay by heat loss from PSII. NPQ was an essential part of the plant response to stress, as indicated by their slower growth [52]. Decreased F_v/F_m was often observed when plants were exposed to abiotic and biotic stresses [53]. The effect of salt stress on the chlorophyll fluorescence of *C. reinhardtii* is demonstrated in Figure 8. F_v/F_m values decreased from 0.695 ± 0.027 to 0.354 ± 0.038 in salt-treated cells during the 48 h cultivation, while it slightly increased from 0.695 ± 0.027 to 0.812 ± 0.022 in the control (Figure 8a). Meanwhile, NPQ values in cells detected under salt stress and in the control, both increased in this culture period. However, NPQ values under salinity increased more significantly (from 0.030 ± 0.002 to 0.199 ± 0.011) compared to the control (from 0.030 ± 0.002 to 0.061 ± 0.010) (Figure 8b). Analogously, Mou et al. [54] also observed lower F_v/F_m and associated induction of NPQ when *Chlamydomonas* sp. ICE-L was cultured under stress. In addition, they also summarized that the lower F_v/F_m indicated stress conditions, and the higher NPQ showed energy dissipation.

Taken together, salt stress affected various physiological and photosynthetic mechanisms associated with cell growth and development of *C. reinhardtii*. Microalgae cells altered their metabolism to adapt to the adverse environment by reducing biomass production and chlorophyll content, and simultaneously by increasing lipid and carotenoid contents. Moreover, salt stress impeded photosynthesis in *C. reinhardtii*, reflected by decreased F_v/F_m and increased NPQ. Salinity may induce excess production of reactive oxygen species (ROS) which caused cell damage. ROS also participated in regulating the expression of many genes and signal transduction pathways. Under stress, microalgae cells subsequently changed the physiological and photosynthetic status by: (1) Enhancing NPQ to reduce light energy absorption and to accelerate energy dissipation, thus mitigating the damage caused by excess excitation energy. (2) Accumulating lipid at the cost of reduced growth and carbohydrate storage as a response to the oxidative stress [55,56].

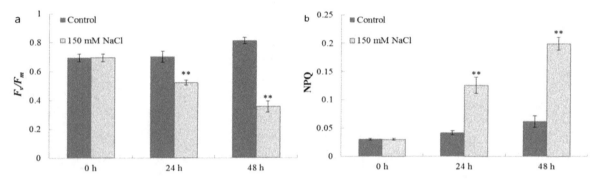

Figure 8. The effects of salt stress on (a) the maximum photo-chemical efficiency (F_v/F_m) and (b) non-photochemical quenching co-efficiency (NPQ) of *C. reinhardtii*. Cell samples harvested after 0, 24, and 48 h cultivation from both the salt treatment and the control, were used for measurements of F_v/F_m and NPQ. Each value is the mean ± SD of three biological replicates. Asterisks indicate statistical significance (* $0.01 < p < 0.05$, ** $p < 0.01$) between control and salt-treated cells according to the Tukey's test.

2.6. The Expression of CrebZIP Genes under Salt Stress

To investigate whether any CrebZIP functioned in regulating stress responses and oil accumulation in *C. reinhardtii*, expression analysis of all 17 *CrebZIP* genes following exposure to salt stress by quantitative real-time (qRT)-PCR were performed. Among the 17 *CrebZIP* genes detected, only six genes including *CrebZIP4, 5, 10, 11, 13*, and *16* showed significant expression changes during salt stress (Figure 9). The rest of 11 *CrebZIP* genes exhibited no obvious expression changes between the salt treatment and the control during 48 h cultivation. The *CrebZIP10, 11*, and *16* genes were up-regulated, whereas *CrebZIP4, 5*, and *13* genes were down-regulated under salt stress compared to the control (Figure 9), indicating that these CrebZIP TFs may be involved in the *C. reinhardtii* defense response to salt stress. Similarly, Zhu et al. also observed the obvious expression changes of several *bZIP* genes in tomato (*Solanum lycopersicum*) plants under salt and drought stresses. Furthermore, they concluded that SlbZIP1 mediates the stress tolerance in tomato plants through regulating an ABA-mediated pathway revealed by silencing of the gene and RNA-seq analysis [57]. Consequently, it was speculated that the six CrebZIP TFs may participate in regulation of photosynthesis and oil accumulation in *C. reinhardtii* under stress conditions, based on the association of *CrebZIP*'s expression and physiological phenotypes of the cells. Nevertheless, to verify whether the six CrebZIPs directly mediate the regulation of *C. reinhardtii* cell growth, photosynthesis, and lipid accumulation in response to stresses, mutants of gain-of-function and loss-of-function of these six CrebZIPs should be generated, and correspondingly, phenotypic analysis needs to be conducted on these mutants under stress in future study. The data from this study was the first evidence showing that CrebZIPs may mediate the regulation of stress responses, particularly the oil accumulation under salt stress, although a few CrebZIPs may be involved in regulating N-starving stress responses in *C. reinhardtii* [58].

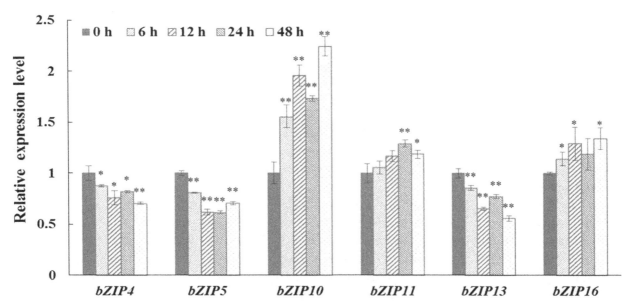

Figure 9. Expression profiles of six *CrebZIP* genes in *C. reinhardtii* under salt stress. Expression of *CrebZIP* genes was determined by qRT-PCR using total RNA from microalgae cells sampled at different time points of salt treatment. The α-tubulin gene was used as the internal reference gene. Each value is the mean $\pm$ SD of six biological replicates. Asterisks indicate statistical significance (* $0.01 < p < 0.05$, ** $p < 0.01$) between control and salt-treated cells according to the Tukey's test.

Although bZIP TFs in higher plants have been intensively explored, the study of bZIP proteins of microalgae have not, limiting the functional analysis of bZIP members in microalgae. The study of bZIP TFs in green plant evolution suggested that the ancestor of green plants possessed four bZIP genes functionally involved in oxidative stress, unfolded protein responses, and light-dependent regulations [15]. Takahashi et al. found that in algae *Vaucheria frigida* and *Fucus distichus*, each of two bZIP proteins, chromoproteins AUREO1 and AUREO2, contained one bZIP domain and one light–oxygen–voltage (LOV)-sensing domain, representing the blue light (BL) receptor. It was hypothesized that because bZIP proteins typically bind DNA by forming heterodimer, AUREO1 and AUREO2 may cooperatively regulate different kinds of BL responses by forming homo- and hetero-dimers [59]. Marie et al. reported that the two bZIP TFs AUREO1a and bZIP10 were likely to be involved in the blue light-dependent transcription of a cyclin gene that regulated the onset of the cell cycle in diatom (*Phaeodactylum tricornutum*) after a period of darkness [60]. Fischer et al. identified a singlet oxygen resistant 1 (SOR1), which was a putative bZIP protein in *C. reinhardtii*, that could stimulate the tolerance of *C. reinhardtii* to high O_2 formation by activating a reactive electrophile species (RES)-induced defense response, thereby enhancing the tolerance of this organism to photo-oxidative stress [61]. In addition to light-dependent regulation and oxidative stress resistance, bZIP TFs were also proved to participate in TAG accumulation in microalgae. Hu et al. revealed that bZIP family members were dominant (five such TFs) among 11 TFs that were potentially involved in the transcriptional regulation of TAG biosynthesis pathways in *Nannochloropsis* [14]. Consistent with these previous reports, the present study provides new data to show that some bZIP TFs may mediate the regulation of photosynthesis and lipid synthesis in microalgae, especially under stress conditions.

3. Materials and Methods

3.1. Microalgae Strains and Growth Conditions

The microalgae *C. reinhardtii* was purchased from the Freshwater Algae Culture Collection of the Institute of Hydrobiology, Chinese Academy of Science, China, and maintained in Tris-Acetate-Phosphate (TAP) medium [62]. The culture of *C. reinhardtii* was inoculated in 250 mL flasks under a continuous illumination of 100 μmol m^{-2} s^{-1} and a temperature of 25 $\pm$ 1 °C. A certain amount of NaCl was added into the TAP medium to a final concentration of 150 mmol L^{-1} in salt stress treatment, while algal cells were cultivated in normal TAP medium as the control.

3.2. Genome-Wide Identification of C. reinhardtii bZIP Gene Family

The bZIP protein sequence of *C. reinhardtii* was identified and downloaded through the following means: Using HMMMER V3.0 software with the Hidden Markov Model (HMM) PF00170 (from Pfam database, http://pfam.xfam.org/) of bZIP domain, from the Phytozome database (http://phytozome. jgi.doe.gov/pz/portal.html) and Plant Transcription Factor Database (PlnTFDB) with the keyword of bZIP. The screening results of the three means were merged, followed by manually removing the redundant or repetitive sequences.

The prediction of the bZIP structure domain was carried out within the amino acid sequence of the selected candidate bZIP protein family member in *C. reinhardtii* with the help of the Conserved Domain Database (CDD, http://www.ncbi.nlm.nih.gov/Structure/cdd/wrpsb.cgi) in the National Center for Biotechnology Information (NCBI, http://www.ncbi.nlm.nih.gov/), and Simple Modular Architecture Research Tool software (SMART, http://smart.embl-heidelberg.de). The candidate genes without bZIP structure domain were removed. The physicochemical properties of the bZIP protein of *C. reinhardtii*, including hydropathicity, molecular mass, instable index, and so forth, were predicted via ProtParam software from the Expasy database (http://web.expasy.org/protparam/).

3.3. Motif Recognition and Phylogenetic Analysis of C. reinhardtii bZIP Gene Family

Motifs of the selected genes were analyzed using the MEME suite (http://meme-suite.org/tools/ meme). Multiple alignments of bZIP protein sequences were performed by ClustalW software and the phylogenetic tree was constructed with MEGA 7.0 software (https://www.megasoftware.net/home).

3.4. bZIP Protein Secondary Structure Prediction

The secondary structure of bZIP proteins in *C. reinhardtii* was predicted using the PBIL LYON-GERLAND database (https://npsa-prabi.ibcp.fr/cgi-bin/npsa_automat.pl?page=/NPSA/npsa_hnn.html).

3.5. bZIP Gene Structure and Chromosomal Assignment Analysis

The *bZIP* gene structure diagram was constructed by Gene Structure Display Server (GSDS) 2.0 software (http://gsds.cbi.pku.edu.cn/), and then quantitative analysis of introns and exons followed. The chromosome map of *CrebZIP* gene family members was depicted by MapInspect software according to the *CrebZIP* genes location on chromosomes of *C. reinhardtii*.

3.6. Measure of Microalgae Biomass Concentration

The biomass concentration of *C. reinhardtii* was indicated by optical cell density, which was measured with a UV-Visible spectrophotometer (UV-1200, Shanghai Jingke Instrument Co., Ltd., Shanghai, China) at 750 nm (OD_{750}). When necessary, the sample was diluted to give an absorbance in the range of 0.1–1.0. The experiments were conducted in triplicate.

3.7. Analysis of Total Lipid Content in Microalgae Cells

The total lipid content was determined by gravimetric analysis according to the method of Chen et al. [63]. The microalgae cells were collected by centrifugation and then lyophilized. About 50 mg of lyophilized microalgae sample was triturated in a mortar, and then the cell disruption was added into 7.5 mL chloroform/methanol (1:2, v/v) mixture. The mixture was placed at 37 °C overnight and then centrifuged to collect the supernatant. Residual biomass was extracted at least once more. All the supernatants were combined, and added into a chloroform and 1% sodium chloride solution to a final volume ratio of 1:1:0.9 (chloroform/methanol/water). The new mixture was centrifuged afterwards, and the subnatant was transferred to a pre-weighted vitreous vial. The sample solution was dried to constant weight at 60 °C under nitrogen flow. Finally, the total lipid content was obtained as a percentage of the dry weight (DW) of the microalgae. The experiments were conducted in triplicate.

3.8. Nile Red Staining and Microscopy for Assessment of Oil Accumulation in Microalgae Cells

Lipid droplets were visualized by fluorescent microscopy using Nile Red staining [64]. Microalgae cells under salt stress and control conditions were collected after 48 h cultivation by centrifugation. The cells pellets were washed with physiological saline solution three times. After the collected cells were re-suspended in the same solution, 10 μL Nile Red stain (0.1 g L^{-1} in acetone) and 200 μL dimethyl sulfoxide were added into an 800 μL microalgae suspension, with final concentration at about 1×10^6 cells mL^{-1}. The stained cells were incubated in the dark for 20 min at room temperature, and then immediately observed by fluorescent microscopy.

3.9. Determination of the Pigment Contents in Microalgae Cells

Contents of microalgae pigments, including chlorophyll (a and b) and carotenoid, were determined according to the methods described by Wellburn [65]. A certain amount of microalgae culture was centrifuged to collect cells. The cell pellets were mixed with methanol at 60 °C for 12 h in darkness, until the microalgae cells whitened completely. The mixture was centrifuged and supernatant extraction was collected. The optical densities of the extraction were measured with a spectrophotometer (UV-1601, Beijing Beifen-Ruili Aanlytical Instrument Co., Ltd., Beijing, China) at 666, 653, and 470 nm. The concentrations of chlorophyll a (Chla), chlorophyll b (Chlb), and carotenoid (Car) of the extraction were calculated as follows (mg L^{-1}):

$$Chla = 15.65 \times OD_{666} - 7.34 \times OD_{653} \tag{1}$$

$$Chlb = 27.05 \times OD_{653} - 11.21 \times OD_{666} \tag{2}$$

$$Car = (1000 \times OD_{470} - 2.86 \times Chla - 129.2 \times Chlb)/221 \tag{3}$$

The cell density of the microalgae culture was obtained using cytometry with a hemacytometer. The experiments were conducted in triplicate.

3.10. Chlorophyll Fluorescence Measurements

The chlorophyll fluorescence was measured using an Imaging-PAM (Pulse Amplitude Modulation) fluorescence monitor (Walz, Effeltrich, Germany) according to the method described by Mou et al. [54]. Samples were dark adapted for 15 min before the fluorescence measurement. Then the dark-level fluorescence yield (F_0), the maximum fluorescence yield (F_m), and the maximum light-adapted fluorescence yield (F'_m) were measured.

The maximum quantum yield of PSII was calculated as:

$$F_v/F_m = (F_m - F_0)/F_m \tag{4}$$

The non-photochemical quenching (NPQ) was calculated as:

$$NPQ = (F_m - F'_m)/F'_m \tag{5}$$

The experiments were conducted in triplicate.

3.11. RNA Isolation and Quantitative Real-Time (qRT)-PCR Analysis

CrebZIP genes expression under salt stress were analyzed by qRT-PCR. Total RNA of C. reinhardtii cells sampled after 0, 6, 12, 24, and 48 h cultivation under salt stress were extracted. RNA from each sample was used to synthesize the cDNAs with the cDNA Synthesis Kit (TaKaRa, Kusatsu, Japan). The primers for the 17 CrebZIP genes and one reference gene (α-tubulin) were designed using

Primer Premier 5.0 software. Sequences of all the primers used in this study are shown in Table 3. The qRT-PCR was performed in an ABI 7500 qRT-PCR system (Applied Biosystems, Foster City, CA, USA) with following reaction conditions: 95 °C for 3 min followed by 40 cycles of 95 °C for 15 s, 60 °C for 30 s, and 72 °C for 30 s. The experiments were repeated six times using independent RNA samples.

Table 3. Primer information of *CrebZIP* and reference genes for quantitative real-time (qRT)-PCR.

Gene Number	Forward Primer (5′-3′)	Reverse Primer (5′-3′)
CrebZIP1	CGGTCGATGACGCTAAGGC	TGGTCGGTGTCGGAGGAGT
CrebZIP2	GAGCGCAAGAAGCAGTACGTGACCT	GATGTTCCGCAGTGCCTCGTTCT
CrebZIP3	ATGCACCAAACGCCAAATCG	ATCTGCTGTAGGTCCGCCAGGGTA
CrebZIP4	GACAGCGGAGACTCAGACAT	CCCTTCTTACGCTGCCTGTA
CrebZIP5	CCGTCATCCATGCACCTACT	GCTGAAGATCAAGACCGCTG
CrebZIP6	CCCGTTTCCAGCCCATCAGAC	TGCAAGGCCGAGGGTGTTGATGAC
CrebZIP7	TGGGAGGCTCTGGCTTTGG	TGGCTGCTGCTGCTGCTGATG
CrebZIP8	GCAAAGGCAAGGGCAAGG	CGCAGGAGTTCTCAGCCGATT
CrebZIP9	GCGCTTCTTCCTCGCTACCA	CGCAAGCCCTTGTGCTGTTA
CrebZIP10	CCCTGACGTCCACCTTAACT	TCCCTCAGACAGCAACTCAG
CrebZIP11	ACGGACGTATGACATGAGCA	ACTGCTGGAACTGGTGGAA
CrebZIP12	CACAGCGACCGCAGCCATAA	GCCCAGCTTGTCCGAGAAGGA
CrebZIP13	GAGCTGCCTTCGACAAGC	CTGTGCTAGGCGATTCAGC
CrebZIP14	GATGGCGGGCTTATGTCGG	CAAGGCGTCCACGTTGTG
CrebZIP15	TGACTCATCCACGCACTTCCTC	TGATGTTGCACCAGCCCTGA
CrebZIP16	GGTTTCCTTCCCTACCCA	ACGGCACGGTTGTCAGCA
CrebZIP17	AAGCGCATTGTTGACGGAG	CGCAGCAGGTCTAATAAGTCG
Creα-tubulin	CTCGCTTCGCTTTGACGGTG	CGTGGTACGCCTTCTCGGC

4. Conclusions

The first genome-wide analysis of the bZIP TF family members in *C. reinhardtii* was performed in this study, with a total of 17 CrebZIP proteins identified. Detailed information was obtained for their evolutionary relationship, exon-intron organization and chromosome assignment, protein structural features, and conserved motifs. Moreover, expression profiling of *CrebZIP* genes by qRT-PCR indicated that six CrebZIPs might be involved in stress response and lipid accumulation in microalgae cells. Salt stress led to the reduction in biomass production, chlorophyll content, and F_v/F_m, but the enhancement in NPQ, carotenoid content, and oil accumulation in *C. reinhardtii*. Collectively, integration of findings in the present study provided new data to indicate that some CrebZIP TFs could play important roles in mediating regulation of cell growth, photosynthesis, and oil accumulation in microalgae, particularly under stress conditions. These *CrebZIP* genes could be utilized to further functionally characterize them, laying the foundation for elucidating their specific regulatory mechanisms and ultimately applying them in genetic improvement programs.

Author Contributions: R.L. designed the experiments and co-wrote the manuscript. C.J. performed the experiments and wrote the manuscript. C.J., X.M., J.H., X.W., J.X., and H.C. analyzed the data. All authors read and approved the final manuscript.

References

1. Wijffels, R.H.; Barbosa, M.J. An outlook on microalgal biofuels. *Science* **2010**, *329*, 796–799. [CrossRef] [PubMed]
2. Fan, J.H.; Cui, Y.B.; Wan, M.X.; Wang, W.L.; Li, Y.G. Lipid accumulation and biosynthesis genes response of the oleaginous *Chlorella pyrenoidosa* under three nutrition stressors. *Biotechnol. Biofuels* **2014**, *7*, 17. [CrossRef] [PubMed]

3.   Wang, Z.T.; Ullrich, N.; Joo, S.; Waffenschmidt, S.; Goodenough, U. Algal lipid bodies: Stress induction, purification, and biochemical characterization in wild-type and starchless *Chlamydomonas reinhardtii*. *Eukaryot. Cell* **2009**, *8*, 1856–1868. [CrossRef] [PubMed]

4.   Cakmak, T.; Angun, P.; Demiray, Y.E.; Ozkan, A.D.; Elibol, Z.; Tekinay, T. Differential effects of nitrogen and sulfur deprivation on growth and biodiesel feedstock production of *Chlamydomonas reinhardtii*. *Biotechnol. Bioeng.* **2012**, *109*, 1947–1957. [CrossRef] [PubMed]

5.   He, Q.; Yang, H.; Wu, L.; Hu, C. Effect of light intensity on physiological changes, carbon allocation and neutral lipid accumulation in oleaginous microalgae. *Bioresour. Technol.* **2015**, *191*, 219–228. [CrossRef] [PubMed]

6.   Merchant, S.S.; Prochnik, S.E.; Vallon, O.; Harris, E.H.; Karpowicz, S.J.; Witman, G.B.; Terry, A.; Salamov, A.; Fritz-Laylin, L.K.; Marechal-Drouard, L.; et al. The *Chlamydomonas* genome reveals the evolution of key animal and plant functions. *Science* **2007**, *318*, 245–250. [CrossRef] [PubMed]

7.   Lv, H.; Qu, G.; Qi, X.; Lu, L.; Tian, C.; Ma, Y. Transcriptome analysis of *Chlamydomonas reinhardtii* during the process of lipid accumulation. *Genomics* **2013**, *101*, 229–237. [CrossRef] [PubMed]

8.   Adrián, L.G.D.L.; Sascha, S.; Jacob, V.; Saheed, I.; Warren, C.; Bilgin, D.D.; Yohn, C.B.; Serdar, T.; Reiss, D.J.; Orellana, M.V. Transcriptional program for nitrogen starvation-induced lipid accumulation in *Chlamydomonas reinhardtii*. *Biotechnol. Biofuels* **2015**, *8*, 207. [CrossRef]

9.   Rolland, N.; Atteia, A.; Decottignies, P.; Garin, J.; Hippler, M.; Kreimer, G.; Lemaire, S.D.; Mittag, M.; Wagner, V. *Chlamydomonas* proteomics. *Curr. Opin. Microbiol.* **2009**, *12*, 285–291. [CrossRef] [PubMed]

10.   Wienkoop, S.; Weiss, J.; May, P.; Kempa, S.; Irgang, S.; Recuenco-Munoz, L.; Pietzke, M.; Schwemmer, T.; Rupprecht, J.; Egelhofer, V.; et al. Targeted proteomics for *Chlamydomonas reinhardtii* combined with rapid subcellular protein fractionation, metabolomics and metabolic flux analyses. *Mol. Biosyst.* **2010**, *6*, 1018–1031. [CrossRef] [PubMed]

11.   Mastrobuoni, G.; Irgang, S.; Pietzke, M.; Aßmus, H.E.; Wenzel, M.; Schulze, W.X.; Kempa, S. Proteome dynamics and early salt stress response of the photosynthetic organism *Chlamydomonas reinhardtii*. *BMC Genom.* **2012**, *13*, 215. [CrossRef] [PubMed]

12.   May, P.; Wienkoop, S.; Kempa, S.; Usadel, B.; Christian, N.; Rupprecht, J.; Weiss, J.; Recuenco-Munoz, L.; Ebenhöh, O.; Weckwerth, W.; et al. Metabolomics- and proteomics-assisted genome annotation and analysis of the draft metabolic network of *Chlamydomonas reinhardtii*. *Genetics* **2008**, *179*, 157–166. [CrossRef] [PubMed]

13.   Valledor, L.; Furuhashi, T.; Recuencomuñoz, L.; Wienkoop, S.; Weckwerth, W. System-level network analysis of nitrogen starvation and recovery in *Chlamydomonas reinhardtii* reveals potential new targets for increased lipid accumulation. *Biotechnol. Biofuels* **2014**, *7*, 171. [CrossRef] [PubMed]

14.   Hu, J.; Wang, D.; Jing, L.; Jing, G.; Kang, N.; Jian, X. Genome-wide identification of transcription factors and transcription-factor binding sites in oleaginous microalgae *Nannochloropsis*. *Sci. Rep.* **2014**, *4*, 5454. [CrossRef] [PubMed]

15.   Corrêa, L.G.; Riaño-Pachón, D.M.; Schrago, C.G.; dos Santos, R.V.; Mueller-Roeber, B.; Vincentz, M. The role of bZIP transcription factors in green plant evolution: Adaptive features emerging from four founder genes. *PLoS ONE* **2008**, *3*, e2944. [CrossRef] [PubMed]

16.   Riechmann, J.L.; Heard, J.; Martin, G.; Reuber, L.; Jiang, C.; Keddie, J.; Adam, L.; Pineda, O.; Ratcliffe, O.J.; Samaha, R.R. *Arabidopsis* transcription factors: Genome-wide comparative analysis among eukaryotes. *Science* **2000**, *290*, 2105–2110. [CrossRef] [PubMed]

17.   Jakoby, M.; Weisshaar, B.; Drögelaser, W.; Vicentecarbajosa, J.; Tiedemann, J.; Kroj, T.; Parcy, F. bZIP transcription factors in *Arabidopsis*. *Trends Plant Sci.* **2002**, *7*, 106–111. [CrossRef]

18.   Zou, M.; Guan, Y.; Ren, H.; Zhang, F.; Chen, F. A bZIP transcription factor, OsABI5, is involved in rice fertility and stress tolerance. *Plant Mol. Biol.* **2008**, *66*, 675–683. [CrossRef] [PubMed]

19.   Yanez, M.; Caceres, S.; Orellana, S.; Bastias, A.; Verdugo, I.; Ruiz-Lara, S.; Casaretto, J.A. An abiotic stress-responsive bZIP transcription factor from wild and cultivated tomatoes regulates stress-related genes. *Plant Cell Rep.* **2009**, *28*, 1497–1507. [CrossRef] [PubMed]

20.   Ying, S.; Zhang, D.F.; Fu, J.; Shi, Y.S.; Song, Y.C.; Wang, T.Y.; Li, Y. Cloning and characterization of a maize bZIP transcription factor, ZmbZIP72, confers drought and salt tolerance in transgenic *Arabidopsis*. *Planta* **2012**, *235*, 253–266. [CrossRef] [PubMed]

21. Wang, J.; Zhou, J.; Zhang, B.; Vanitha, J.; Ramachandran, S.; Jiang, S.Y. Genome-wide expansion and expression divergence of the basic leucine zipper transcription factors in higher plants with an emphasis on sorghum. *J. Integr. Plant Biol.* **2011**, *53*, 212–231. [CrossRef] [PubMed]

22. Que, F.; Wang, G.L.; Huang, Y.; Xu, Z.S.; Wang, F.; Xiong, A.S. Genomic identification of group A bZIP transcription factors and their responses to abiotic stress in carrot. *Genet. Mol. Res.* **2015**, *14*, 13274–13288. [CrossRef] [PubMed]

23. Fukazawa, J.; Sakai, T.; Ishida, S.; Yamaguchi, I.; Kamiya, Y.; Takahashi, Y. Repression of shoot growth, a bZIP transcriptional activator, regulates cell elongation by controlling the level of gibberellins. *Plant Cell* **2000**, *12*, 901–915. [CrossRef] [PubMed]

24. Abe, M.; Kobayashi, Y.; Yamamoto, S.; Daimon, Y.; Yamaguchi, A.; Ikeda, Y.; Ichinoki, H.; Notaguchi, M.; Goto, K.; Araki, T. FD, a bZIP protein mediating signals from the floral pathway integrator FT at the shoot apex. *Science* **2005**, *309*, 1052–1056. [CrossRef] [PubMed]

25. Silveira, A.B.; Gauer, L.; Tomaz, J.P.; Cardoso, P.R.; Carmelloguerreiro, S.; Vincentz, M. The *Arabidopsis* AtbZIP9 protein fused to the VP16 transcriptional activation domain alters leaf and vascular development. *Plant Sci.* **2007**, *172*, 1148–1156. [CrossRef]

26. Shen, H.; Cao, K.; Wang, X. A conserved proline residue in the leucine zipper region of AtbZIP34 and AtbZIP61 in *Arabidopsis thaliana* interferes with the formation of homodimer. *Biochem. Biophys. Res. Commun.* **2007**, *362*, 425–430. [CrossRef] [PubMed]

27. Baena-González, E.; Rolland, F.; Thevelein, J.M.; Sheen, J. A central integrator of transcription networks in plant stress and energy signalling. *Nature* **2007**, *448*, 938–942. [CrossRef] [PubMed]

28. Lara, P.; Oñatesánchez, L.; Abraham, Z.; Ferrándiz, C.; Díaz, I.; Carbonero, P.; Vicentecarbajosa, J. Synergistic activation of seed storage protein gene expression in *Arabidopsis* by ABI3 and two bZIPs related to OPAQUE2. *J. Biol. Chem.* **2003**, *278*, 21003–21011. [CrossRef] [PubMed]

29. Thurow, C.; Schiermeyer, A.S.; Butterbrodt, T.; Nickolov, K.; Gatz, C. Tobacco bZIP transcription factor TGA2.2 and related factor TGA2.1 have distinct roles in plant defense responses and plant development. *Plant J.* **2005**, *44*, 100–113. [CrossRef] [PubMed]

30. Kaminaka, H.; Näke, C.; Epple, P.; Dittgen, J.; Schütze, K.; Chaban, C.; Holt, B.F.; Merkle, T.; Schäfer, E.; Harter, K. bZIP10-LSD1 antagonism modulates basal defense and cell death in *Arabidopsis* following infection. *EMBO J.* **2006**, *25*, 4400–4411. [CrossRef] [PubMed]

31. Nieva, C.; Busk, P.K.; Domínguez-Puigjaner, E.; Lumbreras, V.; Testillano, P.S.; Risueño, M.C.; Pagès, M. Isolation and functional characterisation of two new bZIP maize regulators of the ABA responsive gene *rab28*. *Plant Mol. Biol.* **2005**, *58*, 899–914. [CrossRef] [PubMed]

32. Uno, Y.; Furihata, T.; Abe, H.; Yoshida, R.; Shinozaki, K.; Yamaguchi-Shinozaki, K. *Arabidopsis* basic leucine zipper transcription factors involved in an abscisic acid-dependent signal transduction pathway under drought and high-salinity conditions. *Proc. Natl. Acad. Sci. USA* **2000**, *97*, 11632–11637. [CrossRef] [PubMed]

33. Wellmer, F.; Kircher, S.; Rügner, A.; Frohnmeyer, H.; Schäfer, E.; Harter, K. Phosphorylation of the parsley bZIP transcription factor CRPF2 is regulated by light. *J. Biol. Chem.* **1999**, *274*, 29476–29482. [CrossRef] [PubMed]

34. Ulm, R.; Baumann, A.; Oravecz, A.; Máté, Z.; Adám, E.; Oakeley, E.J.; Schäfer, E.; Nagy, F. Genome-wide analysis of gene expression reveals function of the bZIP transcription factor HY5 in the UV-B response of *Arabidopsis*. *Proc. Natl. Acad. Sci. USA* **2004**, *101*, 1397–1402. [CrossRef] [PubMed]

35. Liu, C.; Mao, B.; Ou, S.; Wang, W.; Liu, L.; Wu, Y.; Chu, C.; Wang, X. OsbZIP71, a bZIP transcription factor, confers salinity and drought tolerance in rice. *Plant Mol. Biol.* **2014**, *84*, 19–36. [CrossRef] [PubMed]

36. Riano-Pachon, D.M.; Correa, L.G.; Trejos-Espinosa, R.; Mueller-Roeber, B. Green transcription factors: A *Chlamydomonas* overview. *Genetics* **2008**, *179*, 31–39. [CrossRef] [PubMed]

37. Liao, Y.; Zou, H.F.; Wei, W.; Hao, Y.J.; Tian, A.G.; Huang, J.; Liu, Y.F.; Zhang, J.S.; Chen, S.Y. Soybean GmbZIP44, GmbZIP62 and GmbZIP78 genes function as negative regulator of aba signaling and confer salt and freezing tolerance in transgenic *Arabidopsis*. *Planta* **2008**, *228*, 225–240. [CrossRef] [PubMed]

39. Mikami, K.; Sakamoto, A.; Iwabuchi, M. The HBP-1 family of wheat basic/leucine zipper proteins interacts with overlapping cis-acting hexamer motifs of plant histone genes. *J. Biol. Chem.* **1994**, *269*, 9974–9985. [PubMed]

40. Cheong, Y.H.; Park, J.M.; Yoo, C.M.; Bahk, J.D.; Cho, M.J.; Hong, J.C. Isolation and characterization of STGA1, a member of the TGA1 family of bZIP transcription factors from soybean. *Mol. Cells* **1994**, *4*, 405–412.

41. Nijhawan, A.; Jain, M.; Tyagi, A.K.; Khurana, J.P. Genomic survey and gene expression analysis of the basic leucine zipper transcription factor family in rice. *Plant Physiol.* **2008**, *146*, 333–350. [CrossRef] [PubMed]

42. Zhang, M.; Liu, Y.; Shi, H.; Guo, M.; Chai, M.; He, Q.; Yan, M.; Cao, D.; Zhao, L.; Cai, H. Evolutionary and expression analyses of soybean basic leucine zipper transcription factor family. *BMC Genom.* **2018**, *19*, 159. [CrossRef] [PubMed]

43. Wang, X.L.; Chen, X.; Yang, T.B.; Cheng, Q.; Cheng, Z.M. Genome-wide identification of bZIP family genes involved in drought and heat stresses in strawberry (*Fragaria vesca*). *Int. J. Genom.* **2017**, *2017*, 1–14. [CrossRef]

44. Kato, Y.; Ho, S.H.; Vavricka, C.J.; Chang, J.S.; Hasunuma, T.; Kondo, A. Evolutionary engineering of salt-resistant *Chlamydomonas* sp. strains reveals salinity stress-activated starch-to-lipid biosynthesis switching. *Bioresour. Technol.* **2017**, *245*, 1484–1490. [CrossRef] [PubMed]

45. Kalita, N.; Baruah, G.; Chandra, R.; Goswami, D.; Talukdar, J.; Kalita, M.C. *Ankistrodesmus falcatus*: A promising candidate for lipid production, its biochemical analysis and strategies to enhance lipid productivity. *J. Microbiol. Biotechnol. Res.* **2011**, *1*, 148–157.

46. Lu, N.; Wei, D.; Chen, F.; Yang, S.T. Lipidomic profiling and discovery of lipid biomarkers in snow alga *Chlamydomonas nivalis* under salt stress. *Eur. J. Lipid Sci. Technol.* **2012**, *114*, 253–265. [CrossRef]

47. Mohan, S.V.; Devi, M.P. Salinity stress induced lipid synthesis to harness biodiesel during dual mode cultivation of mixotrophic microalgae. *Bioresour. Technol.* **2014**, *165*, 288–294. [CrossRef] [PubMed]

48. Zhang, T.; Gong, H.; Wen, X.; Lu, C. Salt stress induces a decrease in excitation energy transfer from phycobilisomes to photosystem II but an increase to photosystem I in the cyanobacterium *Spirulina platensis*. *J. Plant Physiol.* **2010**, *167*, 951–958. [CrossRef] [PubMed]

49. Santos, C.V. Regulation of chlorophyll biosynthesis and degradation by salt stress in sunflower leaves. *Sci. Hortic.* **2004**, *103*, 93–99. [CrossRef]

50. Rao, A.R.; Dayananda, C.; Sarada, R.; Shamala, T.R.; Ravishankar, G.A. Effect of salinity on growth of green alga *Botryococcus braunii* and its constituents. *Bioresour. Technol.* **2007**, *98*, 560–564. [CrossRef] [PubMed]

51. Fedina, I.S.; Grigorova, I.D.; Georgieva, K.M. Response of barley seedlings to UV-B radiation as affected by NaCl. *J. Plant Physiol.* **2003**, *160*, 205–208. [CrossRef] [PubMed]

52. Finazzi, G.; Johnson, G.N.; Dall'Osto, L.; Zito, F.; Bonente, G.; Bassi, R.; Wollman, F.A. Nonphotochemical quenching of chlorophyll fluorescence in *Chlamydomonas reinhardtii*. *Biochemistry* **2006**, *45*, 1490–1498. [CrossRef] [PubMed]

53. Baker, N.R. Chlorophyll fluorescence: A probe of photosynthesis in vivo. *Annu. Rev. Plant Biol.* **2008**, *59*, 89–113. [CrossRef] [PubMed]

54. Mou, S.; Zhang, X.; Ye, N.; Dong, M.; Liang, C.; Qiang, L.; Miao, J.; Dong, X.; Zhou, Z. Cloning and expression analysis of two different *LhcSR* genes involved in stress adaptation in an Antarctic microalga, *Chlamydomonas* sp. ICE-L. *Extremophiles* **2012**, *16*, 193–203. [CrossRef] [PubMed]

55. Kan, G.; Shi, C.; Wang, X.; Xie, Q.; Wang, M.; Wang, X.; Miao, J. Acclimatory responses to high-salt stress in *Chlamydomonas* (Chlorophyta, Chlorophyceae) from Antarctica. *Acta Oceanol. Sin.* **2012**, *31*, 116–124. [CrossRef]

56. Wang, T.; Ge, H.; Liu, T.; Tian, X.; Wang, Z.; Guo, M.; Chu, J.; Zhuang, Y. Salt stress induced lipid accumulation in heterotrophic culture cells of *Chlorella protothecoides*: Mechanisms based on the multi-level analysis of oxidative response, key enzyme activity and biochemical alteration. *J. Biotechnol.* **2016**, *228*, 18–27. [CrossRef] [PubMed]

57. Zhu, M.; Meng, X.; Cai, J.; Li, G.; Dong, T.; Li, Z. Basic leucine zipper transcription factor SlbZIP1 mediates salt and drought stress tolerance in tomato. *BMC Plant Biol.* **2018**, *18*, 83. [CrossRef] [PubMed]

58. Miller, R.; Wu, G.; Deshpande, R.R.; Vieler, A.; Gärtner, K.; Li, X.; Moellering, E.R.; Zäuner, S.; Cornish, A.J.; Liu, B. Changes in transcript abundance in *Chlamydomonas reinhardtii* following nitrogen deprivation predict diversion of metabolism. *Plant Physiol.* **2010**, *154*, 1737–1752. [CrossRef] [PubMed]

59. Takahashi, F.; Yamagata, D.; Ishikawa, M.; Fukamatsu, Y.; Ogura, Y.; Kasahara, M.; Kiyosue, T.; Kikuyama, M.; Wada, M.; Kataoka, H. AUREOCHROME, a photoreceptor required for photomorphogenesis in stramenopiles. *Proc. Natl. Acad. Sci. USA* **2007**, *104*, 19625–19630. [CrossRef] [PubMed]

60. Huysman, M.J.; Fortunato, A.E.; Matthijs, M.; Costa, B.S.; Vanderhaeghen, R.; Van den Daele, H.; Sachse, M.; Inzé, D.; Bowler, C.; Kroth, P.G.; et al. AUREOCHROME1a-mediated induction of the diatom-specific cyclin dsCYC2 controls the onset of cell division in diatoms (*Phaeodactylum tricornutum*). *Plant Cell* **2013**, *25*, 215–228. [CrossRef] [PubMed]

61. Fischer, B.B.; Niyogi, K.K. *SINGLET OXYGEN RESISTANT 1* links reactive electrophile signaling to singlet oxygen acclimation in *Chlamydomonas reinhardtii*. *Proc. Natl. Acad. Sci. USA* **2012**, *109*, E1302–E1311. [CrossRef] [PubMed]

62. Harris, E. *The Chlamydomonas sourcebook: Introduction into Chlamydomonas and Its Laboratory Use*, 2nd ed.; Elsevier Science Publishing Co. Inc.: New York, NY, USA, 2008; ISBN 978-012-370-874-8.

63. Chen, L.; Liu, T.Z.; Zhang, W.; Chen, X.L.; Wang, J.F. Biodiesel production from algae oil high in free fatty acids by two-step catalytic conversion. *Bioresour. Technol.* **2012**, *111*, 208–214. [CrossRef] [PubMed]

64. Greenspan, P.; Mayer, E.P.; Fowler, S.D. Nile red: A selective fluorescent stain for intracellular lipid droplets. *J. Cell Biol.* **1985**, *100*, 965–973. [CrossRef] [PubMed]

65. Wellburn, A.R. The spectral determination of chlorophylls a and b, as well as total carotenoids, using various solvents with spectrophotometers of different resolution. *J. Plant Physiol.* **1994**, *144*, 307–313. [CrossRef]

11

Variations in Physiology and Multiple Bioactive Constituents under Salt Stress Provide Insight into the Quality Evaluation of Apocyni Veneti Folium

Cuihua Chen [1], Chengcheng Wang [1], Zixiu Liu [1], Xunhong Liu [1,2,3,*], Lisi Zou [1], Jingjing Shi [1], Shuyu Chen [1], Jiali Chen [1] and Mengxia Tan [1]

[1] College of Pharmacy, Nanjing University of Chinese Medicine, Nanjing 210023, China; cuihuachen2013@163.com (C.C.); ccw199192@163.com (C.W.); liuzixiu3221@126.com (Z.L.); zlstcm@126.com (L.Z.); shijingjingquiet@163.com (J.S.); 18305172513@163.com (S.C.); 18994986833@163.com (J.C.); 18816250751@163.com (M.T.)

[2] Collaborative Innovation Center of Chinese Medicinal Resources Industrialization, Nanjing 210023, China

[3] National and Local Collaborative Engineering Center of Chinese Medicinal Resources Industrialization and Formulae Innovative Medicine, Nanjing 210023, China

* Correspondence: liuxunh1959@163.com

Abstract: As one of the major abiotic stresses, salinity stress may affect the physiology and biochemical components of *Apocynum venetum* L. To systematically evaluate the quality of Apocyni Veneti Folium (AVF) from the perspective of physiological and the wide variety of bioactive components response to various concentrations of salt stress, this experiment was arranged on the basis of ultra-fast liquid chromatography tandem triple quadrupole mass spectrometry (UFLC-QTRAP-MS/MS) technology and multivariate statistical analysis. Physiological characteristics of photosynthetic pigments, osmotic homeostasis, lipid peroxidation product, and antioxidative enzymes were introduced to investigate the salt tolerance mechanism of AVF under salinity treatments of four concentrations (0, 100, 200, and 300 mM NaCl, respectively). Furthermore, a total of 43 bioactive constituents, including 14 amino acids, nine nucleosides, six organic acids, and 14 flavonoids were quantified in AVF under salt stress. In addition, multivariate statistical analysis, including hierarchical clustering analysis, principal component analysis (PCA), and gray relational analysis (GRA) was employed to systematically cluster, distinguish, and evaluate the samples, respectively. Compared with the control, the results demonstrated that 200 mM and 100 mM salt stress contributed to maintain high quality of photosynthesis, osmotic balance, antioxidant enzyme activity, and the accumulation of metabolites, except for total organic acids, and the quality of AVF obtained by these two groups was better than others; however, under severe stress, the accumulation of the oxidative damage and the reduction of metabolite caused by inefficiently scavenging reactive oxygen species (ROS) lead to lower quality. In summary, the proposed method may provide integrated information for the quality evaluation of AVF and other salt-tolerant Chinese medicines.

Keywords: Apocyni Veneti Folium; salt stress; multiple bioactive constituents; physiological changes; multivariate statistical analysis

1. Introduction

As a major abiotic factors constraint on agriculture, salinity affects about 20% of the cultivated lands in the world and nearly 50% of all irrigated lands [1,2]. In China, about 34.6 million hectares lands are suffering from salinity interference. The medicinal plants among the major and vital groups of crops that exert a significant role in disease prevention and treatment are also being threatened by this constraint [3].

The effects of salt stress on plant growth are mainly revealed in ion toxicity, osmotic stress, and secondary oxidative stress. Plants subjected to salt stress form a series of physiological and molecular mechanisms that respond to salt stress, including ion transport and distribution to maintain ion balance [4,5], osmotic adjustment substances, and metabolites formation to maintain osmotic balance, antioxidant enzyme accumulation, and activity enhancement to resist oxidative stress, signal transduction factors, and salt stress-related genes regulation [6–10].

Apocyni Veneti Folium (AVF), Luobumaye in Chinese, is the dried leaf of *Apocynum venetum* L. (Apocynaceae) [11]. For centuries, AVF has been used to treat cardiac disease, hypertension, nephritis, and neurasthenia, and processed into tea due to its health benefits. Documentaries demonstrated that AVF has functions of antihypertension, antidepression, hepatoprotection, antianxiety, antioxidation, and diuresis [12,13]. It is well established that metabolites of medicinal plants, as sources of natural antioxidant and immune enhancer, are involved in the treatment of human diseases and health disorders [14]. However, their synthesis and accumulation depend on the growing conditions and are vitalized under abiotic stresses [5]. Therefore, a systematic quality assessment method is required for the quality control of herbal medicines.

For the assessment of contents of bioactive components in AVF, many analytical methods have been established by using high performance liquid chromatography (HPLC) coupled to an UV detector [15] or high performance capillary electrophoresis method with diode array detection (HPCE-DAD) [16]. However, these methods have been limited to the quantification of only a few flavones. Additionally, ion trap-time-of-flight (IT-TOF) MS system has been applied to sensitively detect phenolic acids and flavonoids in AVF [17], but not simultaneously detect a variety of bioactive compounds. Recently, ultra-fast liquid chromatography tandem triple quadrupole mass spectrometry (UFLC-QTRAP-MS/MS) is useful for the qualitative and quantitative analysis of bioactive compounds with the advantages of great separation efficiency, high peak capacity, and high sensitivity [18,19] and has emerged as a good tool for the sensitive and selective analysis of various constituents [20].

Apocynum venetum L., as a medicinal halophytic plant, is able to grow in most regions of China, but widely distributed among saline-alkali wasteland, desert edge, and the Gobi desert. According to the Chinese Pharmacopoeia (2015), hyperoside is used as a basis for assessing the quality of AVF and its content should be no less than 0.3% [13]. However, documents have shown that the content of particular class of components is not efficient to evaluate AVF, considering its growing environment [17,21], and there is no published comparative study reported for the quantification of bioactive constituents combined with physiological changes simultaneously.

Choosing of optimum environmental condition to elevate the metabolites has been reported on medicinal plants [22]. In this study, the effects of salt stress on the physiological characteristics and multiple bioactive constituents of AVF were studied, and quantitative results were then further interpreted by multivariate statistical analysis to evaluate its quality. Our investigation may provide a theoretical basis for the quality evaluation of AVF in respect of the increasing in metabolites under superior salt stressed condition and the mechanism of salt tolerance.

2. Results

2.1. Physiological Changes Affected by NaCl

2.1.1. Effects of Salt Stress on Photosynthetic Pigments

Salt-treated plant did not show significant change from the perspective of plant phenotypes even at the end of the 40 days experiment (Figure S1), as compared to the control. Chlorophyll a, total chlorophyll, chlorophyll a/b, and carotenoids were significantly increased in the presence of 100 and 200 mM salt, but the changes were not significant under severe stress (Table 1). However, chlorophyll b had little change compared to the control throughout the salt stressed experiments with value ranging from 0.43 to 0.55 mg g^{-1} FW.

Table 1. Effects of different NaCl concentrations on the chlorophyll (Chl) and carotenoids contents of AVF.

Treatments	Pigment Content				
	Chl a (mg g^{-1} FW)	Chl b (mg g^{-1} FW)	total Chl (mg g^{-1} FW)	Chl a/b	Carotenoids (mg g^{-1} FW)
0 mM	2.35 ± 0.11 c	0.49 ± 0.06 ab	2.84 ± 0.05 c	4.77 ± 0.82 c	0.86 ± 0.07 c
100 mM	3.66 ± 0.19 a	0.55 ± 0.04 a	4.21 ± 0.22 a	6.63 ± 0.14 a	1.25 ± 0.01 a
200 mM	3.00 ± 0.20 b	0.48 ± 0.01 ab	3.47 ± 0.20 b	6.30 ± 0.49 ab	1.04 ± 0.02 b
300 mM	2.31 ± 0.12 c	0.43 ± 0.01 b	2.74 ± 0.11 c	5.33 ± 0.40 bc	0.83 ± 0.01 c

Data are the mean ± SD ($n = 3$). Different letters following values in the same column indicate significant difference among salt treatments using Duncan's multiple-range test at $p < 0.05$.

2.1.2. Effects of Salt Stress on Osmolytes and Lipid Peroxidation

Compared to the control, the content of soluble sugars, proline and soluble proteins under salt treatment showed significant changes, which increased first and then decreased throughout the experiment (Figure 1). It is worth noting that the first two increased about 3.76 and 2.11-fold, respectively, and the osmolytes were affected significantly under 200 mM NaCl treatment. The accumulation of malondiadehyde (MDA) was significantly increased with the elevated salt treatments ranging from 74.42 to 91.21 nmol g^{-1} FW (Table S1).

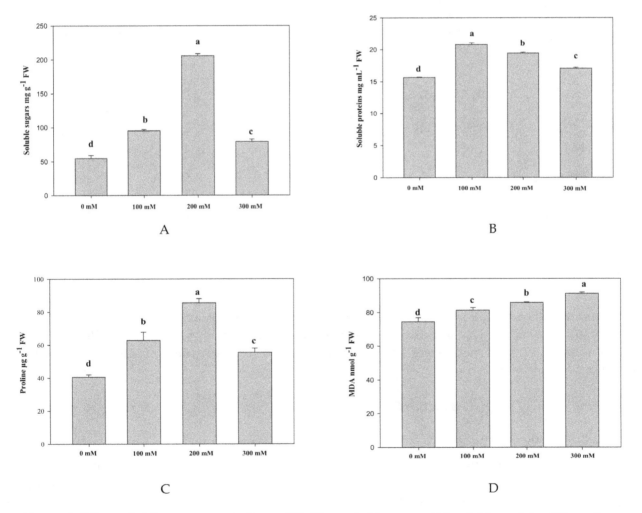

Figure 1. Effects of different concentrations of NaCl on soluble sugars (**A**), soluble proteins (**B**), proline (**C**), and MDA (**D**). Bars are expressed as the mean ± SD ($n = 3$). Bars carrying different letters are significantly different at $p < 0.05$.

2.1.3. Effects of Salt Stress on Antioxidant Enzyme and Ascorbic Acid

It can be observed from Figure 2 that the activity of superoxide dismutase (SOD) was significantly increased under moderate stress and severe stress with respect to the control. Peroxidase (POD) activity was noticeably enhanced, specifically under severe stress, increased to 38.66 U mg^{-1} prot, about 2.2-fold compared with the control (Table S1). However, catalase (CAT) activity in salt-treated AVF was significantly declined compared to the control, but it was elevated with the increasing salt concentrations from 100 to 300 mM (Table S1). Significant change was shown between any two groups of ascorbic acid ranging from 980.9 to 1095 µg mL^{-1}, but it was moderately higher in the salt-treated groups than in the control.

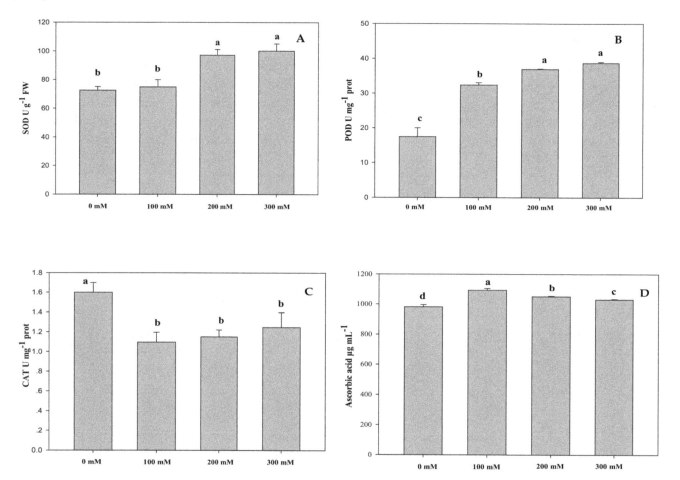

Figure 2. Effects of different NaCl concentrations on the activities of superoxide dismutase (**A**), peroxidase (**B**) catalase (**C**), and ascorbic acid contents (**D**) in Apocyni Veneti Folium (AVF). Bars are expressed as the mean ± SD ($n = 3$). Bars carrying different letters are significantly different at $p < 0.05$ among NaCl treatments.

2.2. Determination of Multiple Bioactive Components

2.2.1. Optimization of Sample Preparation and UFLC-QTRAP-MS/MS Conditions

In order to obtain the optimal extraction efficiency, extraction methods were optimized. Relatively speaking, ultrasonic extraction of samples with a ratio of water volume (mL) to sample weight (g) in accordance with 100:1 for 45 min under 30 °C was the appropriate condition. Then, four kinds of standard compounds with low and high contents, uracil, phenylalanine, neochlorogenic acid, and hyperoside were used to optimize the UFLC-QTRAP-MS/MS conditions. By using the UFLC system with a XBridge® C$_{18}$ column (100 mm × 4.6 mm, 3.5 µm) in the case of the mobile phase of

0.1% formic acid in water—0.1% formic acid in acetonitrile, the flow rate of 0.8 mL min^{-1}, and the column temperature of 30 °C; the analytes were well separated.

MRM (multiple reaction monitoring) technology mainly targeted selection of data for mass spectrometry signal acquisition, recorded signal of the regular ion pairs, and removed the interference ion signal. Only the MS/MS^2 ions selected for mass spectrometry acquisition, in order to achieve more specific, sensitive and accurate analysis of the target molecules [20]. Representative extracted ion chromatograms of 43 analytes in the MRM mode were presented in Figure S2, the detailed information about MS/MS condition for each analyte was listed in Table S2, and the characteristic total ion chromatograms (TIC) was displayed in Figure S3.

2.2.2. Method Validation

Quantitative analysis was performed using the UFLC-QTRAP-MS/MS technique. Calibration curves were constructed by injecting each analyte three times and over six suitable concentrations into UFLC-QTRAP-MS/MS system. As listed in Table S3, the limits of detection (LODs) and limits of quantitation (LOQs) were measured in the range of 0.91–6.15 ng mL^{-1} and 3.03–20.5 ng mL^{-1}, respectively. Each RSD was less than 5%, and the recovery ranged from 95.19 to 103.91%.

2.2.3. Sample Determination

Compared with the control, there was no significant change in asparagine, a higher content of amino acids in AVF, except for 200 mM of salinity stressed group (Table S4). The content of glutamine, proline, and glutamic acid changed notably between any two groups, while the change in the total amino acids was non-significant (Figure 3). Little change was detected on the total nucleosides except for the 300 mM NaCl stressed group. By comparison with the control, the accumulation of inosine and thymidine were considerably reduced. The content of neochlorogenic acid and the total organic acids under salinity stress was significantly reduced, but the change in caffeic acid was different. Compared to the control, kaempferol 3-O-rutinoside, gallocatechin, and epigallocatechin were significantly increased under salt stress, and significant differences were shown from each other. The accumulations of hyperoside, isoquercitrin, astragalin, trifolin, and total flavonoids were changed similarly, increasing first and then decreasing with the increasing salt concentrations.

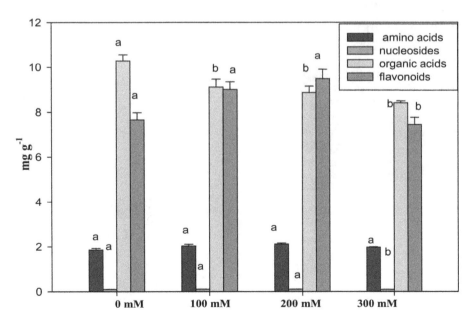

Figure 3. The accumulation of four kinds of constituents in AVF under salt tolerance. Bars are expressed as the mean ± SD ($n = 3$). Bars carrying different letters are significantly different at $p < 0.05$ among NaCl treatments.

2.2.4. Multivariate Statistical Analysis of Samples

A heat map derived from hierarchical clustering analysis intuitively displayed the changes of the accumulation of 43 bioactive components under salinity stress (Figure 4A), and on the other hand, the clustering of samples. In detail, 0 and 100 mM salt treated samples, and 200 and 300 mM salt treated ones were clustered separately and then gathered together. Principal component analysis (PCA) scores plot (Figure 4B) exhibited a statistical distinction based on 43 compositions under salinity stress with R2X [1] and R2X [2] accounted for 47.0% and 16.6% of the total variance, respectively [11]. In the PCA loading plot, chemical markers possessing large loading values of ions, such askaempferol 3-O-rutinoside, hypoxanthine, and thymidine strongly contribute to sample classification. Additionally, gray relational analysis (GRA) is part of the grey system theory and is suitable for solving problems with complicated interrelationships between multiple factors and variables. It provides a reliable guarantee for the quality evaluation of traditional Chinese medicines. The relative correlation degree (r_i) derived from GRA is proportional to the sample quality. Thus, the quality order of AVF under different NaCl treatments was: 200 mM salinity stressed group > 100 mM salinity stressed group > 300 mM salinity stressed group > control, and the corresponding values of r_i were 0.6363, 0.5253, 0.4827, and 0.3984, respectively. These directly revealed that the accumulations of metabolites were affected under saline condition in AVF (Table 2).

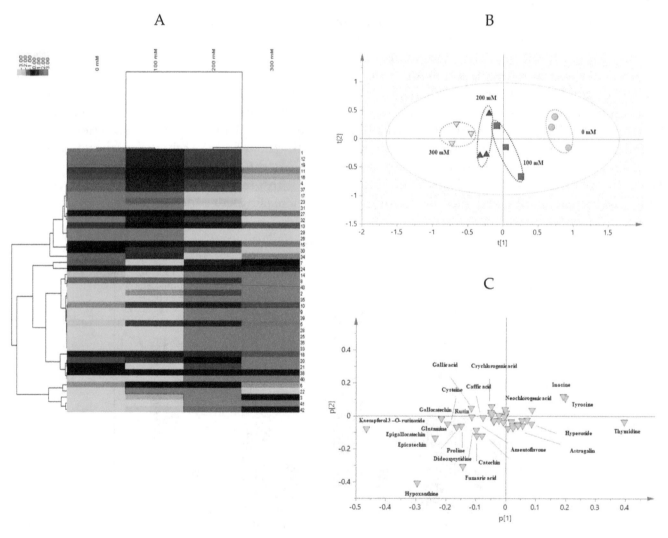

Figure 4. Multivariate statistical analysis of AVF under different salt treatments. Heat map derived from hierarchical clustering analysis (**A**), principal component analysis (PCA) scores plot (**B**), and PCA loading plot (**C**) of AVF.

Table 2. Quality sequencing of the tested samples affected by NaCl.

Treatments	r_i	Quality-Ranking
0 mM	0.3984	4
100 mM	0.5253	2
200 mM	0.6363	1
300 mM	0.4827	3

3. Discussion

Plants exposed to salt stress undergo physiological and biochemical adaptations to help maintain protoplasmic viability in response to salinity. It has been reported that salt stress can cause oxidative stress through increased reactive oxygen species (ROS). Though ROS molecules are the by-products of vital metabolisms, the built-in antioxidant system maintains the ROS under the controlled level. Temporal and spatial-localization of ROS is vital for the regulation of signaling mechanisms [23]. Highly-accumulated ROS, generated as a result of the decreased gas exchange processes and impairment in protective mechanisms, could damage the cellular components, such as lipids, proteins, and nucleic acids [24]. The increased salinity affects primary carbon metabolism, plant growth, and development by ion toxicity, and induces nutritional deficiency, water deficits, and oxidative stress [5,25]. Moreover, it modulates the levels of secondary metabolites, which are physiologically important particularly under stress tolerance [26]. To mitigate ROS-mediated oxidative damage, plants have developed a complex antioxidant defense system that includes osmotic homeostasis, antioxidant enzymes, and metabolites [27–29].

Photosynthesis is involved in the energy metabolism of all the plant systems. When higher plants suffer from salt stress, a growth disorder normally occurs, such as membrane damage and toxic compound accumulation. It can lead to a reduction of the chlorophyll content, disintegration of chloroplast membranes, disruption of photosystem biochemical reactions, and the reduction of photosynthetic activity [30]. The decrease in the chlorophyll content under environmental stress could be attributed to the enhancement of chlorophyll degradation [31]. Carotenoids are accessory light harvesting pigments, preventing the photosynthetic pigments from photo-damage, stabilizing the phospholipids, and scavenging various ROS generated during stressful salinity.

Compared with the control, the content of pigments shown in this study was increased under low and moderated stress, but decreased under severe stress. It indicates that salt stress at specific concentrations might promote the photosynthesis of AVF and severe stress might have a certain inhibitory effect, probably due to the impact of salt on disturbing photosynthesis process, photosynthetic enzymes, chlorophylls, and carotenoids, respectively [5]. The change of chlorophyll a/b reflects the photosynthetic activity of the leaves, and a reasonable value can prevent the excessive light energy in the leaves from inducing the generation of free radicals and photo oxidation of pigment molecules. In summary, it indicated a full utilization of light energy and enhancement of metabolic activity of AVF exposed to low concentrations of NaCl.

The accumulation of osmotic adjustments is one of salt-tolerant mechanisms. It may increase cellular concentrations, help maintain ion homeostasis and water relations, alleviate the negative effects of high ion concentrations on the enzymes and proteins under stressed conditions [32]. Soluble sugars are involved in biosynthetic process and can balance the osmotic strength of cytosol with that of vacuole [33]. The fluctuation in this study under salt stress might be caused by the changes of CO_2 assimilation, source-sink carbon partitioning, and/ or the activity of related enzymes [34]. Soluble proteins in AVF clearly increased depending on the rising of NaCl, and then significantly reduced, indicating that cell turgor maintaining and water acquisition regulation perhaps was affected by salt stress [35,36]. The results were consistent with the literature reported on *Salvia miltiorrhiza* [37]. The increased osmolytes in AVF might be responsible for maintaining homeostasis under low and moderate salt stress, but under severe salt stress, it was notably affected, possibly due to the inability to effectively keep osmotic balance.

Proline accumulation is one of the adaptations of plants to salinity. It has a wide range of biological functions in plants, such as scavenging free-radical by quenching of singlet oxygen, protecting macromolecules against denaturation [38–40], reducing the acidity in the cell, and helping rapid growth after stress [41]. In this study, proline alleviated NaCl stressed induction, but it was significantly reduced caused by osmotic tolerance due to the severe salt stress [42].

Under severe salinity stress, loss of the membrane integrity and stability is a common symptom developed in plants [43] due to excessively formation of free radicals and lipid peroxidation. As one of the lipid peroxidation products, MDA plays an important role in modifying core proteins and in many stressed plants, and it is considered as a useful oxidative marker to indicate the chloroplast lipid peroxidation [44]. A significant increase in the MDA content with salt stress elevated in our study might indicate the oxidative degradation of chloroplast membranes [33,45].

SOD, CAT, and POD, which are involved in antioxidation processes, protect plants from oxidative damage caused by abiotic stresses [46]. SOD catalyzes the dismutation of O_2^- into O_2 and H_2O_2, whereas CAT dis-mutates mostly photorespiratory/respiratory H_2O_2 into H_2O and O_2, and POD is responsible for the removal of H_2O_2 by oxidation of co-substrate, such as phenolic compounds [47]. Integral coordination of antioxidant enzymes could be vital for the redox homeostasis mechanism under the oxidative stress, as previously reported in wheat [48] and barley [49]. In the present study, elevated NaCl increased the activities of SOD and POD, but not CAT. Furthermore, POD might be efficient in clearing the excess H_2O_2 with significantly increased activity and slightly increased SOD activity but significantly decreased in CAT activity appeared under 50 mM NaCl stress. Besides H_2O_2 stress, salt stress might simultaneously enhance superoxide production in cells [50].

As a non-enzymatic free radical scavenger and a key substance in the network of antioxidants, ascorbic acid has been shown to play multiple roles in plant growth. It has also been seen in regulating the normal reactive oxygen species in plant cells together with other small molecules [51,52]. A significant increase was shown in ascorbic acid to protect the body from endogenous damage of oxygen free radicals and then a decrease with the increasing NaCl concentrations, which might be due to the consumption as an enzyme-catalyzed substrate for scavenging ROS.

The increased level of free amino acids in the cell cytoplasm plays an important role in osmotic adjustments, which are also involved in the stability and integrity of cellular membranes in saline environment [53]. The amino acid content was found increased in *Aloe vera* during salinity stress [54]. In this study, the elevated level of amino acids, such as glutamic acid, cysteine, and proline helped provide osmotic protection for AVF. Nucleosides and their derivatives have significant physiological functions. Higher salinity had been reported to induce changes in protein structure, increase in cytoplasmic RNAase activity, leading to decrease in DNA synthesis and creating many cellular menaces to activity required for development processes in plants [55]. Under adverse conditions, transcription factors associated with stress resistance can regulate the simultaneous expression of multiple stress-tolerant genes and the transmission of stress signals [56]. Lower content of nucleosides was shown under severe stress due to the salt stress.

Abiotic stress promotes the synthesis of various secondary metabolites possessing antioxidant activity. Organic acids and flavonoids are ubiquitous in plants and are generally accumulated in response to salinity stress [57]. As mentioned above, phenolic metabolites can cooperate with POD in H_2O_2 scavenging. The phenylpropanoid pathway is the main metabolic route for the synthesis of phenolics and flavonoids [58]. The accumulation of organic acids may vary in different plant species in response to salinity tolerance. In previous reports, they were observed increased in buckwheat sprout [59], but decreased in baby Romaine lettuce [60]. In this experiment, they were significantly depressed under salinity conditions, possibly because of the consumption of scavenging or detoxifying excess free radicals [61].

Flavonoids have a wide array of physiological functions in plants, e.g., involvement in UV filtration and symbiotic nitrogen fixation, and acting as chemical signal messengers for initiating plant-microbe symbiotic associations [62,63], and they also contribute significantly for human by virtue of antioxidative, antiviruses, antiangiogenic, and neuropharmacological effects [64,65]. Flavonoid biosynthesis can be stimulated by the variation of the cellular redox homeostasis and lipid peroxidation of membranes of the plant cell [33,66]. In our study, the increase in flavonoid content under low and moderate might be associated with the increases in chlorophyll, and this enhanced synthesis of secondary metabolites under stressful conditions was believed to protect the cellular structures from oxidative damage and osmotic stress [67]. However, it decreases under severe salt stress, which might be due to the oxidative damage caused by the imbalance between antioxidants formation and ROS scavenging.

According to the results of hierarchical clustering analysis showed in the heat map, the control and low concentrations of salt stressed AVF, and the moderate and high concentrations of salt stressed were aggregated together orderly. This was consistent with PCA results that AVF samples exposed to salt stress (0, 100, 200, and 300 mM, respectively) were sequentially distributed and could be distinguished from each other from the positive to the negative axis of PC1. Moreover, chemical markers in the PCA loading plot provided the possibility of sample differentiation. In all, the results of multivariate statistical analysis indicated that it was not only provided relevant basis for the classification of various salt treatment of AVF, but also comprehensively evaluated the quality of it, that is, moderate salt treated samples were superior to others, and salt treated ones were better than the control.

Generally speaking, in plants under salt stress, photosynthesis, osmotic balance, and metabolic processes are deeply affected [5]. Impairment in the photosynthetic process leads to the higher lipid peroxidation and excessive accumulation of reactive oxygen species (ROS) [24]. The productions of ROS are much higher than its detoxification in abiotic stress conditions. Osmolytes play key roles in maintaining normal osmotic potential, and antioxidant systems of enzymes and metabolites protect plants from oxidative damage and efficiently retain more tolerance against abiotic stresses. In the subsequent experiments, we will use multi-omics approaches to comprehensively interpret the effects of salt stress on the quality of AVF and the salt tolerance mechanism at the transcription, protein, and metabolic levels [68].

4. Materials and Methods

4.1. Plant Materials and Salinity Treatments

The experimental samples of AVF were obtained by the following steps. Firstly, the site for salt stress test was selected in Medicinal Botanical Garden of Nanjing University of Traditional Chinese Medicine (latitude 118°57′1″, East longitude 32°6′5″). The experiment was carried out in the shelter covered by a transparent film that blocked rainwater, while other conditions were similar to the open-air environment. Secondly, the materials and methods of salt stress were given as follow: The botanical origins of the materials were identified by Professor Xunhong Liu (Department for Authentication of Chinese Medicines, Nanjing University of Chinese Medicine); the main root of *Apocynum Venetum* L., two years old from the same plant, and the number of bud and head close to each other, was excavated from the garden in December 2016, and then planted in pots (50 cm height, 34 cm of top diameter, and 26 cm of bottom diameter). Each pot was filled with 25 kg of dry soil and 3 roots, and was placed in the open air before the salt-treated experiments.

Salt stress tests had been conducted since 20 May 2017 when *Apocynum venetum* L. was in normal growth (about 30 cm height). Four levels of salt treatment concentrations, 0 (control, watering), 100 (low stress), 200 (moderate stress), and 300 mM (severe stress) NaCl treatments were designed with 3 replicates at each concentration level and 3 pots per replicate. According to the previous research, by calculating the amount of water, the final determination of the solution per pot was 2 L. In order to prevent osmotic shock, salt concentrations increased gradually by 50 mM NaCl every four days until the designated concentration was reached and lasted 6 times. The photograph of changes of plant phenotypes upon treatment of different concentrations of NaCl was seen in Figure S1. At last, experimental samples were harvested on 30 June 2017. The collected samples of four groups were immediately frozen at $-80\ °C$ for subsequent experiments, some for physiological experiments, some for quantitative analysis, and the rest for the voucher specimens deposited at the Herbarium in School of Pharmacy, Nanjing University of Chinese Medicine.

4.2. Physiological Experiment

4.2.1. Extraction and Assay of Pigment

Four groups of fresh AVF samples (0.2 g) under salt stress were homogenized with ethanol (95%, v/v), filtered, and made up to 2 mL, respectively. Photosynthetic pigments (chlorophyll a, b, total chlorophyll and carotenoids) concentrations were calculated from the absorbance of extract at 665, 649 and 470 nm using the formula [37,69], given as follow: Chlorophyll a (mg g^{-1} FW) = (13.95 × A665 − 6.88 × A649) × 2/(1000 × 0.2); Chlorophyll b (mg g^{-1} FW) = (24.96 × A649 − 7.32 × A665) × 2/(1000 × 0.2); Carotenoids (mg g^{-1} FW) = ((1000 × A470) − (2.05 × Chl a) − (114.8 × Chl b)) × 2/(245 × 1000 × 0.2)

4.2.2. Osmolytes and MDA Assay

Osmolytes, including soluble proteins, soluble sugars and proline, were assayed in this study to measure the salt tolerance of AVF. The content of soluble proteins was determined according to the ultraviolet absorption method. 0.5 g of fresh sample of AVF under salt stress was homogenized and extracted by 8 mL of PBS (phosphate buffer saline, 0.1 mM Na_2HPO_4 and NaH_2PO_4, pH 7.4), respectively. After centrifugation, 1 μL of the supernatant was subjected to detect using UV-Vis spectrophotometer (DENOVIX DS-11, Wilmington, DE, USA), which can directly display the concentration of protein in the solution by detecting the absorbance of the solution at 260 nm and 280 nm with 1 μL of PBS as a control. The content of soluble sugars and proline was quantified by colorimetric method. To measure the content of soluble sugars, 0.5 g of fresh leaves was homogenized with 4.5 mL of PBS in an ice water bath and centrifuge the homogenate at 3500 rpm for 10 min. 0.5 mL of extract was mixed with 3 mL of anthrone solution (75 mg anthrone in 50 mL of 72% sulphuric acid (w/w)), and was immediately placed in a boiling water bath for 10 min. The light absorption was estimated at 620 nm. The content of soluble sugars was determined by using glucose as a standard and expressed as mg g^{-1} FW. With regard to the free proline assay, the procedure was as follows: 0.5 g of harvested leaf fragments were extracted with 4.5 mL of aqueous sulfosalicylic acid (3%, w/v) in boiled water for 10 min and then centrifuged at 3500 rpm for 10 min. After that, 2 mL of glacial acetic acid and 4 mL of acidninhydrin agent (1.25 g acidninhydrin in 30 mL glacial acetic acid and 20 mL of 6 mol L^{-1} H_3PO_4) were added to the homogenate in a test tube. The mixture was incubated in boiling water for 30 min, and then the test tube was placed in the cold water to terminate the reaction. Each test tube was added to 4 mL of toluene and vortexed for 30 s. The supernatant was taken and centrifuged at 3000 rpm for 5 min. Proline content was quantified by using the Bate's method at 520 nm [69,70].

Lipid peroxidation was measured in terms of MDA by thiobarbituric acid (TBA) method [69,71]. Fresh leaf (0.5 g) fragment were homogenized with 5 mL of trichloroacetic acid (3%, m/v), and then centrifuged at 3000 rpm for 10 min. Took 2.0 mL aliquot of the supernatant to the test tube and added 2.0 mL of thiobarbituric acid (0.67%, m/v). The mixture was heated in boiling water for 30 min and then quickly cooled in an ice bath. After centrifugation at 3000 rpm for 10 min, the absorbance of the samples was recorded at 530 nm.

4.2.3. Enzyme Activities and Ascorbic Acid Assay

To determine the antioxidant enzyme activities, 0.5 g of fresh AVF under the stress of different salt concentrations were homogenized with 4.5 mL of PBS in an ice water bath and centrifuged at 3500 rpm for 10 min. The supernatant was collected to determine the antioxidant enzyme activities. SOD activity was assayed by hydroxylamine method, CAT activity was determined by ammonium molybdate method, and POD activity was measured according to the colorimetric method [69,70].

As for the determination of ascorbic acid, it was assayed based on the oxidation of ascorbic acid by iron (III) in the presence of 1,10-phenanthroline with subsequent formation of ferroin and a suitable anion associate according to the Zenki et al. method [72]. All of them were tested by assay kits bought from Nanjing Jiancheng Bioengineering Institute (Nanjing, China). UV-visible absorptions were measured by multi-mode microplate reader (SpectraMax M5, San Jose, CA, USA) and the detection amount of the reaction solution was 200 µL.

4.3. Multiple Bioactive Constituents Assay

4.3.1. Chemicals and Reagents

Ultrapure water was prepared using a Milli-Q purifying system (Millipore, Bedford, MA, USA). Methanol and acetonitrile of HPLC grade were purchased from Merck (Damstadt, Germany). Standard compounds of histidine (1), arginine (3), cysteine (4), asparagine (5), serine (6), lysine (7), glutamine (8), proline (9), cytidine (10), hypoxanthine (11), deoxycytidine (12), uridine (13), tyrosine (14), guanine (15), guanosine(16), inosine(17), deoxyguanosine (19), isoleucine (20), leucine (21), thymidine (23), phenylalanine (24), tryptophan (27), epicatechin (33), rutin (34), hyperoside (35), and quercitrin (37) were purchased from Shanghai Yuanye Biotechnology (Shanghai, China); glutamic acid (2), gallic acid (22) and apigenin (43) were obtained from Chinese National Institute of Control of Pharmaceutical and Biological Products (Beijing, China); fumaric acid (18), gallocatechin (25), epigallocatechin (28), cryptochlorogenic acid (31), kaempferol 3-O-rutinoside (39) and amentoflavone (42) were acquired from Chengdu Chroma Biotechnology (Chengdu, China); neochlorogenic acid (26), chlorogenic acid (29), catechin (30), caffeic acid (32), isoquercitrin (36), avicularin (38), trifolin (40), and astragalin (41) were bought from Baoji Chenguang Biotechnology Co., LTD. (Baoji, China) with the purity greater than 98% and their structures were presented in Figure S4.

4.3.2. Sample Preparation

Four groups of fresh AVF harvested under salt treatments were naturally dried, and then powdered and passed through a 60-mesh sieve. 0.3 g of sample was weighed accurately and ultrasonically extracted with 30 mL of water for 45 min, supplemented with water to compensate for the lost weight, and centrifuged at 12000 rpm for 15 min [11,73]. The supernatant was stored at 4 °C and filtered through a 0.22 µm membrane (Jinteng laboratory equipment Co., Ltd., Tianjin, China) before being subjected to UFLC-MS/MS analysis.

4.3.3. Chromatographic and Mass Spectrometric Conditions

The mobile phase of AB Sciex QTRAP® 4500 UFLC-MS/MS spectrometry consisted of water containing 0.1% formic acid (v/v, A) and acetonitrile containing 0.1% formic acid (v/v, B). The analytes were eluted using a linear gradient program: 1–3 min, 5% B; 3–6 min, 5–15% B; 6–15 min, 15–20% B; 15–17 min, 20–70% B, 17–17.5 min, 70–5% B, and 17.5–23 min, 5% B. The flow rate was 0.80 mL/min. The column temperature was 30 °C. The injection volume was 1 µL. According to our previous reports [11], the standard solution of each analyte was injected separately into the electrospray ionization (ESI) source in the direct infusion mode of MS to acquire the fragmentor voltage and collision energy in both positive and negative modes. Next, ESI source operates in both ion modes using the MRM transition acquiring the spectra and the Analyst 1.6.3 software analyzing data, respectively. In the same ion mode, isomers with the same ion pairs, such as catechin/epicatechin, chlorogenic acid/neochlorogenic acid/cryptochlorogenic acid, gallocatechin/epigallocatechin, hyperoside/isoquercetin, and leucine/isoleucine were separately and injected into UFLC-QTRAP-MS/MS to find the accurate t_R for identification and quantification. The operating parameters were set as follows: GS1 flow, 65 L min^{-1}; GS2 flow, 65 L min^{-1}; and CUR flow, 30 L min^{-1}; gas temperature, 650 °C; pressure of the nebulizer, 5500 V for the positive ion mode, and −4500 V for the negative ion mode, respectively.

4.3.4. Method Validation and Sample Determination

The standard solution containing 43 reference substances was prepared and diluted with water to appropriate concentrations for the construction of calibration curves. The concentrations of 43 analytes in mixed solution were seen in Table S3. The LODs and LOQs of constituents were measured at signal-to-noise (S/N) ratios of 3 and 10, respectively. Precision of the intra and inter-day was expressed as relative standard deviation (RSD). Repeatability was achieved by six different analytical sample solutions prepared by the same sample, and stability was performed by analyzing the variations at 0, 2, 4, 8, 12, and 24 h, respectively. While the recovery test was performed by adding a known amount of corresponding constituents in triplicate at low, medium, or high levels to 0.5 g of 100 mM NaCl treated samples, respectively. The quantitative determination of the bioactive compounds of AVF under different sat stress was performed under the optimal condition by UFLC-QTRAP-MS/MS.

4.3.5. Multivariate Statistical Analysis

Hierarchical cluster analysis is a method of cluster analysis, which seeks to build a hierarchy of clusters. PCA is a statistical procedure that uses an orthogonal transformation to convert a set of observations of possibly correlated variables (entities each of which takes on various numerical values) into a set of values of linearly uncorrelated variables called principal components [11]. Hierarchical clustering analysis and PCA were introduced to cluster and classify samples based on the content of constituents by Java Treeview 3.0 software and SIMCA-P 13.0 software, respectively. Then, GRA was performed according to the contents of 43 bioactive components by Microsoft Excel 2010 for Window 10 to evaluate the quality of AVF under different concentration of salt stress. Specifically, through the establishment of sample dataset and normalization treatment of raw data, the optimal and the worst reference sequences were conducted. After establishing dimension of the differences between comparing sequences and reference sequences, correlation coefficient and correlation degree were calculated, followed by the weight value of the evaluation samples (r_i).

4.4. Data Processing

The mean values of all parameters were taken from the measurements of three replicates with the standard deviation calculated. One-way ANOVA followed by Duncan's multiple-range test was used to compare the means with the significance level set as 0.05 by SPSS 19.0

5. Conclusions

In this study, our aim was to use the changes in physiology and biochemical components as references to study the quality control of AVF response to salt stress. Thus, an efficient analytical method of simultaneous determination of multiple bioactive constituents combined with physiological analysis was established for the quality evaluation based on the multivariate statistical analysis. Investigations into the physiological changes of photosynthetic pigments, osmotic homeostasis, lipid peroxidation, antioxidative enzymes, and ascorbic acid could provide comprehensive insights into the response mechanisms induced by salt stress. Furthermore, a total of 43 bioactive constituents, including amino acids, nucleosides, organic acids, and flavonoids, were successfully identified and quantified in different salinity-treated AVF with the application of UFLC-QTRAP-MS/MS technology. Multivariate statistical analysis was performed for the group classification and quality evaluation. Overall, the quality of AVF subjected to NaCl was superior to the control and AVF treated with 200 mM NaCl had the best quality. In general, this study was conducted to the quality evaluation of AVF concerning the impacts caused by salinity on the physiology and bioactive constituents. The results might provide a valuable reference for the quality assessment of other herbal medicines and the development of salt-tolerant plants in saline soils.

Author Contributions: C.C. and X.L. conceived and designed the experiments. C.C., C.W. and Z.L. performed the experiments. C.C., L.Z., J.S., S.C., J.C., and M.T. analyzed the data and drafted the manuscript. All authors contributed to the revision of this manuscript and approved the final manuscript.

Acknowledgments: This research was supported by the Priority Academic Program Development of Jiangsu Higher Education Institutions of China (NO. ysxk-2014) and Postgraduate Research & Practice Innovation Program of Jiangsu Province (KYCX18_1606).

References

1. Zhao, G.M.; Han, Y.; Sun, X.; Li, S.H.; Shi, Q.M.; Wang, C.H. Salinity stress increases secondary metabolites and enzyme activity in safflower. *Ind. Crop. Prod.* **2015**, *64*, 175–181.

2. Zhou, Y.; Tang, N.Y.; Huang, L.J.; Zhao, Y.J.; Tang, X.Q.; Wang, K.C. Effects of Salt Stress on Plant Growth, Antioxidant Capacity, Glandular Trichome Density, and Volatile Exudates of Schizonepeta tenuifolia Briq. *Int. J. Mol. Sci.* **2018**, *19*, 252. [CrossRef] [PubMed]

3. Aghaei, K.; Komatsu, S. Crop and medicinal plants proteomics in response to salt stress. *Front. Plant Sci.* **2013**, *4*, 8. [CrossRef] [PubMed]

4. Liu, A.L.; Xiao, Z.X.; Li, M.; Wong, F.; Yung, W.S.; Ku, Y.S.; Wang, Q.W.; Wang, X.; Xie, M.; Yim, A.K.; et al. Transcriptomic reprogramming in soybean seedlings under salt stress. *Plant Cell Environ.* **2018**, 1–17.

5. Munns, R.; Tester, M. Mechanisms of salinity tolerance. *Annu. Rev. Plant Biol.* **2008**, *59*, 651–681. [CrossRef] [PubMed]

6. Arbona, V.; Manzi, M.; Ollas, C.; Gómez-Cadenas, A. Metabolomics as a Tool to Investigate Abiotic Stress Tolerance in Plants. *Int. J. Mol. Sci.* **2013**, *14*, 4885–4911. [CrossRef] [PubMed]

7. Zhang, B.; Liu, K.; Zheng, Y.; Wang, Y.; Wang, J.; Liao, H. Disruption of AtWNK8 Enhances Tolerance of Arabidopsis to Salt and Osmotic Stresses via Modulating Proline Content and Activities of Catalase and Peroxidase. *Int. J. Mol. Sci.* **2013**, *14*, 7032–7047. [CrossRef] [PubMed]

8. Jaleel, C.A.; Riadh, K.; Gopi, R.; InèsHameed, M.; Inès, J.; Al-Juburi, H.; Zhao, C.X.; Shao, H.B.; Rajaram, P. Antioxidant defense responses: Physiological plasticity in higher plants under abiotic constraints. *Acta Physiol. Plant.* **2009**, *31*, 427–436. [CrossRef]

9. Tran, L.S.; Urao, T.; Qin, F.; Maruyama, K.; Kakimoto, T.; Shinozaki, K.; Yamaguchi-Shinozaki, K. Functional analysis of AHK1/ATHK1 and cytokinin receptor histidine kinases in response to abscisic acid, drought, and salt stress in Arabidopsis. *Proc. Natl. Acad. Sci. USA* **2007**, *104*, 20623–20638. [CrossRef] [PubMed]

10. Tang, X.L.; Mu, X.M.; Shao, H.B.; Wang, H.; Brestic, M. Global plant-responding mechanisms to salt stress: Physiological and molecular levels and implications in biotechnology. *Crit. Rev. Biotechnol.* **2015**, *35*, 425–437. [CrossRef] [PubMed]

11. Chen, C.H.; Liu, Z.X.; Zou, L.X.; Liu, X.H.; Chai, C.; Zhao, H.; Yan, Y.; Wang, C.C. Quality evaluation of Apocyni Veneti Folium from different habitats and commercial herbs based on simultaneous determination of multiple bioactive constituents combined with multivariate statistical analysis. *Molecules* **2018**, *23*, 573. [CrossRef] [PubMed]

12. The Pharmacopoeia Committee of the Health Ministry of People's Republic of China. *Pharmacopoeia of People's Republic of China*; Guangdong Scientific Technologic Publisher: Guangzhou, China, 1995; p. 182.

13. Pharmacopoeia Commission of the Ministry of Health of the People's Republic of China. *Pharmacopoeia of the People's Republic of China*; Part I; Medical Science and Technology Press: Beijing, China, 2015; pp. 211–212.

14. Ksouri, R.; Megdiche, W.; Debez, A.; Falleh, H.; Grignon, C.; Abdelly, C. Salinity effects on polyphenol content and antioxidant activities in leaves of the halophyte *Cakile maritima*. *Plant Physiol. Biochem.* **2007**, *45*, 244–249. [CrossRef] [PubMed]

15. Zhou, C.Z.; Gao, G.H.; Zhou, X.M.; Yu, D.; Chen, X.H.; Bi, K.S. Simultaneous determination of five active components in traditional Chinese medicine *Apocynum venetum* L. by RP-HPLC–DAD. *J. Med. Plants Res.* **2011**, *5*, 735–742.

16. Liu, X.H.; Zhang, Y.C.; Li, S.J.; Wang, M.; Wang, L.J. Simultaneous Determination of Four Flavonoids in Folium Apocyni Veneti by HPCE-DAD. *Chin. Pharmacol. J.* **2010**, *45*, 464–467.

17. An, H.J.; Wang, H.; Lan, Y.X.; Hashi, Y.; Chen, S.Z. Simultaneous qualitative and quantitative analysis of phenolic acids and flavonoids for the quality control of *Apocynum venetum* L. leaves by

HPLC–DAD–ESI–IT–TOF–MS and HPLC–DAD. *J. Pharmaceut. Biomed.* **2013**, *85*, 295–304. [CrossRef] [PubMed]

18. Chen, F.; Zhang, F.S.; Yang, N.Y.; Liu, X.H. Simultaneous Determination of 10 Nucleosides and Nucleobases in *Antrodia camphorata* Using QTRAP LC–MS/MS. *J. Chromatogr. Sci.* **2014**, *52*, 852–861. [CrossRef] [PubMed]

19. Yuan, M.; Breitkopf, S.B.; Yang, X.; Asara, J.M. A positive/negative ion-switching, targeted mass spectrometry-based metabolomics platform for bodily fluids, cells, and fresh and fixed tissue. *Nat. Protoc.* **2012**, *7*, 872–881. [CrossRef] [PubMed]

20. March, R.E. An introduction to quadrupole ion trap mass spectrometry. *J. Mass Spectrom.* **1997**, *32*, 351–369. [CrossRef]

21. Shi, J.Y.; Li, G.L.; Zhang, R.; Zheng, J.; Suo, Y.R.; You, J.M.; Liu, Y.J. A validated HPLC-DAD-MS method for identifying and determining the bioactive components of two kinds of luobuma. *J. Liq. Chromatogr. Relat. Technol.* **2011**, *34*, 537–547. [CrossRef]

22. Wahid, A.; Ghazanfar, A. Possible involvement of some secondary metabolites in salt tolerance of sugarcane. *J. Plant Physiol.* **2006**, *163*, 723–730. [CrossRef] [PubMed]

23. Mateos-Naranjo, E.; Andrades-Moreno, L.; Davy, A.J. Silicon alleviates deleterious effects of high salinity on the halophytic grass Spartina densiflora. *Plant Physiol. Biochem.* **2013**, *63*, 115–121. [CrossRef] [PubMed]

24. Kim, Y.H.; Khan, A.L.; Kim, D.H.; Lee, S.Y.; Kim, K.M.; Waqas, M.; Jung, H.Y.; Shin, J.H.; Kim, J.G.; Lee, I.J. Silicon mitigates heavy metal stress by regulating P-type heavy metal ATPases, Oryza sativa low silicon genes, and endogenous phytohormones. *BMC Plant Biol.* **2014**, *14*, 13. [CrossRef] [PubMed]

25. Flowers, T.J.; Colmer, T.D. Salinity tolerance in halophytes. *New Phytol.* **2008**, *179*, 945–963. [CrossRef] [PubMed]

26. Parihar, P.; Singh, S.; Singh, R.; Singh, V.P.; Prasad, S.M. Effect of salinity stress on plants and its tolerance strategies: A review. *Environ. Sci. Pollut. Res. Int.* **2015**, *22*, 4056–4075. [CrossRef] [PubMed]

27. Deinlein, U.; Stephan, A.B.; Horie, T.; Luo, W.; Xu, G.; Schroeder, J.I. Plant salt-tolerance mechanisms. *Trends Plant Sci.* **2014**, *19*, 371–379. [CrossRef] [PubMed]

28. Golldack, D.; Li, C.; Mohan, H.; Probst, N. Tolerance to drought and salts tress in plants: Unraveling the signaling networks. *Front. Plant Sci.* **2014**, *5*, 151. [CrossRef] [PubMed]

29. Roy, S.J.; Negrão, S.; Tester, M. Salt resistant crop plants. *Curr. Opin. Biotechnol.* **2014**, *26*, 115–124. [CrossRef] [PubMed]

30. Gururani, M.A.; Venkatesh, J.; Tran, L.S.P. Regulation of photosynthesis during abiotic stress-induced photoinhibition. *Mol. Plant* **2015**, *8*, 1304–1320. [CrossRef] [PubMed]

31. Gururani, M.A.; Mohanta, T.K.; Bae, H. Current understanding of the interplay between phytohormones and photosynthesis under environmental stress. *Int. J. Mol. Sci.* **2015**, *16*, 19055–19085. [CrossRef] [PubMed]

32. Flowers, T.J.; Munns, R.; Colmer, T.D. Sodium chloride toxicity and the cellular basis of salt tolerance in halophytes. *Ann. Bot.* **2015**, *115*, 419–431. [CrossRef] [PubMed]

33. D'Souza, M.R.; Devaraj, V.R. Biochemical responses of Hyacinth bean (*Lablab purpureus*) to salinity stress. *Acta Physiol. Plant.* **2010**, *32*, 341–353.

34. Rosa, M.; Prado, C.; Podazza, G.; Interdonato, R.; Gonzalez, J.A.; Hilal, M.; Prado, F.E. Soluble sugars: Metabolism, sensing and abiotic stress. A complex network in the life of plants. *Plant Signal. Behav.* **2009**, *4*, 388–393. [CrossRef] [PubMed]

35. Abbaspour, H.; Afshari, H.; Abdel-Wahhab, A. Influence of salt stress on growth, pigments, soluble sugars and ion accumulation in three pistachio cultivars. *J. Med. Plants Res.* **2012**, *6*, 2468–2473. [CrossRef]

36. Mittal, S.; Kumari, N.; Sharma, V. Differential response of salt stress on *Brassica juncea*: Photosynthetic performance, pigment, proline, D1 and antioxidant enzymes. *Plant Physiol. Biochem.* **2012**, *54*, 17–26. [CrossRef] [PubMed]

37. Gengmao, Z.; Quanmei, S.; Yu, H.; Shihui, L.; Changhai, W. The physiological and biochemical responses of a medicinal plant (*Salvia miltiorrhiza* L.) to stress caused by various concentrations of NaCl. *PLoS ONE* **2014**, *9*, e89624. [CrossRef] [PubMed]

38. Sharma, S.; Verslues, P.E. Mechanisms independent of abscisic acid (ABA) or proline feedback have a predominant role in transcriptional regulation of proline metabolism during low water potential and stress recovery. *Plant Cell Environ.* **2010**, *33*, 1838–1851. [CrossRef] [PubMed]

39. Szabados, L.; Savoure, A. Proline: A multifunctional amino acid. *Trends Plant Sci.* **2010**, *15*, 89–97. [CrossRef] [PubMed]

40. Kumar, S.G.; Reddy, A.M.; Sudhakar, C. NaCl effects on proline metabolism in two high yielding genotypes of mulberry (*Morus alba* L.) with contrasting salt tolerance. *Plant Sci.* **2003**, *165*, 1245–1251. [CrossRef]

41. Cuin, T.A.; Shabala, S. Compatible solutes reduce ROS-induced potassium efflux in Arabidopsis roots. *Plant Cell Environ.* **2007**, *30*, 875–885. [CrossRef] [PubMed]

42. Çoban, Ö.; Baydar, N.G. Brassinosteroid effects on some physical and biochemical properties and secondary metabolite accumulation in peppermint (*Mentha piperita* L.) under salt stress. *Ind. Crops Prod.* **2016**, *86*, 251–258. [CrossRef]

43. Bajji, M.; Kinet, J.M.; Lutts, S. The use of the electrolyte leakage method for assessing cell membrane stability as a water stress tolerance test in durum wheat. *Plant Growth Regul.* **2002**, *36*, 61–70. [CrossRef]

44. Yamauchi, Y.; Sugimoto, Y. Effect of protein modification by malondialdehyde on the interaction between the oxygen-evolving complex 33 kDa protein and photosystem II core proteins. *Planta* **2010**, *231*, 1077–1088. [CrossRef] [PubMed]

45. Wang, Q.H.; Liang, X.; Dong, Y.J.; Xu, L.L.; Zhang, X.W.; Kong, J.; Liu, S. Effects of exogenous salicylic acid and nitric oxide on physiological characteristics of perennial ryegrass under cadmium stress. *J. Plant Growth Regul.* **2013**, *32*, 721–731. [CrossRef]

46. Tasgin, E.; Atici, O.; Nalbantoglu, B.; Popova, L.P. Effects of salicylic acid and cold treatments on protein levels and on the activities of antioxidant enzymes in the apoplast of winter wheat leaves. *Phytochemistry* **2006**, *67*, 710–715. [CrossRef] [PubMed]

47. Kang, G.Z.; Wang, C.H.; Sun, G.C.; Wang, Z.X. Salicylic acid changes activities of H2O2-metabolizing enzymes and increases the chilling tolerance of banana seedlings. *Environ. Exp. Bot.* **2003**, *50*, 9–15. [CrossRef]

48. Nwugo, C.C.; Huerta, A.J. The effect of silicon on the leaf proteome of rice (*Oryza sativa* L.) Plants under Cadmium-Stress. *J. Proteome Res.* **2011**, *10*, 518–528. [CrossRef] [PubMed]

49. Melo, A.M.P.; Roberts, T.H.; Moller, I.M. Evidence for the presence of two rotenone-insensitive NAD(P)H dehydrogenases on the inner surface of the inner membrane of potato tuber mitochondria. *Biochim. Biophys. Acta* **1996**, *1276*, 133–139. [CrossRef]

50. Apel, K.; Hirt, H. Reactive oxygen species: Metabolism, oxidative stress, and signal transduction. *Annu. Rev. Plant Biol.* **2004**, *55*, 373–399. [CrossRef] [PubMed]

51. Wang, Z.Y.; Xiong, L.; Li, W.; Zhu, J.K.; Zhu, J. The plant cuticle is required for osmotic stress regulation of abscisic acid biosynthesis and osmotic stress tolerance in Arabidopsis. *Plant Cell* **2011**, *23*, 1971–1984. [CrossRef] [PubMed]

52. Moradi, F.; Ismail, A.M. Responses of photosynthesis, chlorophyll fluorescence and ROS-Scavenging systems to salt stress during seedling and reproductive stages in rice. *Ann. Bot.* **2007**, *99*, 1161–1173. [CrossRef] [PubMed]

53. Mishra, A.; Patel, M.K.; Jha, B. Non-targeted metabolomics and scavenging activity of reactive oxygen species reveal the potential of *Salicornia brachiata* as a functional food. *J. Funct. Foods* **2015**, *13*, 21–31. [CrossRef]

54. Murillo-Amador, B.; Córdoba-Matson, M.V.; Villegas-Espinoza, J.A.; Hernández-Montiel, L.G.; Troyo-Diéguez, E.; García-Hernández, J.L. Mineral content and biochemical variables of *Aloe vera* L. under salt stress. *PLoS ONE* **2014**, *9*, e9487. [CrossRef] [PubMed]

55. Niu, X.; Bressan, R.A.; Hasegawa, P.M.; Pardo, J.M. Ion homeostasis in NaCl stress environments. *Plant Physiol.* **1995**, *109*, 735–742. [CrossRef] [PubMed]

56. Qi, Z.; Xiong, L. Characterization of a Purine Permease Family Gene OsPUP7 Involved in Growth and Development Control in Rice. *Chin. Bull. Botany* **2013**, *55*, 1119–1135.

57. Zhao, X.; Wang, W.; Zhang, F.; Deng, J.; Li, Z.; Fu, B. Comparative metabolite profiling of two rice genotypes with contrasting salt stress tolerance at the seedling stage. *PLoS ONE* **2014**, *29*, e108020. [CrossRef] [PubMed]

58. Yokozawa, T.; Kashiwada, Y.; Hattori, M.; Chung, H.Y. Study on the components of Luobuma with peroxynitrite-scavenging activity. *Biol. Pharm. Bull.* **2002**, *25*, 748–752. [CrossRef] [PubMed]

59. Lim, J.H.; Park, K.J.; Kim, B.K.; Jeong, J.W.; Kim, H.J. Effect of salinity stress on phenolic compounds and carotenoids in buckwheat (*Fagopyrum esculentum* M.) sprout. *Food Chem.* **2012**, *135*, 1065–1070. [CrossRef] [PubMed]

60. Chisari, M.; Todaro, A.; Barbagallo, R.N.; Spagna, G. Salinity effects on enzymatic browning and antioxidant capacity of fresh-cut baby Romaine lettuce (*Lactuca sativa* L. cv. Duende). *Food Chem.* **2010**, *119*, 1502–1506. [CrossRef]

61. Nichenametla, S.N.; Taruscio, T.G.; Barney, D.L.; Exon, J.H. A review of the effects and mechanisms of polyphenolics in cancer. *Crit. Rev. Food Sci. Nutr.* **2006**, *46*, 161–183. [CrossRef] [PubMed]

62. Chen, C.H.; Xu, H.; Liu, X.H.; Zou, L.S.; Wang, M.; Liu, Z.X.; Fu, X.S.; Zhao, H.; Yan, Y. Site-specific accumulation and dynamic change of flavonoids in Apocyni Veneti Folium. *Microsc. Res. Tech.* **2017**, *80*, 1315–1322. [CrossRef] [PubMed]

63. Lee, B.H.; Jeong, S.M.; Lee, J.H.; Kim, J.H.; Yoon, I.S.; Lee, J.H.; Choi, S.H.; Lee, S.M.; Chang, C.G.; Kim, H.C.; et al. Quercetin inhibits the 5-hydroxytryptamine type 3 receptor-mediated ion current by interacting with pre-transmembrane domain I. *Mol. Cells* **2005**, *20*, 69–73. [PubMed]

64. Kim, Y.H.; Lee, Y.J. RAIL apoptosis is enhanced by quercetin through Akt dephosphorylation. *J. Cell Biochem.* **2007**, *100*, 998–1009. [CrossRef] [PubMed]

65. Winkel-Shirley, B. Biosynthesis of flavonoids and effects of stress. *Curr. Opin. Plant Biol.* **2002**, *5*, 218–223. [CrossRef]

66. Bettaieb, I.; Knioua, S.; Hamrouni, I.; Limam, F.; Marzouk, B. Water-deficit impact on fatty acid and essential oil composition and antioxidant activities of cumin (*Cuminum cyminum* L.) aerial parts. *J. Agr. Food Chem.* **2011**, *59*, 328–334. [CrossRef] [PubMed]

67. Close, D.C.; McArthor, C. Rethinking the role of many plant phenolics—Protection against photodamage not herbivores? *OIKOS* **2002**, *99*, 166–172. [CrossRef]

68. Hirayama, T.; Shinozaki, K. Research on plant abiotic stress responses in the post-genome era: Past, present and future. *Plant J.* **2010**, *61*, 1041–1052. [CrossRef] [PubMed]

69. Wang, X.H.; Huang, J.L. *Principles and Techniques of Plant Physiological Biochemical Experiment*, 3rd ed.; Higher Education Press: Beijing, China, 2015.

70. Zeng, J.W.; Chen, A.M.; Li, D.D.; Yi, B.; Wu, W. Effects of Salt Stress on the Growth, Physiological Responses, and Glycoside Contents of *Stevia rebaudiana* Bertoni. *J. Agric. Food Chem.* **2013**, *61*, 5720–5726. [CrossRef] [PubMed]

71. Wu, F.B.; Zhang, G.P.; Dominy, P. Four barley genotypes respond differently to cadmium: Lipid peroxidation and activities of antioxidant capacity. *Environ. Exp. Bot.* **2003**, *50*, 67–78. [CrossRef]

72. Zenki, M.; Tanishita, A.; Yokoyama, T. Repetitive determination of ascorbic acid using iron(III)-1.10-phenanthroline-peroxodisulfate system in a circulatory flow injection method. *Talanta* **2004**, *64*, 1273–1277. [CrossRef] [PubMed]

73. Hua, Y.J.; Wang, S.N.; Chai, C.; Liu, Z.S.; Liu, X.H.; Zou, L.S.; Wu, Q.N.; Zhao, H.; Yan, Y. Quality Evaluation of Pseudostellariae Radix Based on Simultaneous Determination of Multiple Bioactive Components Combined with Grey Relational Analysis. *Molecules* **2016**, *22*, 13. [CrossRef] [PubMed]

12

Overexpression of Transglutaminase from Cucumber in Tobacco Increases Salt Tolerance through Regulation of Photosynthesis

Min Zhong [1], Yu Wang [1], Yuemei Zhang [1], Sheng Shu [1], Jin Sun [1,2] and Shirong Guo [1,2,*]

[1]   Key Laboratory of Southern Vegetable Crop Genetic Improvement, Ministry of Agriculture,
     College of Horticulture, Nanjing Agricultural University, Nanjing 210095, China;
     2016204040@njau.edu.cn (M.Z.); ywang@njau.edu.cn (Y.W.); zym941128@163.com (Y.Z.);
     shusheng@njau.edu.cn (S.S.); jinsun@njau.edu.cn (J.S.)
[2]   Suqian Academy of Protected Horticulture, Nanjing Agricultural University, Suqian 223800, China
*   Correspondence: srguo@njau.edu.cn

Abstract: Transglutaminase (TGase) is a regulator of posttranslational modification of protein that provides physiological protection against diverse environmental stresses in plants. Nonetheless, the mechanisms of TGase-mediated salt tolerance remain largely unknown. Here, we found that the transcription of cucumber *TGase* (*CsTGase*) was induced in response to light and during leaf development, and the CsTGase protein was expressed in the chloroplast and the cell wall. The overexpression of the *CsTGase* gene effectively ameliorated salt-induced photoinhibition in tobacco plants, increased the levels of chloroplast polyamines (PAs) and enhanced the abundance of D1 and D2 proteins. TGase also induced the expression of photosynthesis related genes and remodeling of thylakoids under normal conditions. However, salt stress treatment reduced the photosynthesis rate, PSII and PSI related genes expression, D1 and D2 proteins in wild-type (WT) plants, while these effects were alleviated in *CsTGase* overexpression plants. Taken together, our results indicate that TGase-dependent PA signaling protects the proteins of thylakoids, which plays a critical role in plant response to salt stress. Thus, overexpression of TGase may be an effective strategy for enhancing resistance to salt stress of salt-sensitive crops in agricultural production.

Keywords: TGase; photosynthesis; salt stress; polyamines; cucumber

1. Introduction

Photosynthesis is the basic manufacturing process in plants; it can increase carbon gains and improve crop yield and quality [1]. Salinity, along with other environmental stresses such as drought and chilling, induces inhibition of photosynthetic activity by disruption of the chloroplast structure and reduction in CO_2 assimilation [2]. The effects of environmental stresses on photosynthesis in cucumber have been extensively studied. We have described several defense systems that protect the photosynthetic apparatus by exogenous polyamines (PAs) application. For example, exogenous spermidine (Spd) delays chlorophyll degradation under heat stress, and the regulation of fatty acids and accumulation of PAs in thylakoid membranes are induced by exogenous putrescine (Put) under salt stress [3,4]. PAs are low molecular weight aliphatic amines; the biochemical properties of PAs are quite simple, but their regulation of processes is strikingly complex and wide processes [5]. The majority of PAs in higher plants are Spd, spermine (Spm), and their precursor Put, which derive from arginine in chloroplasts [6,7]. On the other hand, as the polycationic nature of PAs at physiological pH, PAs could be free molecules conjugated with organic acids or bound to negatively charged macromolecules such as proteins, nucleic acids, and chromatin through transglutaminases (TGases) enzymatic activity, stabilizing their structures [8,9]. These interactions are essential for the effects of PAs on plant cell growth and developmental processes and plant response to various stresses.

Moreover, PAs, as organic cations and permeant buffers, have been reported to protect the photosynthetic apparatus by regulating the size of the antenna proteins of light harvesting chlorophyll a/b protein complexes (LHCII) and the larger subunit of ribulose bisphosphate carboxylase-oxygenase during stresses, such as UV-B radiation and salt stress [10,11]. PAs are also synthesized and oxidized in chloroplasts, while the addition of PAs inhibits the destruction of thylakoids and prevents the loss of pigment during salt stress [12,13]. The levels of endogenous PAs are also related to chlorophyll biosynthesis and the rate of photosynthesis during stresses [14]. PA accumulation in the lumen promotes an increase in ATP and the electric field in *vivo* and in *vitro* [15]. In addition, endogenous PAs might be involved in the assembly of photosynthetic membrane complexes such as thylakoid membranes [16].

TGases are crucial factors of the thylakoid system but are often rather ignored. TGases catalyze proteins by establishing ε-(γ-glutamyl) links, then regulate proteins post-translational modification and the covalent binding of PAs to protein substrates [17,18]. TGases are widely distributed in microorganisms, animals, and plants. However, research on these enzymes in plants is more rarely reported than in animal systems, in which was detected for the first time in *Arabidopsis thaliana* the presence of only one gene, *AtPng1p*, which encodes a putative N-glycanase containing the Cys-His-Asp triad of the TGase catalytic domain and was expressed ubiquitously [19]. Although TGases are found in several organs in lower and higher plants, they are activated in a Ca^{2+} dependent manner and are involved in fertilization, abiotic and biotic stresses, senescence, and programmed cell death, under different light environment conditions, including natural habitats, but the function of TGases in chloroplasts has received the most attention [18,20]. The activity of TGases has been shown to be light sensitive, and some proteins of the photosystems (LHCII, CP29, CP26, and CP24) have been shown to be endogenous substrates of TGase in chloroplasts [21]. Meanwhile, TGase not only localized in the chloroplast grana and close to LHCII but also localized in the walls of the bulliform cells of leaves. Its activity was light dependent and its abundance depended on the degree of grana development [22]. Moreover, the expression and activity of TGase was involved in length of light exposure in maize [23].

In earlier works, we described the effects of salt and heat stresses on changes of free PA contents in leaves of cucumber and tomato [24,25]. Specifically, the content of PAs decreased under stress conditions, resulting in severe damage to photosynthetic organs such as chloroplasts, which were severely deformed into irregular shapes, and increased starch granules. TGase was induced by salt stress and involved in the protection of the photosynthetic apparatus [26]. In this article, we reveal that TGase has a positive role in PAs accumulation and induce the transcript of photosynthetic genes in chloroplast, which may play crucial roles in the regulation of photosynthetic organ stability. To our knowledge, this is the first report to show that the TGase positively regulates plant's photosynthetic through accumulation of PAs to enhance salt tolerance.

2. Results

2.1. Expression Profile Analysis of TGase

To elucidate the molecular function of TGase, we analyzed the gene expression of *TGase* in cucumber. First, to investigate the effect of light on the expression of *TGase* in cucumber plants, we monitored the transcriptional levels of *TGase* in light-treated cucumber seedlings. The transcript level of *TGase* was gradually induced by light and reached the highest level at 16 h (Figure 1A). These results indicate that light plays a vital role in regulating *TGase* expression. To further explore the expression of *TGase* at different developmental stages, we investigated the transcript levels of *TGase* in leaves ranging from 1 to 8 weeks old by quantitative real-time PCR (qPCR). The expression level of *TGase* increased with the growth and development of the leaves (Figure 1B). *TGase* transcript levels in 4 and 8 week-old plants was significantly higher than those in 1-week-old plants (Figure 1B). Furthermore, *TGase* transcript levels increased during leaf development from young to mature leaves (Figure 1C). Some reports showed that TGase was widely present in plant tissue [17]. We extracted RNA from roots, stems, leaves, flowers, and fruits, and then analyzed the transcript levels of *TGase* in these tissues via qPCR. Our results also showed that *TGase* was present in all investigated tissues and was highly expressed in leaves and flowers, but minimally expressed in roots and stems (Figure 1D).

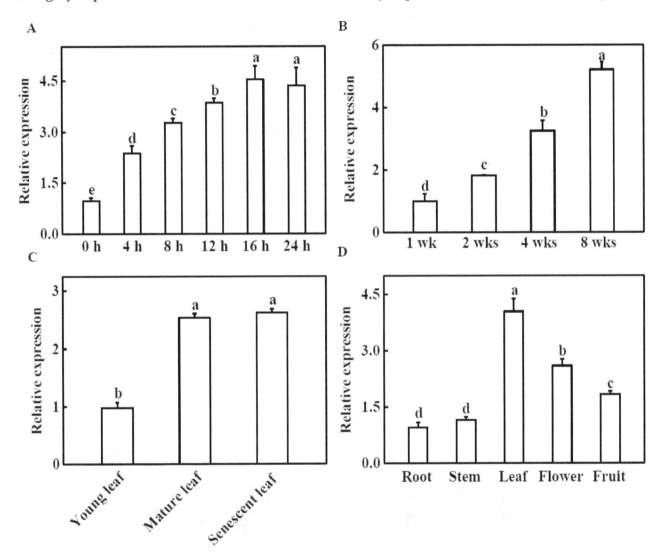

Figure 1. Expression profiles of *TGase*. (**A**) Light-induced *TGase* expression in cucumber. The cucumber seedlings were exposed to light 0, 4, 8, 12, 16 and 24 h and *TGase* transcript levels were analyzed by quantitative real-time PCR (qPCR). (**B**) Transcript levels of *TGase* in cucumber leaves at 1, 2, 4, 8 weeks of development. (**C**) Transcript levels of *TGase* in young, mature and old leaves of cucumber plants. (**D**) qPCR analysis of *TGase* transcript in roots, stems, leaves, flowers and fruit of cucumber. Each histogram represents a mean $\pm$ SE of four independent experiments ($n = 4$). Different letters indicate significant differences between treatments ($p < 0.05$) according to Duncan's multiple range test.

2.2. Immunolocalization of TGase Protein in Cucumber Leaves

Subcellular immunolocalization in cucumber leaf mesophyll cells provided details on the presence of TGase. The signal was localized in the chloroplasts and near the chloroplast grana (Figure 2A). The presence of TGase spots in the cell wall was detected (Figure 2B). These were not significantly localized to TGase in other cell organelles.

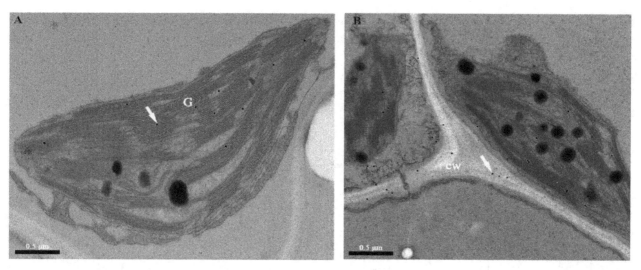

Figure 2. TEM immunolocalization of TGase in cucumber leaf chloroplasts using the monoclonal antibody (1:1000). (**A**) Signal in the granary of the chloroplasts. (**B**) Signal in the cell well. G, grana; cw, cell wall.

2.3. Effects of TGase on the Biomass and Photosynthetic Characteristics of Transgenic Tobacco Lines

To analyze the role of cucumber TGase in salt tolerance, we overexpressed the *TGase* gene in tobacco plants. As shown in Figure 3A, as compared with WT plants, the biomass was higher in the *CsTGase*-overexpressing (*CsTGase*OE) plants after salt stress. On the other hand, the *CsTGase*OE plants had a higher biomass relative to WT plants under normal conditions. As shown in Figure 3B, the proline content in all the plants increased after salt treatment, but this increase was much greater in *CsTGase*OE plants than in the WT plants. In WT plants, the chlorophyll a and chlorophyll b contents decreased by 62.0% and 51.3%, respectively, after salt treatment, whereas they maintained higher levels in *CsTGase*OE plants compared with those of the WT (Figure 3C,D).

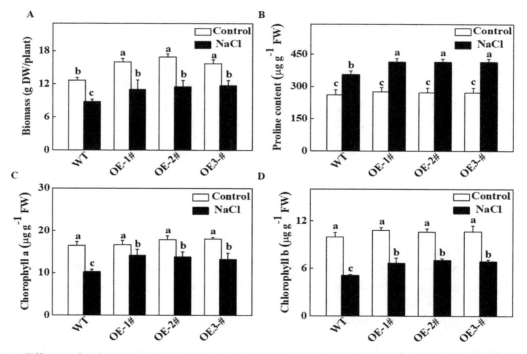

Figure 3. Effects of salt stress on biomass, proline and chlorophyll content in wild-type (WT) and *CsTGase*OE plants. (**A**) Biomass. (**B**) Proline content in leaves. (**C,D**) Chlorophyll a and b content. Each histogram represents a mean ± SE of four independent experiments ($n = 4$). Different letters indicate significant differences between treatments ($p < 0.05$) according to Duncan's multiple range test.

We also evaluated the effects of salt stress on photosynthetic gas exchange parameters. As shown in Figure 4A, under normal conditions, the net photosynthesis rate (Pn) of *CsTGase*OE plants was significantly higher than that of the WT plants. Salt stress resulted in a significant decrease in Pn, but this value was still maintained at a higher level in *CsTGase*OE plants than in the WT (Figure 4A). Similar results were observed for stomatal conductance (Gs) (Figure 4B) and transpiration rate (Tr) (Figure 4D). However, there were no significant differences in intercellular CO_2 concentration (Ci) in WT and *CsTGase*OE plants after salt stress (Figure 4C). These results suggest that TGase plays a critical role in the tobacco response to salt stress, especially in maintaining photosynthetic properties.

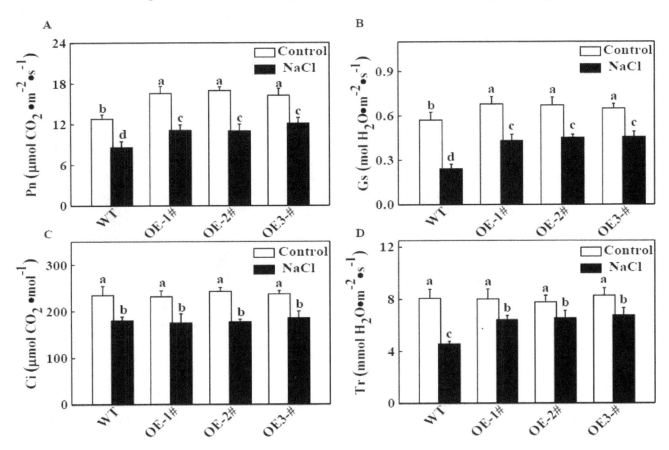

Figure 4. Photosynthetic parameters of WT and *CsTGase*OE plants in response to salt stress. (**A**) Net photosynthetic rate (Pn). (**B**) Stomatal conductance (Gs). (**C**) Intercellular CO_2 concentration (Ci). (**D**) Transpiration rate (Tr). Each histogram represents a mean ± SE of four independent experiments ($n = 6$). Different letters indicate significant differences between treatments ($p < 0.05$) according to Duncan's multiple range test.

2.4. Effects of TGase on Endogenous PA Content in Thylakoid Membranes

To determine whether PAs were involved in *TGase*-induced salt tolerance by protecting photosynthetic properties, we first measured the endogenous concentration of PAs in thylakoid membranes using a sensitive HPLC method. Under normal conditions, in *CsTGase*OE plants, the thylakoid associated PAs (Put, Spd and Spm) showed significantly higher levels compared with those of WT plants. Salt-induced bound Put, Spd, and Spm increased in comparison to the WT (Figure 5A–C). Meanwhile, the PA concentration increased by 161.8%, 155.9%, and 167.4% in the three *CsTGase*OE lines, respectively, compared with the WT under normal conditions (Figure 5D). PA accumulation rose by 28.7% in WT plants after salt treatment, but was still lower than in *CsTGase*OE plants.

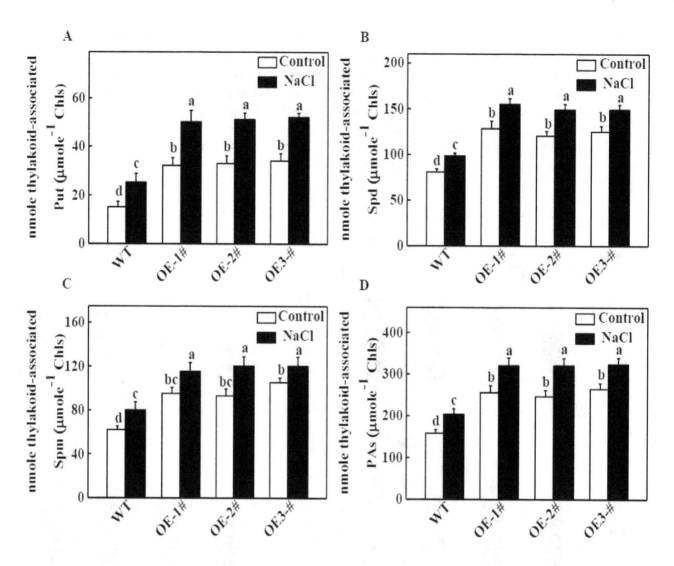

Figure 5. Effects of salt stress on the contents of endogenous putrescine (Put), spermidine (Spd), spermine (Spm) and total polyamines (PAs) in chloroplast of WT and *CsTGase*OE plants. (**A**) Thylakoid-associated Put content. (**B**) Thylakoid-associated Spd content. (**C**) Thylakoid-associated Spm content. (**D**) Total thylakoid-associated PAs content. Each histogram represents a mean ± SE of three independent experiments (*n* = 4). Different letters indicate significant differences between treatments (*p* < 0.05) according to Duncan's multiple range test.

2.5. Effects of TGase on the Ultrastructure of Thylakoids

PAs are a major positive factor in chloroplast ultrastructure [14]. To determine whether TGase regulates the ultrastructure of chloroplasts, we assayed the architecture of the thylakoid network using transmission electron microscopy (TEM). Under normal conditions, TEM revealed that the chloroplasts of WT plants had well-structured thylakoid membranes composed of grana connected by stroma lamellae (Figure 6A). Interestingly, overexpression of *TGase* resulted in chloroplasts having more grana and a larger size than those of the WT plants (Figure 6B–D). And in *CsTGase*OE plants, chloroplasts grana stacks reached up to 600 nm, whereas in the WT plants, chloroplast grana stacks were a maximum of 200 nm (Figure S1). These results suggest that TGase plays an important role in chloroplast development. Furthermore, under salt stress, chloroplasts were severely deformed into irregular shapes, and starch granules accumulated; moreover, a separation between cell membranes and chloroplasts was observed in WT plants (Figure 6E). However, the disintegration of grana thylakoids was significantly lower in *CsTGase*OE plants compared to the salt-treated WT plants (Figure 6E–H).

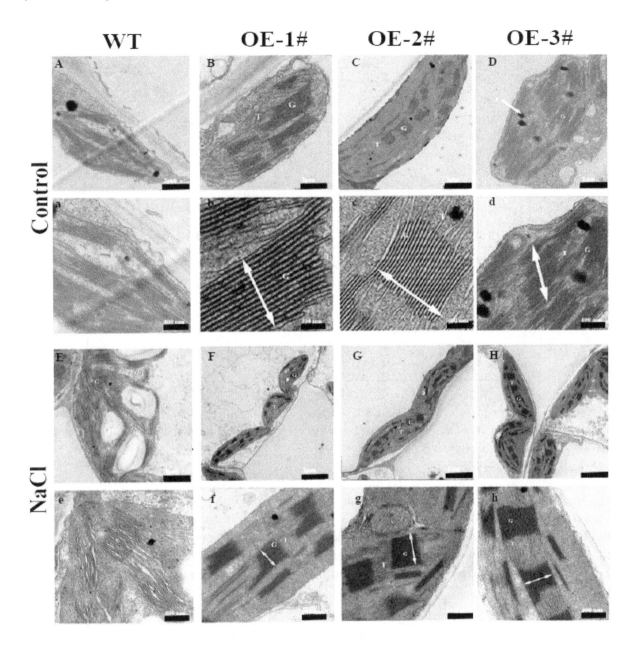

Figure 6. Electron microscopy in chloroplast of WT and *CsTGase*OE plants after salt stress. (**B–D**, **b–d**, **F–H** and **f–h**) shows an increased grana appression and a reduced stroma thylakoid network with respect to the WT (**A**,**a**,**E**,**e**) under normal conditions and salt stress. G, grana; T, thylakoid; SG, starch grana, P, plastoglobule. Grana height is indicated by white arrows. Scale bars for chloroplast and thylakoid are indicated. Three biological replicates were performed, and similar results were obtained.

2.6. Effect of TGase on Chl a Fluorescence Transients (OJIP)

A kinetic comparison was made of the raw OJIP transients measured in *CsTGase*OE and WT plants after salt stress. No significant difference was observed under the absence of NaCl in any of the plants. However, there were significant differences in *CsTGase*OE and WT plants under salt stress (Figure 7). In WT plants, salt stress resulted in a significant decrease in the intensities of fluorescence at J, I, and P levels with no major change in the minimal fluorescence (F_0) (Figure 7). Compared to the WT plants, the OJIP fluorescence transient and F_m values were near the normal levels in *CsTGase*OE plants after salt stress (Figure 7).

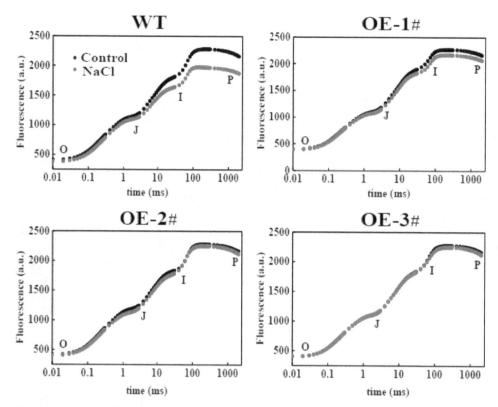

Figure 7. Change in Chl *a* fluorescence transient curves (OJIP) (log time scale) in leaves of WT and *CsTGase*OE plants under salt stress.

2.7. Effect of TGase on Nonphotochemical Quenching (NPQ) Induction

NPQ induction during light and dark transition periods was monitored. Transformed tobacco illuminated with 1 min light (500 μmol m^{-2} s^{-1}) and treat with 4 min dark showed NPQ values proximal to ~0.8, whereas there was little activation of photoprotection in the WT, with an NPQ of ~0.4 (Figure 8A). At a light intensity level of 500 μmol m^{-2} s^{-1} the NPQ value of the WT was significantly decreased by salt stress compared to that of the *CsTGase*OE plants (Figure 8B).

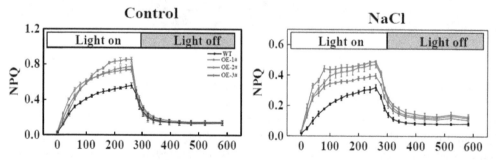

Figure 8. Nonphotochemical quenching (NPQ) induction and relaxation kinetics of WT and *CsTGase*OE plants under salt stress.

2.8. Effect of TGase on Quantum Yield of Energy Conversion in PSII and PSI

The F_v/F_m values were not significantly different in all plants under normal conditions (Figure 9A). However, the F_v/F_m values of the WT plants were significantly lower than those of *CsTGase*OE plants under salt stress (Figure 9A), indicating that photoinhibition was more severe in WT plants and the TGase has a positive role in photoprotection. The Y(II) in *CsTGase*OE plants was 19.8–24.0% higher than in WT plants, mainly due to an extremely higher photochemical quenching coefficient (qP) under normal conditions. After salt stress, the Y(NO) and Y(NPQ) values of *CsTGase*OE plants were lower compared with those of WT plants (Figure 9C–E). In *CsTGase*OE plants, the quantum yield of regulated energy dissipation (Y(NPQ)), an important process that consumes excess absorbed light energy and protects the photosynthetic apparatus, was higher than that in WT plants after salt stress (Figure 9G).

Similar to F_v/F_m and Y(II), P_m and Y(I) increased under normal conditions and were notably higher after salt stress in *CsTGase*OE plants compared with WT plants. Y(ND) showed a significant decrease in *CsTGase*OE plants compared to that of the WT plants after salt stress (Figure 9F). In addition, Y(NA) showed a significant increase in *CsTGase*OE leaves compared with those of the WT plants after salt stress (Figure 9H).

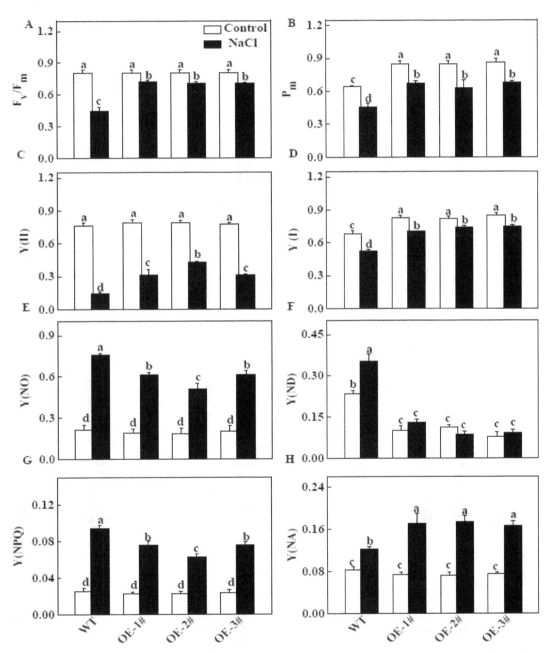

Figure 9. Changes in PSII and PSI in leaves of WT and *CsTGase*OE plants under salt stress. (**A**) The maximal quantum efficiency of PSII (F_v/F_m). (**B**) The maximum fluorescence of PSI (P_m). (**C**) The effective quantum efficiency of PSII (Y(II)). (**D**) The effective quantum efficiency of PSI (Y(I)). (**E**) The nonregulated energy dissipation (Y(NO)). (**F**) The oxidation status of PSI donor side (Y(ND)). (**G**) The regulated energy dissipation (Y(NPQ)). (**H**) The reduction status of PSI accept or side (Y(NA)). Each histogram represents a mean ± SE of four independent experiments ($n = 4$). Different letters indicate significant differences between treatments ($p < 0.05$) according to Duncan's multiple range test.

2.9. Effects of TGase on the Regulation of Photosynthesis Related Gene Expression

To analyze whether TGase is involved in affecting photosynthesis-related genes, we first examined the expression of photosynthesis-related genes in the WT and *TGase*OE plants, such as PSII-related genes (*NtpsbA/B/C/D/E*), PSI-related genes (*NtpsaA/B*), ATP synthesis-related genes (*NtatpA/B*), Calvin cycle-related genes (*NtrbcL/NtrbcS/NtFBPase*) and cytochrome-related genes (*NtpetA/B/D*). As shown in Figure 10, the transcript levels of photosynthesis-related genes were higher in *CsTGase*OE plants than the WT plants under normal conditions. We further analyzed the expression patterns of those genes in WT and *CsTGase*OE plants after 7 days of salt stress. Except for *NtpsbD*, the transcript levels of these genes in WT and *CsTGase*OE plants were suppressed under salt stress, but its levels in *CsTGase*OE plants were still higher than those in the WT (Figure 10).

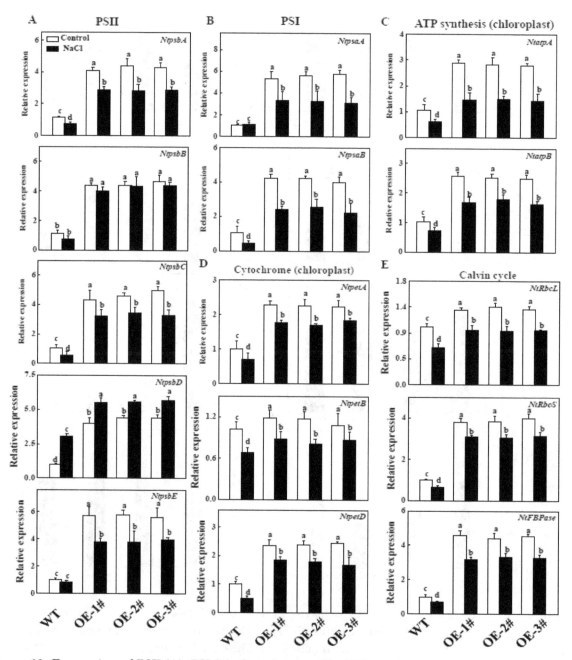

Figure 10. Expression of PSII (**A**); PSI (**B**); Cytochrome (**C**); ATP synthesis (**D**); Calvin cycle-related genes (**E**) in WT and *CsTGase*OE plants under salt stress. Each histogram represents a mean ± SE of four independent experiments (*n* = 4). Different letters indicate significant differences between treatments (*p* < 0.05) according to Duncan's multiple range test.

To confirm these results, we monitored the differential changes of some photosynthesis proteins using western blot (WB). As shown in Figure 11, in agreement with the results of gene expression, the levels of D2 proteins in *CsTGase*OE plants were significantly higher than that of WT plants under normal conditions. After 7 days of salt stress treatment, a significant reduction of D1, D2, and Cytf proteins was detected in the leaves of WT plants, but higher levels of these proteins were still present in the leaves of *CsTGase*OE plants (Figure 11 and Figure S2). In addition, LHCA1, and LHCB1 had showed no significant differences in WT and *CsTGase*OE plants before and after salt stress (Figure 11 and Figure S2).

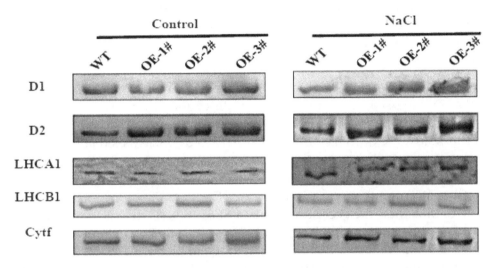

Figure 11. Analysis of thylakoid membrane protein changes in WT and *CsTGase*OE plants under salt stress. Thylakoid membrane proteins were separated by 12% SDS-urea-PAGE, transferred to PVDF membranes and probed with antisera against known thylakoid membrane proteins obtained from Agrisera company.

3. Discussion

CsTGase expression was dependent on light induction (Figure 1A), and *CsTGase* transcript levels increased with leaf development and aging, especially in mature and senescent leaves (Figure 1B,C), supporting the notion that TGase mainly functions in the senescence process [27]. Furthermore, cucumber TGase (CsTGase) was located not only in the chloroplast grana but also in the cell wall (Figure 2). These results indicate that TGase plays a vital role in early plant development.

Salt stress and other abiotic stresses can decrease photosynthetic capacity. Under salt conditions, plants overexpressing *CsTGase* showed enhanced salt tolerance displaying vigorous growth and higher Pn and Gs (Figures 3 and 4), suggesting that this gene might be particularly involved in the salt stress response. Additionally, some photosynthesis-related genes such as *NtpsbA*, *NtpsbB*, *NtpsbC*, *NtpsbD*, *NtpsbE*, *NtpsaA*, *NtpsaB*, *NtpetA*, *NtpetB*, *NtatpA*, and *NtatpB*, had higher transcript levels in *CsTGase* overexpressing plants (Figure 10), suggesting that the improved tolerance of transgenic plants over-expressing *CsTGase* might result from regulation of photosynthetic systems. Further, identification of the regulation of photosynthetic systems and unraveling this regulatory network may shed light on the mechanism underlying *CsTGase*-dependent tolerance to salt stress.

Many reports have shown that PAs play critical roles in regulating plant responses to abiotic stresses such as salinity, high temperature, and cold stresses, which are related to changes in endogenous PA levels and gene modifications [25,28,29]. *CsTGase* overexpressing lines had higher PA contents than the WT under normal conditions (Figure 5). Moreover, the overexpression of *CsTGase* increased endogenous PA levels in chloroplasts compared with those of the WT under salt stress (Figure 5), indicating that TGase may play positive roles in PA-dependent pathways to enhance plant resistance to salt stress.

The salt stress induced serious changes in photochemical efficiency and was often associated with the suppression of PSII and PSI activity [30]. In the present study, salt stress resulted in a significant decline in F_v/F_m and based on the analysis of the OJIP curves, levels at J, I, and P transients were gradually decreased in WT plants under salt stress (Figure 7). This inhibition resulted in strong fluorescence and pronounced suppression of total fluorescence emission, indicating that salt stress reduced PSII and PSI electron transport. However, overexpression of *CsTGase* interrupted the decrease in the OJIP curves and F_m values (Figure 7). These results are consistent with the fact that TGase plays a critical role in photoprotection [31]. Electron microscopy revealed that *CsTGase* overexpression resulted in an increase in grana stacking (Figure 6). Moreover, salt stress induced more severe damage to the ultrastructure of chloroplast and thylakoids in WT plants compared with *CsTGase*OE plants. Taken together, these results indicated that TGase plays a critical role in the protection of chloroplast under salt stress.

It was demonstrated that chlorophyll-a/b proteins (LHCII, CP29, CP26, and CP24) as substrates of TGase, as well as TGase itself, can catalyze the modification of light harvesting complex II by PAs in a light-dependent pathway [32,33]. However, the regulation of PSI through TGase has not been well studied. In this study, the decrease of Y(I) in the treated WT leaves resulted from the increase in the donor side limitation of PSI, as reflected by Y(ND), whereas the Y(ND) was not increased in the *CsTGase*OE plants (Figure 9). This finding indicates that salt stress releases an excessive amount of light energy to PSI in WT plants but not in *CsTGase*OE plants. The proportion of reduced electron carriers cannot be oxidized on the acceptor side of PSI by Y(NA) in WT plants, which is often used as an indicator of PSI photoinhibition. Our results showed that TGase could alleviate the salt stress caused the acceptor side limitation of PSI, as reflected by the higher value of Y(NA) in the *CsTGase*OE plants under salt stress than in the WT plants. An increase in Y(NA) indicates an increasing acceptor side limitation [34]. This phenomenon indicates that TGase may play a positive role in PSI under salt stress.

Apart from the effect on photosystems, we also found that the Calvin cycle is regulated by *CsTGase*. Indeed, the upregulation of some Calvin cycle genes, such as *NtRbcL* and *NtRbcS*, were observed in *CsTGase*OE plants under normal conditions. Furthermore, the expression of the *NtFBPase* gene was slightly enhanced in the leaves of *CsTGase*OE plants (Figure 10). However, the expression of Calvin cycle-related genes was inhibited in WT and *CsTGase*OE plants after salt stress, but these genes still maintained a high level in *CsTGase*OE plants (Figure 10). Rubisco and FBPase are sensitive to oxidative stress, as salt stress has been shown to induce ROS accumulation [35]. Hence, we hypothesize that TGase may be involved in the downregulation of ROS accumulation by enhancing endogenous PA content, especially in chloroplasts, thereby protecting the photosynthetic organs and the Calvin cycle.

A number of previous studies have established the role of TGase in maintaining the activity of chloroplast-related proteins, which is associated with enhancement of PA conjugation to light harvesting complex II (LHCII) proteins [21]. Proteomic analysis revealed that some proteins of photosystems were substrates of TGase, such as, LHCII, CP29, CP26, and CP24 [36]. In this study, the D1 and D2 proteins were strongly accumulated in *CsTGase*OE plants, while their levels were noticeably decreased in WT plants under salt stress. In addition, the LHCA1, LHCB1 and Cytf protein levels were not significantly different in WT and *CsTGase*OE plants (Figure 11 and Figure S2). Taken together, these results suggest a critical role for TGase in maintaining protein stability under salt stress.

In summary, this study provides compelling evidence to support our assumption that TGase participates in the enhancement of salt tolerance by modulating photosynthetic proteins; the manipulation of endogenous TGase activity could increase PA levels to alleviate salt-induced photoinhibition in tobacco plants. Meanwhile, TGase increased the levels of chloroplast proteins and protected these proteins under salt stress. Therefore, TGase mediates the processing of PA biosynthesis and the protection of chloroplast proteins to enhance salt tolerance, thereby increasing the survival of plants under salt stress.

4. Materials and Methods

4.1. Cucumber Plant Materials and Treatments for Expression Analysis

The cucumber *Cucumis sativus* L. cv. 9930 genotype was used in the experiment. Seeds were germinated and grown in 250 cm^3 plastic pots filled with peat. The plants were watered daily with Hoagland's nutrition solution in the chamber. The growing conditions were as follows: 14/10 light/dark cycle, 28/22 °C day/night temperatures and 600 µmol m^{-2} s^{-1} photosynthetic photon flux density (PPFD).

To analyze the possible influence of light on gene expression, 30-day-old cucumber plants growing under the normal 16/8 h photoperiod, were incubated to a 24 h illumination period, with or without being previously subjected to 24 h darkness. Leaf samples were taken at 0, 4, 8, 12, and 24 h and during the continuous illumination period.

To analyze the tissue-specific expression of *CsTGase*, roots, stems, leaves, flowers and fruits were collected. Young, mature, and senescent leaves were collected. We also collected the leaves at 1 week, 2 weeks, 4 weeks, and 8 weeks during plant growth.

4.2. TEM Observations: Immunogold Transmission Electron Microscopy

Cucumber leaf sections were fixed, dehydrated, and embedded in Lowicryl K4M resin (Pelco International, Redding, CA, USA), following the previously described procedures [37]. A monoclonal antibody, anti-plant TGase (produced in rabbit; Univ-bio, Shanghai, China) at 1:1000 dilutions was used. A gold AffiniPure anti rabbit IgG was used as a secondary antibody. For electron microcopy, a Jeol-JEM-1010 transmission electron microscope operating at 80 kV.

4.3. Generation and Selection of Transgenic Plants

To obtain the cucumber *TGase* overexpression (*TGase*OE) construct, the 1836 bp full-length coding DNA sequence (CDS) was amplified with the primer *TGase*OE-F (5′-CGAGCTCATGGATGATC GTGAGGCGTTTAAGA-3′) and *TGase*OE-R (5′-CGGGGTACCACGTTGCATGCAATTCCCGTAG-3′) using cucumber cDNA as the template. The PCR product was digested with *Sac*I and *Kpn*I and inserted behind the CaMV 35S promoter in the binary vector pCAMBIA1301-GUS. The *TGase*OE-GUS plasmid was transformed into *Agrobacterium tumefaciens* strain EHA105. NC89 tobacco plants were used for transformation, as described by Horsch [38].

4.4. Salt Tolerance Analysis of the Transgenic Plants

Three-week-old plants grown in vermiculite were treated with 200 mM NaCl for 7 days, and leaves were used to measure the biomass, proline content, and chlorophyll content. For biomass measurements, plants were dried for 48 h at 75 °C and then weighed. The proline content was measured according to Bates et al. [39]. The chlorophyll content was measured by UV spectrophotometry as described by Yang et al. [40].

Three-week-old plants grown in vermiculite were treated with 200 mM NaCl for 14 days, and the leaves were used to measure the photosynthetic parameters. The net photosynthetic rate (Pn), intercellular CO_2 concentration (Ci), and stomatal conductance (Gs) were measured according to Zhang et al. [41].

To avoid light impact on the TGase activity and PA content, potted plants in the experimental field were exposed to a 14/10 light/dark cycle. Then, we collected the samples and measured photosynthetic parameters at 12 h of illumination.

4.5. Thylakoid Isolation

Thylakoids were isolated as previously described with minor modifications [13]. Intact chloroplasts from the fully expanded leaves were homogenized in 50 mM KCl, 1 mM $MgCl_2$,

1 mM $MnCl_2$, 1 mM EDTA, 0.5 mM KH_2PO_4, 25 mM HEPES, pH 7.6, 330 mM sorbitol, 10 μM sodium ascorbate, and 0.2% (w/w) bovine serum albumin. The homogenates were filtered through a 300-μm and then 100-μm nylon mesh and debris was removed by centrifugation at $300\times g$ for 1 min, followed by centrifugation at $4000\times g$ for 10 min to collect the thylakoids. The pellet was separated from starch, resuspended and washed in 7 mM $MgCl_2$, 10 mM KCl, and 25 mM HEPES, pH 7.6 to break intact chloroplasts and removed free polyamines. Finally, for the polyamine analysis, thylakoids were resuspended in the medium containing 7 mM $MgCl_2$, 50 mM KCl, 25 mM HEPES, pH 7.6 and 330 mM sorbitol.

4.6. Analysis of Endogenous Polyamines in Thylakoid Membranes

The contents of endogenous PAs in thylakoid membranes were analyzed according to a method described by Zhang et al. [41] with some modifications. Briefly, for polyamine analysis, isolated thylakoids were incubated in 1.6 mL of 5% (w/v) cold perchloric acid (PCA) for 1 h on ice. After centrifugation for 20 min at $12,000\times g$, the pellet was used to determine bound PAs. PAs were analyzed using a high-performance liquid chromatography with a 1200 series system (Agilent Technologies, Santa Clara, CA, USA), a C18 reversed-phase column (4.6 mm by 250 mm, 5 μm Kromasil) and a two solvent system including a methanol gradient (36%–64%, v/v) at a flow rate of 0.8 mL min^{-1}.

4.7. Observation of the Ultrastructure of the Chloroplast

Tobacco leaves were collected and cut into pieces of approximately 1 mm^2 and fixed by vacuum infiltration with 3% glutaraldehyde and 1% formaldehyde in a 0.1 M phosphate buffer (pH 7.4) for 2 h (primary fixation). After washing, the sample were fixed for 2 h in osmium tetroxide at room temperature; then, the samples were dehydrated in acetone and embedding in Durcupan ACM, Ultrathin sections of the leaf pieces (70 nm) were cut, stained with uranium acetate and lead citrate in series and examined using a H7650 transmission electron microscope (Hitachi, Tokyo, Japan) at an accelerating voltage of 80 kV. A minimum of 50 chloroplasts of each type of plant were examined.

4.8. Chl a fluorescence Measurement and OJIP Transient Analyses

The chlorophyll a (Chl a) fluorescence induction kinetics were measured at 12 h of illumination using dual portable fluorescence (Dual-PAM-100, Walz, Germany). Measurements were analyzed using the automated induction program provided by the Dual-PAM software. PSII and PSI activities were quantified by chlorophyll fluorescence and P700$^+$ absorbance changes.

The OJIP curves showed a polyphasic rise. The initial fluorescence (F_0) (approx. 50 μs) was set as O, followed by the O to J phase (ends at approx. 2 ms), then the J to I phase (ends at approx. 30 ms) and I to P phase (at the peak of the OJIP curve). The JIP measurement is named after the basic steps in fluorescence transience when plotted on a logarithmic time scale [42]. Leaves were dark adapted for 30 min prior to the measurement or the NPQ measurement leaves were continuously illuminated for 270 s with 500 μmol photons m^{-2} s^{-1} using the Handy-PEA (multihit- mode). Every 30 s was given a 3000 μmol photons m^{-2} s^{-1} (duration 0.8 s) saturating pulse for maximal fluorescence, F_m'. To calculate the NPQ at the end of the actinic light phase we followed the equation, NPQ = $F_m/F_m' - 1$ [43]. The tests were shown in the middle portion of infiltration on the ventral surface of the leaves. Measurements were taken in 6 replications.

4.9. Quantitative Real-Time PCR

Total RNA was extracted from three biological replicates of leaves using an RNA extraction kit (Tiangen, Beijing, China) according to the manufacturer's instructions. The first strand cDNA was synthesized from 1 μg of DNase-treated RNA using reverse transcriptase (Takara, Dalian, China) following the manufacturer's protocol. Quantitative real-time PCR (qPCR) was using gene specific primers (Table S1) in 20 μL reaction system using SYBR Premix Ex Taq II (Takara, Dalian, China).

Tobacco *β-actin* was used as a reference gene for tobacco; cucumber *actin* was used as a reference gene for cucumber. Relative gene expression was calculated according to Livak and Schmittgen [44].

4.10. Protein Extraction and Western Blotting

For the thylakoid membranes, the intact chloroplasts were reputed in low osmotic buff (50 mM HEPES–KOH (pH 7.6) and 2 mM $MgCl_2$) on the ice, then the thylakoid membranes were collected and the protein content was determined by a BCA Protein Assay Kit (Solarbio, Beijing, China). For immunoblot analysis, thylakoid proteins were solubilized and separated on 12% SDS-urea-PAGE gels. After electrophoresis, the proteins were transferred to polyvinylidene difluoride (PVDF) membranes (Millipore, Billerica, MA, USA) and probed using commercial antibodies specific for the PSII subunits (D1 and D2) (AS05084 and AS06146), light-harvesting antenna proteins (LHCA1 and LHCB1) (AS01005 and AS01004) and Cytb6f subunit (Cytf) (AS14169), at 1:5000 dilutions were used. These antibodies were from Agrisera (Vännäs, Sweden). At least three independent replicates were used for each determination. Accumulation of proteins were quantified using Quantity One software (Bio-Rad, Hercules, California, CA, USA).

4.11. Statistical Analysis

At least 4 independent replicates were used for each determination. Statistical analysis of the bioassays was performed using the SPSS 20 statistical package (SPSS Inc., Chicago, IL, USA). Experimental data were analyzed with a Duncan's multiple range test at $p < 0.05$.

Author Contributions: S.G. designed the research and proposed the research proceeding. M.Z. and Y.W. performed the experiments and wrote the main manuscript text. Y.Z. prepared all figures and modified this manuscript until submit. S.S. and J.S. improved the manuscript. All authors reviewed and approved the manuscript.

References

1. Ioannidis, N.E.; Kotzabasis, K. Polyamines in chemiosmosis in vivo: A cunning mechanism for the regulation of ATP synthesis during growth and stress. *Front. Plant Sci.* **2014**, *5*, 71. [CrossRef] [PubMed]
2. Yang, Y.; Guo, Y. Unraveling salt stress signaling in plants. *J. Integr. Plant Biol.* **2018**, *60*, 796–804. [CrossRef] [PubMed]
3. Zhou, H.; Guo, S.; An, Y.; Shan, X.; Wang, Y.; Shu, S.; Sun, J. Exogenous spermidine delays chlorophyll metabolism in cucumber leaves (*Cucumis sativus* L.) under high temperature stress. *Acta Physiol. Plant.* **2016**, *38*, 224. [CrossRef]
4. Shu, S.; Yuan, Y.; Chen, J.; Sun, J.; Zhang, W.; Tang, Y.; Zhong, M.; Guo, S. The role of putrescine in the regulation of proteins and fatty acids of thylakoid membranes under salt stress. *Sci. Rep.* **2015**, *5*, 14390. [CrossRef] [PubMed]
5. Sagor, G.H.M.; Zhang, S.; Kojima, S.; Simm, S.; Berberich, T.; Kusano, T. Reducing cytoplasmic polyamine oxidase activity in *Arabidopsis* increases salt and drought tolerance by reducing reactive oxygen species production and increasing defense gene expression. *Front. Plant Sci.* **2016**, *7*, 214. [CrossRef] [PubMed]
6. Bortolotti, C.; Cordeiro, A.; Alcázar, R.; Borrell, A.; Culiañez-Macià, F.A.; Tiburcio, A.F.; Altabella, T. Localization of arginine decarboxylase in tobacco plants. *Physiol. Plant.* **2004**, *120*, 84–92. [CrossRef] [PubMed]
7. Del Duca, S.; Serafini-Fracassini, D.; Cai, G. Senescence and programmed cell death in plants: Polyamine action mediated by transglutaminase. *Front. Plant Sci.* **2014**, *5*, 120. [CrossRef] [PubMed]
8. Serafini-Fracassini, D.; Del Duca, S.; D'Orazi, D. First evidence for polyamine conjugation mediated by an enzymic activity in plants. *Plant Physiol.* **1988**, *87*, 757–761. [CrossRef]
9. Tiburcio, A.F.; Altabella, T.; Bitrián, M.; Alcázar, R. The roles of polyamines during the lifespan of plants: From development to stress. *Planta* **2014**, *240*, 1–18. [CrossRef] [PubMed]
10. Sfichi, L.; Loannidis, N.; Kotzabasis, K. Thylakoid-associated polyamines adjust the UV-B sensitivity of the photosynthetic apparatus by means of light-harvesting complex II changes. *Photochem. Photobiol.* **2004**, *80*, 499–506. [CrossRef]

11. Demetriou, G.; Neonaki, C.; Navakoudis, E.; Kotzabasis, K. Salt stress impact on the molecular structure and function of the photosynthetic apparatus—The protective role of polyamines. *Biochim. Biophys. Acta* **2007**, *1767*, 272–280. [CrossRef]

12. Dondini, L.; Del Duca, S.; Dall'Agata, L.; Bassi, R.; Gastaldelli, M.; Della Mea, M.; Di Sandro, A.; Claparols, I.; Serafini-Fracassini, D. Suborganellar localisation and effect of light on Helianthus tuberosus chloroplast transglutaminases and their substrates. *Planta* **2003**, *217*, 84–95. [PubMed]

13. Shu, S.; Yuan, L.-Y.; Guo, S.-R.; Sun, J.; Yuan, Y.-H. Effects of exogenous spermine on chlorophyll fluorescence, antioxidant system and ultrastructure of chloroplasts in *Cucumis sativus* L. under salt stress. *Plant Physiol. Biochem.* **2013**, *63*, 209–216. [CrossRef] [PubMed]

14. Hamdani, S.; Yaakoubi, H.; Carpentier, R. Polyamines interaction with thylakoid proteins during stress. *J. Photochem. Photobiol.* **2011**, *104*, 314–319. [CrossRef] [PubMed]

15. Ioannidis, N.E.; Cruz, J.A.; Kotzabasis, K.; Kramer, D.M. Evidence that putrescine modulates the higher plant photosynthetic proton circuit. *PLoS ONE* **2012**, *7*, e29864. [CrossRef]

16. Ioannidis, N.E.; Ortigosa, S.M.; Veramendi, J.; Pintó-Marijuan, M.; Fleck, I.; Carvajal, P.; Kotzabasis, K.; Santos, M.; Torné, J.M. Remodeling of tobacco thylakoids by over-expression of maize plastidial transglutaminase. *Biochim. Biophys. Acta* **2009**, *1787*, 1215–1222. [CrossRef] [PubMed]

17. Serafini-Fracassini, D.; Del Duca, S. Transglutaminases: Widespread cross-linking enzymes in plants. *Ann. Bot.* **2008**, *102*, 145–152. [CrossRef]

18. Lilley, G.R.; Skill, J.; Griffin, M.; Bonner, P.L. Detection of Ca^{2+}-dependent transglutaminase activity in root and leaf tissue of monocotyledonous and dicotyledonous plants. *Plant Physiol.* **1998**, *117*, 1115–1123. [CrossRef]

19. Della Mea, M.; Caparrós-Ruiz, D.; Claparols, I.; Serafini-Fracassini, D.; Rigau, J. *AtPng1p*. The first plant transglutaminase. *Plant Physiol.* **2004**, *135*, 2046–2054. [CrossRef]

20. Del Duca, S.; Faleri, C.; Iorio, R.A.; Cresti, M.; Serafini-Fracassini, D.; Cai, G. Distribution of transglutaminase in pear pollen tubes in relation to cytoskeleton and membrane dynamics. *Plant Physiol.* **2013**, *161*, 1706–1721. [CrossRef]

21. Del Duca, S.; Tidu, V.; Bassi, R.; Esposito, C.; Serafmi-Fracassini, D. Identification of chlorophyll-a/b proteins as substrates of transglutaminase activity in isolated chloroplasts of *Helianthus tuberosus* L. *Planta* **1994**, *193*, 283–289. [CrossRef]

22. Campos, N.; Castañón, S.; Urreta, I.; Santos, M.; Torné, J. Rice transglutaminase gene: Identification, protein expression, functionality, light dependence and specific cell location. *Plant Sci.* **2013**, *205*, 97–110. [CrossRef] [PubMed]

23. Villalobos, E.; Santos, M.; Talavera, D.; Rodrıguez-Falcón, M.; Torné, J. Molecular cloning and characterization of a maize transglutaminase complementary DNA. *Gene* **2004**, *336*, 93–104. [CrossRef] [PubMed]

24. Duan, J.; Li, J.; Guo, S.; Kang, Y. Exogenous spermidine affects polyamine metabolism in salinity-stressed *Cucumis sativus* roots and enhances short-term salinity tolerance. *J. Plant Physiol.* **2008**, *165*, 1620–1635. [CrossRef] [PubMed]

25. Sang, Q.; Shan, X.; An, Y.; Shu, S.; Sun, J.; Guo, S. Proteomic analysis reveals the positive effect of exogenous spermidine in tomato seedlings' response to high-temperature stress. *Front. Plant Sci.* **2017**, *8*, 120. [CrossRef] [PubMed]

26. Tang, Y.-Y.; Yuan, Y.-H.; Shu, S.; Guo, S.-R. Regulatory mechanism of NaCl stress on photosynthesis and antioxidant capacity mediated by transglutaminase in cucumber (*Cucumis sativus* L.) seedlings. *Sci. Hortic.* **2018**, *235*, 294–306. [CrossRef]

27. Sobieszczuk-Nowicka, E.; Zmienko, A.; Samelak-Czajka, A.; Łuczak, M.; Pietrowska-Borek, M.; Iorio, R.; Del Duca, S.; Figlerowicz, M.; Legocka, J. Dark-induced senescence of barley leaves involves activation of plastid transglutaminases. *Amino Acids* **2015**, *47*, 825–838. [CrossRef]

28. Yuan, Y.; Zhong, M.; Shu, S.; Du, N.; Sun, J.; Guo, S. Proteomic and physiological analyses reveal putrescine responses in roots of cucumber stressed by NaCl. *Front. Plant Sci.* **2016**, *7*, 1035. [CrossRef]

29. Zhuo, C.; Liang, L.; Zhao, Y.; Guo, Z.; Lu, S. A cold responsive ethylene responsive factor from Medicago falcata confers cold tolerance by up-regulation of polyamine turnover, antioxidant protection, and proline accumulation. *Plant Cell Environ.* **2018**, *41*, 2021–2032. [CrossRef]

30. Oukarroum, A.; Bussotti, F.; Goltsev, V.; Kalaji, H.M. Correlation between reactive oxygen species production and photochemistry of photosystems I and II in Lemna gibba L. plants under salt stress. *Environ. Exp. Bot.* **2015**, *109*, 80–88. [CrossRef]

31. Ioannidis, N.E.; Malliarakis, D.; Torné, J.M.; Santos, M.; Kotzabasis, K. The over-expression of the plastidial transglutaminase from maize in Arabidopsis increases the activation threshold of photoprotection. *Front. Plant Sci.* **2016**, *7*, 635. [CrossRef] [PubMed]

32. Del Duca, S.; Tidu, V.; Bassi, R.; Serafini-Fracassini, D.; Esposito, C. Identification of transglutaminase activity and its substrates in isolated chloroplast of Helianthus tuberosus. *Planta* **1994**, *193*, 283–289. [CrossRef]

33. Sobieszczuk-Nowicka, E.; Krzesłowska, M.; Legocka, J. Transglutaminases and their substrates in kinetin-stimulated etioplast-to-chloroplast transformation in cucumber cotyledons. *Protoplasma* **2008**, *233*, 187. [CrossRef] [PubMed]

34. Huang, W.; Yang, S.-J.; Zhang, S.-B.; Zhang, J.-L.; Cao, K.-F. Cyclic electron flow plays an important role in photoprotection for the resurrection plant Paraboearufescens under drought stress. *Planta* **2012**, *235*, 819–828. [CrossRef] [PubMed]

35. Miller, G.; Suzuki, N.; Ciftci-Yilmaz, S.; Mittler, R. Reactive oxygen species homeostasis and signalling during drought and salinity stresses. *Plant Cell Environ.* **2010**, *33*, 453–467. [CrossRef] [PubMed]

36. Campos, A.; Carvajal-Vallejos, P.; Villalobos, E.; Franco, C.; Almeida, A.; Coelho, A.; Torné, J.; Santos, M. Characterisation of *Zea mays* L. plastidial transglutaminase: Interactions with thylakoid membrane proteins. *Plant Biol.* **2010**, *12*, 708–716. [CrossRef] [PubMed]

37. Campos, N.; Villalobos, E.; Fontanet, P.; Torné, J.M.; Santos, M. A peptide of 17 aminoacids from the N-terminal region of maize plastidial transglutaminase is essential for chloroplast targeting. *Am. J. Mol. Biol.* **2012**, *2*, 245–257. [CrossRef]

38. Horsch, R.B.; Fry, J.E.; Hoffmann, N.L.; Eichholtz, D.; Rogers, S.G.; Fraley, R.T. A simple and general-method for transferring genes into plants. *Science* **1985**, *227*, 1229–1231.

39. Bates, L.; Waldren, R.; Teare, I. Rapid determination of free proline for water-stress studies. *Plant Soil* **1973**, *39*, 205–207. [CrossRef]

40. Yang, Q.; Chen, Z.-Z.; Zhou, X.-F.; Yin, H.-B.; Li, X.; Xin, X.-F.; Hong, X.-H.; Zhu, J.-K.; Gong, Z. Overexpression of *SOS* (Salt Overly Sensitive) genes increases salt tolerance in transgenic Arabidopsis. *Mol. Plant* **2009**, *2*, 22–31. [CrossRef]

41. Zhang, R.H.; Li, J.; Guo, S.R.; Tezuka, T. Effects of exogenous putrescine on gas-exchange characteristics and chlorophyll fluorescence of NaCl-stressed cucumber seedlings. *Photosynth. Res.* **2009**, *100*, 155–162. [CrossRef] [PubMed]

42. Force, L.; Critchley, C.; van Rensen, J.J. New fluorescence parameters for monitoring photosynthesis in plants. *Photosynth. Res.* **2003**, *78*, 17. [CrossRef] [PubMed]

43. Bilger, W.; Björkman, O. Relationships among violaxanthin deepoxidation, thylakoid membrane conformation, and nonphotochemical chlorophyll fluorescence quenching in leaves of cotton (*Gossypium hirsutum* L.). *Planta* **1994**, *193*, 238–246. [CrossRef]

44. Livak, K.J.; Schmittgen, T.D. Analysis of relative gene expression data using real-time quantitative PCR and the $2^{-\Delta\Delta CT}$ method. *Methods* **2001**, *25*, 402–408. [CrossRef] [PubMed]

13

Transcriptome Sequence Analysis Elaborates a Complex Defensive Mechanism of Grapevine (*Vitis vinifera* L.) in Response to Salt Stress

Le Guan [1], Muhammad Salman Haider [1], Nadeem Khan [1], Maazullah Nasim [1], Songtao Jiu [2], Muhammad Fiaz [1], Xudong Zhu [1], Kekun Zhang [1] and Jinggui Fang [1,*]

[1] College of Horticulture, Nanjing Agricultural University, Nanjing 210095, China; guanle@njau.edu.cn (L.G.); salman.hort1@gmail.com (M.S.H.); 2016104235@njau.edu.cn (N.K.); maazullah.nasim@gmail.com (M.N.); fiaz.m2002@gmail.com (M.F.); zhuxudong@njau.edu.cn (X.Z.); 2006204006@njau.edu.cn (K.Z.)

[2] Department of Plant Science, School of Agriculture and Biology, Shanghai Jiao Tong University, Shanghai 200240, China; 2013104019@njau.edu.cn

* Correspondence: fanggg@njau.edu.cn

Abstract: Salinity is ubiquitous abiotic stress factor limiting viticulture productivity worldwide. However, the grapevine is vulnerable to salt stress, which severely affects growth and development of the vine. Hence, it is crucial to delve into the salt resistance mechanism and screen out salt-resistance prediction marker genes; we implicated RNA-sequence (RNA-seq) technology to compare the grapevine transcriptome profile to salt stress. Results showed 2472 differentially-expressed genes (DEGs) in total in salt-responsive grapevine leaves, including 1067 up-regulated and 1405 down-regulated DEGs. Gene Ontology (GO) and Kyoto Encyclopedia of Genes and Genomes (KEGG) annotations suggested that many DEGs were involved in various defense-related biological pathways, including ROS scavenging, ion transportation, heat shock proteins (HSPs), pathogenesis-related proteins (PRs) and hormone signaling. Furthermore, many DEGs were encoded transcription factors (TFs) and essential regulatory proteins involved in signal transduction by regulating the salt resistance-related genes in grapevine. The antioxidant enzyme analysis showed that salt stress significantly affected the superoxide dismutase (SOD), peroxidase (POD), catalase (CAT) and glutathione S-transferase (GST) activities in grapevine leaves. Moreover, the uptake and distribution of sodium (Na^+), potassium (K^+) and chlorine (Cl^-) in source and sink tissues of grapevine was significantly affected by salt stress. Finally, the qRT-PCR analysis of DE validated the data and findings were significantly consistent with RNA-seq data, which further assisted in the selection of salt stress-responsive candidate genes in grapevine. This study contributes in new perspicacity into the underlying molecular mechanism of grapevine salt stress-tolerance at the transcriptome level and explore new approaches to applying the gene information in genetic engineering and breeding purposes.

Keywords: grapevine; salt stress; ROS detoxification; phytohormone; transcription factors

1. Introduction

Grapevine (*Vitis vinifera*) is an economic fruit crop, primarily categorized into the table (fresh) and wine grapes [1]. Recent shifts in the environment have become the critical limiting factors for yield and grapevine products. Thus, it is indispensable to characterize the salt-tolerant grapevine varieties by screening salt resistance-related genes and genetically transform them to enable plants to withstand high salt concentrations. One-fifth of irrigated agricultural lands are affected by soil salinity, which leads to escalating the salt effects on plant growth investigations in the recent few years [2–4]. High soil salinity affects plants in multiple ways, such as inhibition of water uptake in the root zone, which makes it difficult for the plants to take up water; and results in dehydration of plant cells, leading to cell turgor and in response, plants have to increase osmotic pressure in their cells [5]. Also, due to the decrease in K^+/Na^+ value, the original balance of ions in plant cells might be interrupted, which has a toxic effect on enzymes, chlorophyll degradation and recurrent protein synthesis [6]. Simultaneously, salt induces cellular toxicity, which leads to undue reactive oxygen species (ROS) production and accumulation in different cellular compartments, resulting in lipid peroxidation (LPO) of biological membranes, ions leakage and DNA-strand cleavage [7].

Plants can evolve a complex defensive mechanism to counteract the salinity effects [8], which includes activation of numerous signaling sensors that conclusively excites various transcription factors (TFs) to induce stress-responsive genes, which enable plants to nurture and transcend the adverse conditions. In salinity, factors involved in signaling are: (i) discerning accretion or elimination of ions to stabilize the K^+/Na^+ balance and other ion levels via salt-inducible enzyme Na^+/H^+ antiporter (V-ATPase or PPase) and K^+ and Na^+ transporters (SOS family); (ii) biosynthesis of congenial solutes to adjust the vacuolar ionic balance and restore water in the biochemical reaction (Like polyols and mannitol); (iii) adjust the cell membrane structure; (iv) synthesis of multiple resistance-oriented proteins like ROS and pathogenesis-related proteins (PR family); and (v) induction of plant hormones (ABA, JA and IAA). These biological pathways improve the inclination of salt tolerance are likely to collaborate and may have the synergistic effect [6,9]. Besides, various transcription factors (TFs), such as HD-Zip, ERF, WRKY, bHLH are known to play a vital role in regulating salt resistance mechanism in plants [1,10].

Recently, next-generation sequencing (NGS) technology based high throughput RNA-seq technology has been extensively used to unveil and compare the transcriptome profile under abiotic stresses [1], which provides large-scale data to identify and characterize the DEGs. Previously, extensive studies have been carried out on antioxidant metabolism, ionomic uptake and transport, hormonal metabolism and stress signaling [11,12] but the underlying molecular mechanism of salt stress tolerance remain to be elucidated. Though several studies focusing on morphological variations, biochemical and physiological components are available in grapevine, however, there is no report on transcriptomic studies particularly molecular research associated with salt stress tolerance. Therefore, to comprehend the molecular mechanism of salt tolerance in grapevine, Illumina RNA-seq libraries were constructed from both control and salt-treated grapevine leaves. In addition, gene ontology (GO) enrichment analysis was also performed to investigate biochemical and physiological cues in response to salt stress. qRT-PCR analysis of critical salt stress-responsive genes was also carried out to validate RNA-seq results. The obtained information provides more profound insights into the grapevine molecular mechanism in improving breeding strategies for the development of transgenic plants, which can better resist the abiotic stress.

2. Results

2.1. Global Transcriptome Sequence Analysis

The transcriptomic sequencing of cDNA generated from both control and salt-treated grapevine leaf samples produced 21.2 and 21.4 million raw reads, respectively (Table S1). Following the filtering and trimming process, 20.2 and 20.6 million clean reads were retrieved from control and treatment group, respectively, corresponding to 8.16 Gb data, intimating the tag density from both control and salt-treatment, representing about 20 million reads, which is adequate for quantitative analysis of gene expression. For sequence alignments, SOAPaligner/soap2 software (http://soap.genomics.org.cn) was used as reference genome of grapevine (Version 1.0), suggesting total mapped reads as 67.4% matched complemented with both unique (57.42%) or multiple (9.96%) genomic positions (Table S1).

Transcriptome analysis can compare the number of DEGs and their expression pattern in different tissues. In our transcriptomic study, 21,746 transcripts were obtained from control and 21,541 transcripts from the treatment group. Among these expressed transcripts, 14,767 transcripts showed no significant changes in their expression level ($|\log_2 FC|$ < 1), while 2472 transcripts were differentially expressed in the salt-treatment group ($|\log_2 FC| \geq 1$) at false discovery rate (FDR) <0.001), which includes 1067 (43.16%) up-regulated and 1405 (56.87%) down-regulated transcripts (Table S2). Moreover, 20 DEGs suggested their expression only in the control group and 27 DEGs were only expressed in the treatment group (Table S3).

2.2. GO and KEGG Analysis of DEGs in Response to Salt Stress

GO-based enrichment analysis functionally characterizes and annotated the 1, 591 (64.36% of 2, 472) DEGs into 45 functional groups, of which molecular function contains 15 groups, cellular component (15 groups) and biological process (9) (Figure 1 and Table S4) between control and salt-treated group. In molecular function (MF), "ATPase activity" (GO: 0042623) with 178 transcripts, followed by "phosphatase activity" (GO: 0008138) with 113 transcripts and least transcripts (4) were found in both "ABA binding (GO: 0010427)" and "Hsp90 protein binding (GO: 0010329)". In cellular component (CC), "photosynthetic membrane" possessed the highest number of transcripts (GO: 0034357, 106 transcripts), whereas, "thylakoid membrane" consisted of 97 transcripts (GO: 0042651). Furthermore, in biological process (BP), "response to oxidative stress" (GO: 0006979) harbored 164 transcripts, followed by "salinity response" (GO: 0009651) with 148, while "SOS response (GO: 0009432)", "stomatal closure (GO: 0090332)" and "cytochrome b6f complex (GO: 0010190)" with three transcripts each were the least group.

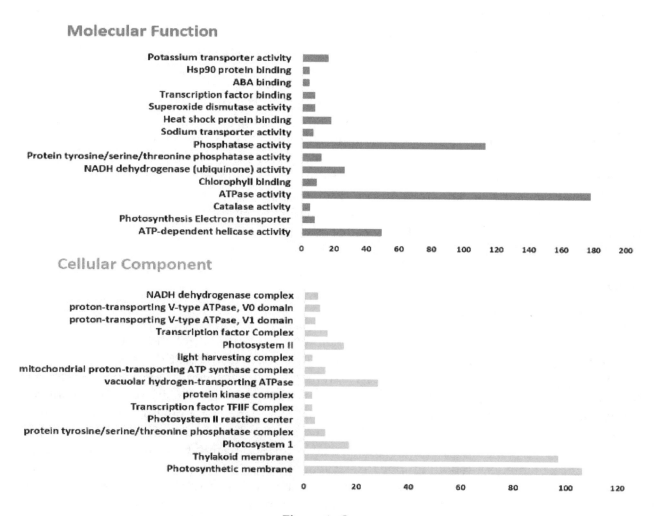

Figure 1. *Cont.*

Transcriptome Sequence Analysis Elaborates a Complex Defensive Mechanism of Grapevine...

223

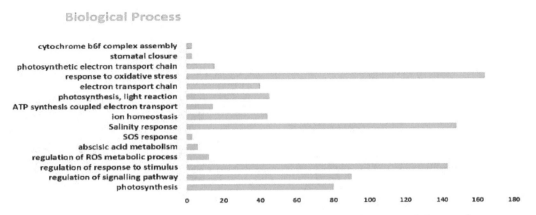

Figure 1. Gene ontology (Go) based annotations of 2472 DEGs. The main GO terms are categorized into "molecular function", "cellular component" and "biological process".

KEGG database simulates the functional annotation of the cells or the organism by sequence similarity and genome information. In this study, 453 (18.32% of 2472) transcripts were allocated to 30 pathways in KEGG database (Figure 2 and Table S5), while "Signal transduction" pathway with 79 transcripts was the most enriched pathway followed by "Folding, sorting and degradation" (65 transcripts) and "Carbohydrate metabolism" (63 transcripts).

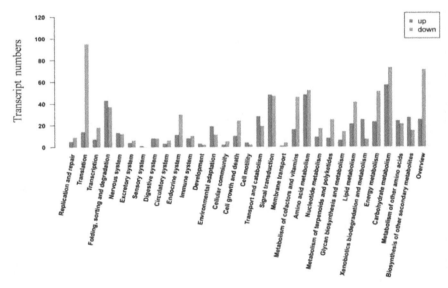

Figure 2. Kyoto Encyclopedia of Genetics and Genomics (KEGG) database analysis of DEGS (up and down-regulated) enriched in different biological pathways. The X-axis represents enriched pathways and Y-axis represents the total number of transcripts.

2.3. Photosynthetic Efficiency of Grapevine in Response to Salt Stress

To verify the extent of salt severity on grapevine physiology, photosynthetic efficiency and related parameters were estimated in the control and treatment group by using a portable Li-COR meter. Results suggested that net photosynthesis rate (A_N) was significantly reduced from 23.98 ± 1.33 (0 h) to 13.42 ± 1.31 (48 h) during the salt stress period. Likewise, an about 2-fold decrease in stomatal conductance significantly inhibited the net CO_2 assimilation rate (Ci; 35.78%) and transpiration rate (E; 51.33%) after 48 h of salt stress as compared to control plants (Figure 3). In the transcriptomic study, the DEGs encoding photosystem II CP47 (psbB) in PSII and photosystem I P700 (psaB) in PSI were down-regulated in salt-treated grapevine leaf samples as compared to control (Table S2), which is consistent with the physiological investigations of decreased net photosynthesis rate. Moreover, six DEGs encoding ATP-synthase and one DEG-related to the cytochrome b6-f complex were also down-regulated in grapevine leaf tissues after exposure to salt stress.

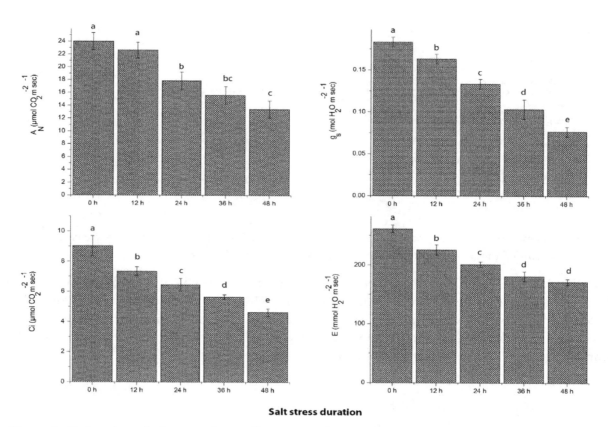

Figure 3. Estimation of photosynthetic efficiency, including net photosynthesis rate (A_N), stomatal conductance (g_S), transpiration rate (E) and net CO_2 assimilation rate (Ci) in grapevine leaves in response to salt stress as compared with control. Values represent mean $\pm$ SE ($n = 3$) and the significance level of 0.05 was used for different letters above bars.

2.4. Production and Scavenging of Reactive Oxygen Species (ROS) in Response to Salt Stress

ROS production is a universal plant response to almost all type of abiotic stresses. In response, plants accumulate various antioxidant enzymes (SOD, POD and CAT) that can quench free radicals, such as H_2O_2 and $O_2^{\bullet-}$ [12]. In this study, 44 DEGs were identified as enzymes in the ROS detoxification and scavenging system. These DEGs were functionally characterized into different ROS enzymes encoding Fe superoxide dismutase (Fe-SODs, 1 transcript), catalase (CAT, 2 transcripts), peroxidase (POD, 8 transcripts), glutathione S-transferase (GST, 16 transcripts), alternative oxidase (AOX, 1 transcript), glutathione-ascorbate (GSH-AsA) cycle (6 transcripts), the peroxiredoxin/thioredoxin (Prx/Trx, 9 transcripts) and polyphenol oxidase (PPO, 1 transcript) (Table 1 and Table S6).

Table 1. List of differentially-expressed genes related to redox metabolism and respiratory chain in grapevine perceived during salt stress.

Trait Name	Description	No. of Up-Regulated	No. of Down-Regulated	Sum
ROS scavenging	Fe-SOD	0	1	1
	POD	8	0	8
	CAT	2	0	2
GSH-AsA cycle	MDAR	1	0	1
	APx	1	0	1
	GR	0	2	2
	Grx	1	1	2
GPX pathway	GST	8	8	16
Prx/Trx	Trx	4	5	9
Cyanide-resistant respiration	AOX	0	1	1
Copper-containing enzymes	PPO	0	1	1

Fe-SOD: Fe superoxide dismutase; POD: peroxidase; CAT: catalase, APX: ascorbate peroxidase; MDAR: monodehydroascorbate reductase; GR: glutathione reductase; Grx: glutaredoxin; GST: glutathione S transferase; Trx: thioredoxin; AOX: alternative oxidase, PPO: polyphenol oxidase.

The metalloenzyme superoxide dismutase (SOD) provides primary defense line against ROS (superoxide radicals, $O_2^{\bullet-}$) and dismutates $O_2^{\bullet-}$ into O_2 and H_2O_2. SODs have three isozymes that are localized in different cellular compartments and vary in their functional properties, including copper-zinc (Cu/Zn-SOD), manganese (Mn-SOD) and iron (Fe-SOD). While only one Fe-SOD with slightly down-regulated expression level ($|\log_2 FC| > 1$) was found in this research, might be due to the severity of salt that suppressed the transcription of Fe-SOD gene in grapevine leaves. These findings are consistent with the previous reports [13–15] and were also confirmed by the SOD activity measurement, in which SOD activity was increased within 36h of salt stress but drastically decreased after 48 h (Figure 4a). In current findings, the activities of CAT and POD were progressively persuaded at 48 h of salt stress treatment (Figure 4b,c). Transcriptomics analysis showed that the DEGs encoding CAT and POD were up-regulated under salt stress, of which two POD transcripts, VIT_13s0067g02360 ($|\log_2 FC| = 3.68$) and VIT_08s0040g02200 ($|\log_2 FC| = 3.51$) were remarkably up-regulated in salt-treated group as compared to control, while remaining six POD and the two CAT transcripts were slightly up-regulated (their $|\log_2 FC|$ values were about 1), which is consistent with the physiological data of increased activities of antioxidant enzymes. These findings suggested a common response of antioxidant enzymes to detoxify ROS effects.

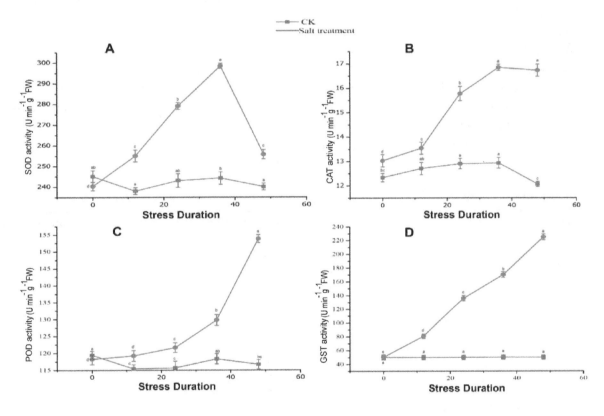

Figure 4. Changes in the enzyme activities of SOD (**A**), CAT (**B**), POD (**C**) and GST (**D**) in grapevine leaves grown for 48 h under control and salt stress. Values represent mean $\pm$ SE ($n = 3$) and the significance level of 0.05 was used for different letters above bars.

GSH and GST also play a crucial defense-related role against ROS caused by salt stress [16,17]. In this study, six transcripts involved in ascorbate-glutathione (AsA-GSH) cycle and 16 Glutathione S-transferase (GST) transcripts were detected, of which salt stress significantly induced two GST transcripts (VIT_05s0049g01070 and VIT_05s0049g01100), when compared with remaining four genes. GST activity was also significantly increased at 48 h of salt stress (Figure 3d), revealing its essential roles in the ROS scavenging process.

2.5. Heat Shock Protein (HSP) and Pathogenesis-Related Proteins (PR) in Response to Salt Stress

Heat shock proteins (HSPs) are the molecular chaperones that act as stress-responsive proteins, thus protecting plants from stress damage, which include mainly HSP100s, HSP90s, HSP70s, HSP60s (cpn60s) and small heat-shock proteins (sHSPs). Overall, 39 HSPs-related DEGs were further divided into high molecular weight HSPs (HMW HSPs; 4 transcripts), low molecular weight HSPs (LMW HSPs; 17 transcripts), heat stress transcription factors (6 transcripts) and other HSPs (12 transcripts) (Table 2 and Table S7). Three transcripts encoding HMW HSPs were down-regulated, while one transcript (VIT_18s0041g01230) was up-regulated. Similarly, 16 of the 17 LMW HSPs were up-regulated and some of them showed very high expression levels compared to the control, for example, VIT_16s0098g01060 ($|\log_2 FC| = 3.755743188$), VIT_13s0019g00860 ($|\log_2 FC| = 3.361853346$) and VIT_12s0035g01910 ($|\log_2 FC| = 3.135505213$), intimating that LMW HSPs play a more important role than the HMW HSPs in response to salt stress in grapevine. Six heat stress TFs (3 up-regulated and 3 down-regulated) showed very diverse transcription levels, suggesting their complex regulatory mechanism over HSPs. Also, 8 chaperone protein DnaJ (6 up-regulated and 2 down-regulated) transcripts identified in the current research, as DnaJ is a vital cofactor plays a central role in transducing stress-induced protein damage to induce heat shock gene transcription, while the up-regulation may suggest the extensive cellular protein damage by salt severity.

Plants can enhance tolerance mechanism against salt stress through over-expression of pathogenesis-related proteins. In the grapevine transcriptome, 37 DEGs encoding disease resistance proteins were identified and classified into 4 pathogenesis-related proteins (PR-1; all up-regulated), 2 chitinase (both down-regulated), 1 beta-1, 3-glucanase (down-regulated), 8 lipid transfer proteins (7 up-regulated, 1 down-regulated), 6 thaumatin-like proteins (1 up-regulated, 5 down-regulated), 1 germin protein (down-regulated), 13 disease resistance proteins (9 up-regulated, 4 down-regulated) and 2 snakin were perceived as up-regulated (Table 2 and Table S7).

Table 2. List of differentially-expressed genes related to heat-shock proteins (HSPs) and pathogens resistance (PRs) proteins in grapevine perceived during drought stress.

Trait Name	Description	No. of Up-Regulated	No. of Down-Regulated	Sum
Heat shock proteins	HMW HSPs	1	3	4
	LMW HSPs	16	1	17
	small HSPs	12	6	18
	other HSPs	7	5	12
	heat-stress transcription factors	3	3	6
PR-1	pathogenesis-related protein 1	4	0	4
PR-2	β-1,3-glucanase	0	1	1
PR-3,4,8,11	chitinase	0	2	2
PR-5	Thaumatin-like protein	1	5	6
PR-14	lipid transfer protein	7	1	8
PR-15	germin-like protein	0	1	1
	Disease resistance proteins	9	4	13
	snakin	2	0	2

HMW HSPs: High molecular weight heat shock proteins; LMW HSPs: Low molecular weight heat shock proteins.

2.6. Regulation of Hormonal Signaling in Response to Salt Stress

Hormones are pivotal to plants in stress adaptive signaling cascades and act as a central integrator to connect and reprogram different responses, such as photosynthesis and activities of ROS enzymes, protein structure and gene expression and accumulation of secondary metabolites [18–20]. In this experiment, various DEGs encoding hormone signaling was involved in abscisic acid (ABA), jasmonic acid (JA), auxin (IAA), gibberellin (GA), ethylene (ETH), brassinosteroid (BR) synthesis and signal transduction pathways (Table S8). Under salt stress, ABA is known to play a protective role in plants against LPO by assisting the accumulation of metabolites that act as osmolytes and also tends to close their stomata to reduce water loss by transpiration. Moreover, 8 transcripts encoding protein phosphatase 2C (PP2C) were down-regulated in the salt-treated grapevine leaves, while PP2C is deliberated as a negative regulator of the ABA signaling. Also, 2 ABA receptor PYL (1 up and 1 down-regulated) were also detected, which indicated that salt stress-induced not only the regulators but also the receptors in the ABA transduction pathway, by which ABA signaling pathway was enhanced quickly and then participated in the salt stress defense process.

Other plant hormones, like auxin and ethylene, also have important roles in plants to cope with salt stress. In this experiment, 23 auxin-related transcripts were detected, in which 2 auxin response factors (ARF) and 3 auxin-responsive proteins were down-regulated, while 12 auxin-induced proteins and 3 indole-3-acetic acid-induced proteins were up-regulated. Out of the 12 auxin-induced proteins, 4 transcripts (VIT_04s0023g00530, VIT_03s0038g01100, VIT_03s0038g01090 and VIT_04s0023g00520) were only expressed in the salt-treated samples, suggesting their close interaction with salt stress. In ethylene synthesis, 3 ACC oxidases (ACO) homologs were up-regulated, whereas 23 transcripts encoding ethylene-responsive TFs revealed variation in up-regulation (13 transcripts) and down-regulation (10 transcripts) in grapevine under salt stress.

2.7. Ion Transport Systems Mediating Na$^+$ Homeostasis in Response to Salt Stress

Salt tolerance mechanism works basically by reducing the undue accretion of Na$^+$ in the cytosol of the plant cell. The quantification of ionic concentrations suggested that Na$^+$ concentration increased significantly in leaf and root tissues (Figure 5). Leaves had higher Na$^+$ accumulation (5.51 $\pm$ 0.48), which was 40.47% more than Na$^+$ level of roots (3.28 $\pm$ 0.23). Moreover, Cl$^-$ concentration increased significantly at about 5-folds in leaves and 9-folds in roots as compared to their corresponding controls, respectively. In contrast, K$^+$ showed a decreasing trend in both tissues (leaf and root) after 48h of salt stress as compared with the control group (Figure 5). In the transcriptomic analysis, 14 DEGs were found to be involved in the ion transport systems, which include 2 vacuolar-type H$^+$-ATPase (V-type proton ATPase, 1 up and 1 down-regulated), 2 sodium/hydrogen exchanger (both down-regulated), 3 cyclic nucleotide-gated ion channel (CNGC, 2 up-regulated and one down-regulated), 2 potassium transporter (both down-regulated), 2 K$^+$ efflux antiporter (both down-regulated), 2 sodium-related cotransporter (sodium/pyruvate cotransporter, sodium/bile acid cotransporter (down-regulated) transcripts (Table S9). In the vacuolar membrane, V-ATPase is the central H$^+$ pump, which creates a transmembrane proton gradient and drives the Na$^+$/H$^+$ antiporter to transport the excessive Na$^+$ in the cytoplasm to vacuoles [21].

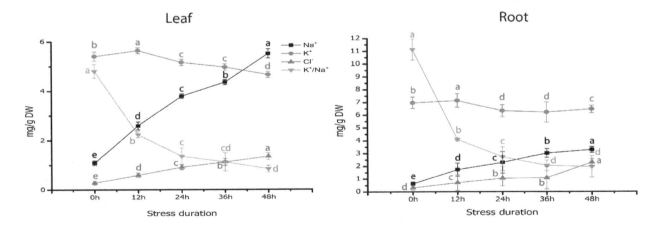

Figure 5. Ion concentrations of sodium (Na$^+$), potassium (K$^+$), chlorine (Cl$^-$) and K$^+$/Na$^+$ ratio in leaf and root samples of grapevine grown for 48 h of salt stress. Values represent means $\pm$ SE (n = 3) and the significance level of 0.05 was used for different letters above bars.

2.8. Transcription Factors in Response to Salt Stress

Transcription factors (TFs) are proteins that cooperate with other transcriptional regulators and bind cis-elements at the promoter region, thus up-regulate the downstream activities of many stress-related genes, results in inducing stress resistance in plants. Almost all the TFs identified in the present transcriptome data have already been reported to play a significant role to counter salt stress (Table S10). Results revealed five MYB transcripts (4 up-regulated, 1 down-regulated), 8 WRKY transcripts (all down-regulated), 1 C2H2 transcript (down-regulated), 4 DOF transcripts (3 up-regulated, 1 down-regulated), 6 HD-zip transcripts (all up-regulated), 5 bHLH transcripts (1 up-regulated and 4 down-regulated), 4 ZAT transcripts (1 up-regulated, 3 down-regulated), 6 NAC transcripts (1 up-regulated and 5 down-regulated), 3 PHD transcripts (all up-regulated) and 23 ERF TFs, intimating their critical roles in the grapevine resistance to salt stress (Figure 6).

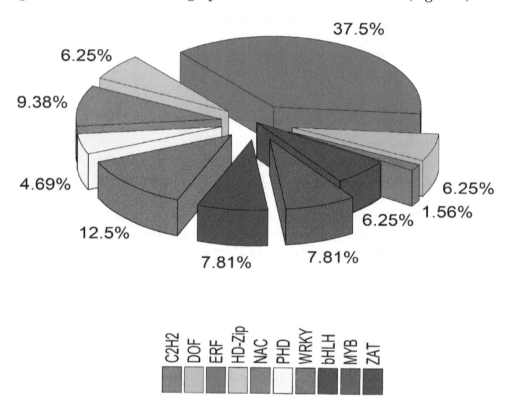

Figure 6. The sector diagram of major TFs identified and the total number of DEGs in grapevine leaf tissues after 48 h of salt stress compared with control.

2.9. qRT-PCR Validation of Illumina RNA-Seq Results

To validate the reliability of RNA-seq transcriptome, 16 DEGs were randomly selected to analyze the gene expression that was correlated with salt stress response and covering almost all the primary functions in various biological pathways, including transcription factors, metabolism, plant hormone signaling, disease resistance and ion transport (Table S11). The result suggested that expression of 16 DEGs treated with 0.8% soil salinity at the interval of 0, 12, 24, 36 and 48 h is inconsistent with the transcriptomic findings, validating the accuracy and reproducibility of the Illumina RNA-seq. Though, out of 16 DEGs, 12 DEGs showed recurrent expression pattern in response to salt stress, in which 8 genes were up-regulated (Figure 7a,c–e,g–j) and 4 genes were down-regulated (Figure 7k,n–p) with prolonged salt stress. Based on the expression patterns of these 12 genes, we selected them as candidate genes to further validate their expressional variations following the different concentrations of salt stress and recovery process.

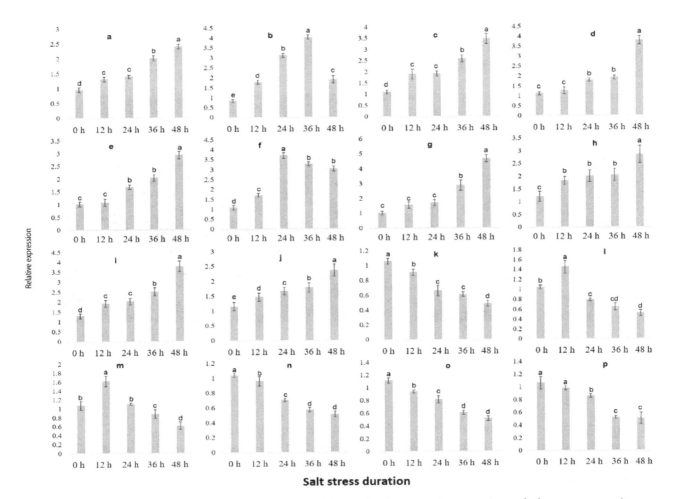

Figure 7. qRT-PCR validation of illumina Hiseq findings and screening of damage severity prediction marker genes. Values represent means ± SE ($n = 3$) and the significance level of 0.05 was used for different letters above bars. Genes have continual increasing or decreasing expression patterns were selected as candidate genes. **a:** VIT_05s0020g03740.t01, **b:** VIT_16s0050g02530.t01 **c:** VIT_19s0015g01070.t01, **d:** VIT_05s0049g00520.t01, **e:** VIT_13s0067g02360.t01, **f:** VIT_12s0035g01910.t01, **g:** VIT_04s0023g00530.t01, **h:** VIT_05s0049g01070.t01, **i:** VIT_06s0004g05670.t01, **j:** VIT_10s0003g01810.t01, **k:** VIT_00s0332g00110.t01, **l:** VIT_00s0201g00080.t01, **m:** VIT_07s0005g00160.t01, **n:** VIT_11s0052g01180.t01, **o:** VIT_14s0128g00020.t01, **p:** VIT_05s0062g00300.t01.

2.10. Salt Stress Recovery and the Selection and Validation of Marker Genes

Herein, 12 marker genes showing regular expression patterns, defined their potential as useful markers to determine the stress severity in grapevine plants. The growth status of grapevine plants was monitored, which indicated that salt severity turned grapevine leaves yellow and brownish blemishes were developed after a prolonged duration of salt stress and eventually die (Figures 8 and 9). At 1.5% salt concentration, grapevine plants can be recovered to normal growth conditions within 10 days of salt treatment by removing the salt stress, though few injured leaves could not survive even after the recovery, might be due to over-accumulation of salt. On the contrary, the plants died after prolonged salt stress duration (15 days), though there was no phenotypic evidence of death before going for recovery. Likewise, the critical time of recovery for 3.0% salt stress is 6 days. Similarly, qRT-PCR analysis of 12 candidate genes showed increased/ decreased expression level following the different doses of salt application. Furthermore, some genes showed unique expression pattern following 10 days of stress, such as VIT_05s0020g03740 showed an increasing trend within 9 days of salt stress but decreased significantly on the 10th day of salt stress (Figure 10a); whereas, the expression level of VIT_05s0049g00520, VIT_05s0049g01070 and VIT_06s0004g05670 was induced within 9 days after stress but sharply induced on the 10th day of salt stress (Figure 10d,h,i). Nevertheless, two transcripts (VIT_00s0332g00110 and VIT_05s0062g00300) showed a gradual decrease in their expression level till the 9th day but the sharp decrease was observed at the 10th day after NaCl application (Figure 10k,p).

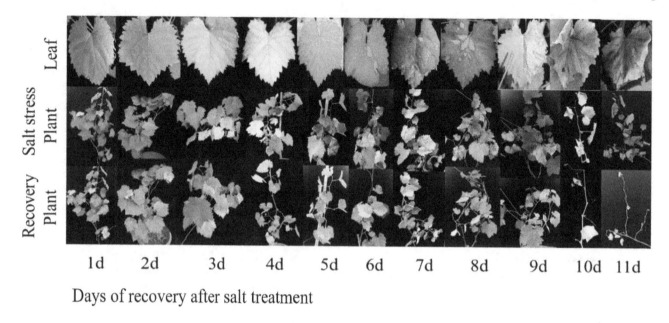

Figure 8. Grapevine growth status under salt stress and after removing salt stress. Grapevine plants were treated by 1.5% SS (salt stress) for 1, 2, 3,4, 5, 6, 7, 8, 9, 10 and 11 days (d), respectively and recovered by washing away the salt in the medium. Recovered plants were photographed 15 days after salt stress was removed.

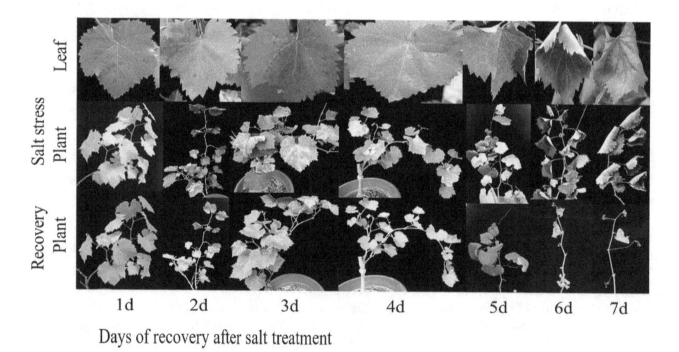

Figure 9. Grapevine growth status under salt stress and after removing salt stress. Grapevine plants were treated by 3.0% SS for 1, 2, 3, 4, 5, 6 and 7 days (d), respectively and recovered by washing away the salt in the medium. Recovered plants were photographed 15 days after salt stress was removed.

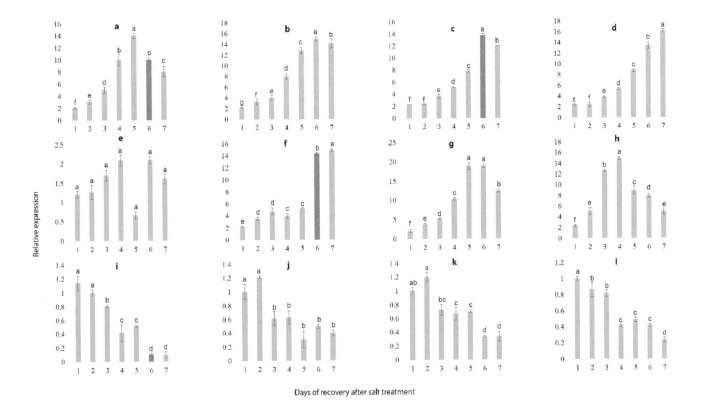

Figure 10. Expression patterns of the 12 candidate genes following the 7 days of 1.5% SS. Red bars indicate the sharp change in gene expression levels. Values represent mean ± SE ($n = 3$) and the significance level of 0.05 was used for different letters above bars. Genes with continual increasing or decreasing expression patterns were selected as candidate genes. **a**: VIT_05s0020g03740.t01, **b**: VIT_19s0015g01070.t01, **c**: VIT_05s0049g00520.t01, **d**: VIT_13s0067g02360.t01, **e**: VIT_04s0023g00530.t01, **f**: VIT_05s0049g01070.t01, **g**: VIT_06s0004g05670.t01, **h**: VIT_10s0003g01810.t01, **i**: VIT_07s0005g00160.t01, **j**: VIT_11s0052g01180.t01, **k**: VIT_14s0128g00020.t01, **l**: VIT_05s0062g00300.t01.

Interestingly, some of these genes showed a similar sharp expression level at the 6th day under 3.0% salt stress, such as transcript VIT_05s0020g03740 kept increasing until 5th day of stress but suddenly decreased on the 6th day of salt stress (Figure 11a), while the expression levels of VIT_05s0049g00520 and VIT_05s0049g01070 kept slow increasing trends up to 5 days of salt stress but showed a sharp increase on the 6th day (Figure 11d,h). Moreover, transcript VIT_00s0332g00110 showed a gradually decreasing trend up to 5 days of salt stress but significantly reduced on the 6th day of salt stress (Figure 11k). Based on above-mentioned findings, grapevine plants cannot be survived by curative processes after 10 days at 0.8% of NaCl and after 6 days at 1.5% of NaCl, which indicates that regardless of high or low concentrations of salt, these four genes with recurrent expression pattern could be used as potential markers to predict the severity imposed by salt stress.

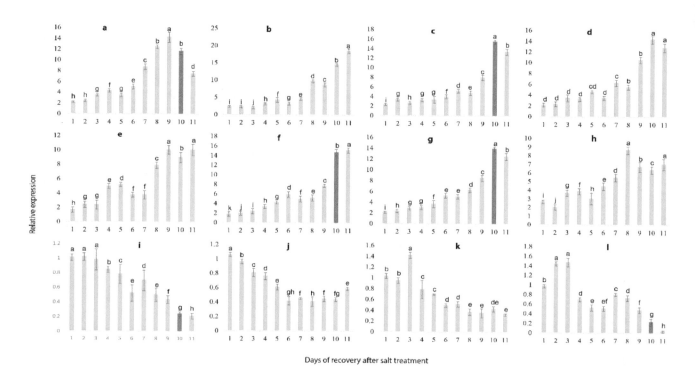

Days of recovery after salt treatment

Figure 11. Expression patterns of the 12 candidate genes following the 11 days of 3.0% SS. Red bars indicate the sharp change in gene expression levels. Values represent mean $\pm$ SE ($n = 3$) and the significance level of 0.05 was used for different letters above bars. Genes with continual increasing or decreasing expression patterns were selected as candidate genes. **a:** VIT_05s0020g03740.t01, **b:** VIT_19s0015g01070.t01, **c:** VIT_05s0049g00520.t01, **d:** VIT_13s0067g02360.t01, **e:** VIT_04s0023g00530.t01, **f:** VIT_05s0049g01070.t01, **g:** VIT_06s0004g05670.t01, **h:** VIT_10s0003g01810.t01, **i:** VIT_07s0005g00160.t01, **j:** VIT_11s0052g01180.t01, **k:** VIT_14s0128g00020.t01, **l:** VIT_05s0062g00300.t01.

3. Discussion

Salt stress is considered as most severe abiotic stress, which impairs all principal physiological functions, including photosynthesis, lipid metabolism and synthesis of proteins [22]. To confront the stress, plants are compelled to initiate protective responses, like restoring cellular ion concentrations and reducing the toxicity of ions like Na^+/H^+, K^+ and Cl^-. Moreover, the accretion of osmoprotectants and hydrophilic proteins, such as sugars, polyols, proline, glycine betaine (GB), amino acids (AA) and amines are crucial for governing the osmotic potential pressure. Also, the accumulation of ROS enzymes and antioxidants is vital to prevent tissue damage by eliminating the free radicals induced by salt stress [22,23].

Grapevine plants alter their physiology to combat salt stress severity. Current findings suggested that reduced stomatal conductance resulted in the inhibition of net photosynthesis rate and CO_2 exchange, which is considered as a primary response of grapevine to reduce transpiration rate to avoid salt accumulation in stomatal apertures of leaves. Zhang et al. [24] depicted that stress factors damage the photosynthetic pigments (chlorophylls and carotenoids) in both photosystems (PSI and PSII), which affect their light-absorbing efficiency, resulting in hindered photosynthetic capability. Our results are consistent with the findings of similar work on peach [25] and grapes [1], proposing that photosynthetic efficiency and CO_2 balance were affected by the reduced stomatal conductance. Moreover, the light-harvesting proteins (CP47) in PSII and chlorophyll binding proteins (P700) in PSI were down-regulated by the salt stress. Similar study intimated that salt stress induces ROS production, which damages the LHCs in PSI and impairs the PSII proteins involved in the evolution of oxygen [26]. Also, the modifications in leaf biochemistry decrease the synthesis of ATP amount, leading to regeneration of the RuBISCO, which results in down-regulation of photosynthetic metabolism [27], favor our findings of down-regulation of RuBISCO and ATP-related transcripts. In *Arabidopsis*, oxidative stress activates SnRK2 in ABA signaling, which regulates the stomatal conductance [27] and up-regulation of SnRK21 in our findings might be the reason for the inhibited photosynthetic activity of grapevine leaves.

Salt stress affects the large-scale metabolic activities that result in excessive ROS accumulation, which include singlet oxygen ($^1O_2$), superoxide radical ($O_2^{\bullet -}$), hydrogen peroxide (H_2O_2) and hydroxyl radical ($\bullet OH$), while similar results were observed in *Medicago truncatula* [28]. The ROS cytotoxicity activates the oxygen species, leading to disruption of optimum metabolic activities, which induce lipid peroxidation in plants [29,30]. Thus, the equilibrium between ROS production and quenching is critical under salt stress. Plants can unfold a complex antioxidative defense system to limit the oxidative damage, which mainly comprised of enzymatic antioxidant (SOD, CAT, POD and GST) and non-enzymatic antioxidants (AsA, GSH, proline and phenolic compounds) [31,32]. In the present study, the antioxidative defense system was activated in salt-treated leaves, although SOD-related transcripts were down-regulated, while CAT, POD and GST-related transcripts were significantly up-regulated. Similar research on *Pyrus pyrifolia* [33] and *Fagopyrum tataricum* [34] depicted that over-expression of GST transcripts significantly enhances salt stress tolerance. Moreover, the enhanced activities of ROS enzymes (CAT, POD and GST) are consistent with the transcriptomic data, which symbolize their vital functions in ROS detoxification. However, CAT activity increased significantly in our findings till 36 h of salt stress but drastically decrease at 48 h, which is in agreement with the down-regulation of SOD-related transcripts in grapevine under salt stress. Zhang et al. [35] reported that salt stress up-regulates the expression of CAT, POD and GST and increases the corresponding enzymes activities. Similarly, the complex accumulation pattern of antioxidant enzyme activities was observed in our findings, which is consistent with the findings of grapevine [36] and soybean [37] under salt stress.

HSPs are the molecular chaperones known to participate in the translocation and degradation of damaged proteins under abiotic stresses [38,39]. In the current study, HSP70 and HSP90 were down-regulated, while various heat stress TFs, small HSPs (sHSPs16–30 kDa) and other HSPs, like DnaJ, were up-regulated by salt stress. This irregular trend of HSPs may suggest that HSPs play an adaptive stress role by altering the growth and development of the plant. Several homologs of HSPs were also found to be activated in *Betula halophila* and *F. tataricum* under salt stress, intimating their regulatory role in various signaling-related pathways [34,40]. The disease resistance proteins can protect plants from pathogens by infection-induced responses of the immune system [41]. In our study, most pathogenesis-related proteins, non-specific lipid-transfer protein and disease resistance proteins were remarkably up-regulated, signifying that these genes not only function in disease resistance but also play essential roles in plant responses to salt stress [42].

Ionic compartmentalization and absorption are essential for growth under saline conditions because stress disrupts ion homeostasis [43]. Plant roots uptake Na^+ and other ions with water from the soil and translocate these ions to the leaves via transpiration stream. With the evaporation of water, a high level of salt gets accumulated in the apoplast and other cellular compartments. Ionic imbalance induces cellular toxicity via replacement of K^+ by Na^+ ions via interfering K^+ channels in the plasma membrane of the root [9,44], while all the potassium transporters were down-regulated in our findings and resulted in lower K^+ concentration in the salt-treated group as compared to control. The plant can resist the cytosolic salt accumulation in the vacuole and other cellular compartments to facilitate their metabolic functions [45,46]. This process involves the regulation of the expressions of some ionic channels and transporters-related genes, which enables the control of Na^+ transport within the plant [4,47]. In *Arabidopsis*, vacuolar $AtNa^+/H^+$ exchanger *SOS1* (Salt Overlay Sensitive) assists Na^+ extrusion from root cells [48,49] but Na^+/H^+ exchanger was down-regulated in our findings, which might be the reason of Na^+ accumulation in the grapevine roots. In addition, NHX1 was down-regulated in our results, while *AtNHX1* cloned plants resulted in high Na^+ in shoot tissues by altering the gene expression of Na^+ transporters [50]. The transcriptional activation of vacuolar-type ATPase (V-ATPase) in our findings suggested that it assists plants in reducing Na^+ accumulation by interacting H^+ pumps to counter salt stress [51]. The high-affinity K^+ transporters (HKTs) can mediate Na^+ transport and Na^+–K^+ symport, while the over-expressed *Arabidopsis AtHKT1* showed high Na^+ in leaves and reduced accumulation in roots [52], favor our findings of higher Na^+ accumulation in grapevine leaf tissues as compared to root. The inhibition of Na^+ influx is the correlative index of cyclic nucleotide-gated ion channels CNGCs that were up-regulated in this study, which is in good agreement with the findings reported in halophyte shrub (*Nitraria sibirica*) [53]. The genetic factors that control the accumulation and transport of Cl^- from root to leaf tissues or enable plants to maintain low leaf Cl^- level are the critical determinant of salt stress tolerance in plants [54], while higher accumulation of Cl^- was observed in roots as compared to leaves in our findings, intimating that grapevine possesses the salt tolerance mechanism. However, in response to higher Cl^- level, plants harness the Cl^-/H^+ transporters (CLCs) to maintain low Cl^- accumulation, especially in aerial parts [55], whereas, no gene related to Cl^- transport was found in our findings.

Phytohormones create a web of signals that are pivotal to plant growth, initiation of flowers, hypocotyl germination and abiotic stress response, which mainly include abscisic acid (ABA), auxin (AUX), jasmonic acid (JA), brassinosteroid (BR) and ethylene (ETH) [56,57]. High salt concentration triggers the ABA level in many plants, which is a well-known fact [25]. In our study, abscisic acid receptor PYL9 was up-regulated and its negative regulator PP2Cs were down-regulated, proposing that the ABA signaling pathway was activated in grapevine in response to salt stress. However, transcriptomic profiling of Jute (*Corchorus* spp.) revealed that DEGs encoding PYL were down-regulated under salt stress [58], which is contradicting with our findings as well as with the basic model of ABA signaling. Additionally, auxin stimulates cell elongation and cell division, also induces sugar and mineral accumulation at the site of application. Under salt stress, all ARFs and their repressors were down-regulated, whereas, most AUX/IAA proteins and IAA synthase were found up-regulated in our findings, suggesting that genes encoding IAA participate significantly in plant development in response to salt stress conditions, while similar results were reported in *V. vinifera* under oxidative stress [59]. JA generally reconciles specific signaling mechanisms involved in senescence, flowering and defense responses, while all the critical enzymes encoding JA were down-regulated, depicting that gene related to JA were suppressed by the salt severity in *Vitis vinifera*. Another study demonstrated that JA level was enhanced in salt-tolerant cultivars as compared to sensitive cultivars [60]. Salt stress inhibits the cell multiplication and expansion by suppressing the activities of growth-promoting hormones, including gibberellins and cytokinins [60], while these results are in favor of current findings.

TFs are regulatory proteins, demonstrated to be involved in regulating the stress-responsive gene expression in many plants responding to abiotic stress. Various MYB genes have been identified and known to induce plant responses to salt stress acclimation, such as *Arabidopsis* [61], rice [62] and wheat [63]. The over-expression of rice MYBs (*OsMYB48-1* and *OsMYB3R-2*) proposed alleviated tolerance to abiotic stresses, such as salt, cold and drought [64,65]. WRKY gene family is regarded as an essential TFs involved in salt stress response, such as *ZmWRKY33* in maize [66] and *GhWRKY39* in cotton [67], but, in our study, all the eight WRKY transcripts, including two WRKY33 were down-regulated under salt stress, which indicates the complexity of the WRKY regulatory mechanism and diverse nature in different stress conditions. Many other TFs with no direct response to salt stress but were triggered by other physiological changes like ROS and endogenous hormones. ERF family interact with ABA signaling pathway (dependent and/or independent) and respond to abiotic stresses [59,60]. In our study, 23 ERF transcripts were detected in the salt-treated grapevine leaves; meanwhile, ten ABA-related transcripts were also identified, intimating their essential roles in ABA-dependent ERF regulatory mechanism in grapevine. PHD finger proteins, especially PHD2, were reported to be involved in the salt stress response, which is known to induce by salt-induced oxidative stress [68]. All the three PHD transcripts detected in our transcriptome were up-regulated, which suggested the significance ROS synthesis caused by salt stress in the tested samples.

Specific genes regulate various plant traits and some of these genes expressed in unique patterns before the emergence of apparent traits, thus by detecting these unique expression signals, we can predict the occurrence of the corresponding phenotype. Hence, to measure the stress severity induced by NaCl stress, grapevine plants were subject to salt treatment with different doses of NaCl and time interval and were recovered by washing off the salt from roots. Results suggested that grapevine plants can be recovered within 10 days at 1.5% of NaCl dose and within 6 days under 3% salt stress. Similar expression patterns of genes (VIT_05s0049g00520, VIT_05s0049g01070, VIT_05s0020g03740 and VIT_00s0332g00110) observed in both NaCl treatments (1.5% and 3%) after 10 and 6 days, respectively, which makes them be the marker genes to estimate the salt severity. Selected genes were involved mainly in the maintenance of cellular structure and functions in plants. For instance, gene VIT_05s0020g03740 (non-specific lipid-transfer protein, LTP) and VIT_05s0049g00520 (proline-rich cell wall protein-like, PRPs) are known to play essential roles in maintaining the stability of the cell wall, membrane and osmotic pressure of the cell [69–71]. VIT_05s0049g01070 (glutathione S-transferase-like, GST) encoded as a critical protein, which has several physiological functions like ROS detoxification and protecting the DNA from damage [72]. Also, VIT_00s0332g00110 (Photosystem II reaction center protein) was involved in the most important physiological function (photosynthesis). Taken together, the transcriptional status of these marker genes reflects the vitality in grapevine plants. Moreover, a similar technique to predict marker genes can also be implicated on other crops where natural environmental disasters, such as temperature (low and high), water-logging and drought prevails occasionally. Taken together, grapevine possesses a complex regulatory mechanism of salt stress-tolerance, which mainly involves the regulation of key genes that are summarized in Figure 12.

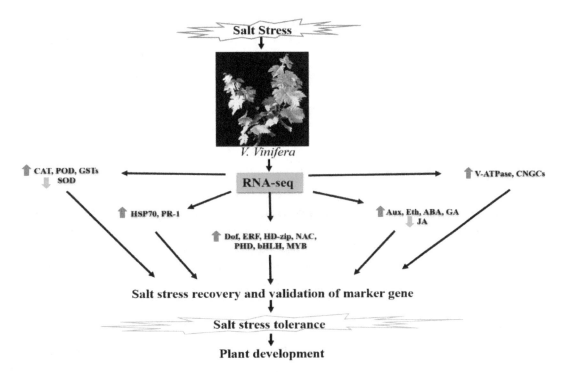

Figure 12. A schematic complex regulatory mechanism of salt stress tolerance in grapevine. Red arrows indicating up-regulated genes and green arrows indicating down-regulated genes. CAT, catalase; POD, peroxidase; GST, glutathione-s-transferase; SOD, superoxide dismutase; HSP70, heat shock protein 70 kDa; PR-1, Pathogenesis-related protein1; Dof, DNA-binding with one finger; ERF, ethylene responsive factor; HD-Zip, homeodomain-leucine zipper; NAC, NAC transcription factor; bHLH, basic helix-loop-helix; MYB, MYB transcription factor; Aux, auxin; Eth, ethylene; ABA, abscisic acid; GA, gibberellic acid; JA, jasmonic acid; V-ATPase, vacuolar-type ATPase; CNGCs, cyclic nucleotide-gated channels.

4. Materials and Methods

4.1. Plant Material and Salt Treatments

Two-year-old grapevine (Summer Black Cv.) pot grown plants obtained from Jiangsu Academy of Agriculture Sciences (JAAS), Nanjing, China and kept in greenhouse conditions (25 ± 5 °C), provided with 65% relative humidity (RH) and 16 h-light and 8 h-dark photoperiod at Nanjing Agricultural University, China. The grapevine plants were kept in a medium of soil-peat-sand at 3:1:1 (v:v:v) and used as experimental materials. Overall, ten grapevine plants were selected and categorized into salt-treated (5 plants) and control (5 plants) groups. NaCl (0.8%) was selected to induce salinity stress in grapevine plants. Fourth-unfolded leaf from both the NaCl-treated and control groups were collected at the interval of 0 (control), 12, 24, 36 and 48 h. Each sample has three replicates. Collected leaf samples were immediately frozen dried in liquid nitrogen and then stored at −80 °C until further analysis.

4.2. RNA Extraction, cDNA Library Construction and Illumina Deep Sequencing

Trizol reagent method was used to extract the total RNA from both salt-treated and control grapevine leaf samples (Invitrogen, Carlsbad, CA, USA). The RNA quantity was determined by using Micro-spectrophotometer (Nano-100, ALLSHENG, Hangzhou, China) and further mRNA purification and cDNA library construction were performed with the Ultra™ RNA Library Prep Kit for Illumina (MA, USA) by following the manufacturer's protocol. The final sampling collected after 48 h from salt treatment was sequenced against control (0 h) on an Illumina HiseqTM2500.

4.3. Mapping of reads, Gene Annotation and Analysis of Gene Expression Level

The raw sequence data were filtered by removing low-quality sequences and adapter reads by using HISAT [1]. After quality trimming, clean reads were mapped to the V. vinifera reference genome using Bowtie (1.1.2) by adapting standard mapping parameters [59]. In this data, >100 bp read length with <2 mismatches were mapped to reference genome. To calculate the gene expression and RPKM (and reads per kilobase per million, SAM tools and BamIndexStats.jar were used. Then, DEGseq2 was used to obtain differentially-expressed genes (DEGs) between Log$_2$ and the stationary phase [59]. The genes with FDR less than 0.001 and 2-fold change were pondered as DEGs.

4.4. Gene Ontology (GO) and Kyoto Encyclopedia of Genes and Genomics (KEGG)

For GO annotations, the DEGs were subjected GO database (http://www.geneontology.org/) by using program Blast2Go (http://www.blast2go.com/Ver.2.3.5). To classify genes or their products into terms (molecular function, biological process and cellular component) GO enrichment analysis by using GO-seq was used to under biological functions of DEGs [59]. For KEGG annotations, all the DEGs were mapped to the KEGG database (https://www.genome.jp/kegg/pathway.html) and looked for enriched pathways compared to the background genome [73].

4.5. Estimation of Photosynthesis Rate and Determination of Several Enzymatic and Ionic Concentrations

For photosynthesis rate (A$_N$), stomatal conductance rate (g$_S$), CO$_2$ exchange (Ci) and transpiration rate (E), 4th unfolded leaves were used from control and salt-treated grapevine plants between 9:00–11:00 AM, on full sunny day, using portable Li-COR (Li-6400XT, NE, USA) as briefly described by Haider et al. [1].

Leaf samples treated with 0.8% NaCl for 0, 12, 24, 36 and 48 h were used to determine the antioxidative enzymes activities, including SOD, CAT, POD and GST. The activity of SOD was measured using NBT at 560 nm; CAT activity was measured by monitoring disappearance of H$_2$O$_2$ at 240 nm, the POD was determined by guaiacol oxidation method following the method briefly explained by Haider et al. [74,75]. GST activity was determined using Glutathione S-transferase (GST) activity determination kit (Shanghai solarbio Bioscience & Technology Co., LTD, Shanghai, China) following the manufacturer's protocol.

For ionic concentrations, 0.5 g of leaf and root sample were first oven dried at 70 °C for 48 h and then ground to powder and digested in HNO3: HClO4 (2:1, v:v). The concentrations of selected ions (e.g., Na^+, K^+ and Cl^-) were determined using ICP-MS (Thermo Electron Corporation, MA, USA) as previously explained by Ma et al. [76]. The data were subjected to one-way analysis of variance (ANOVA) by using three replicates for each sample and expressed mean ± standard error (SE). Statistical analysis was carried out using Minitab (Ver 16) and SPSS (Ver 15.0) at $p < 0.05$ level of significance.

4.6. Quantitative Real-Time PCR (qRT-PCR) Analysis of DEGs and Validation of Illumina RNA-Seq Results

Sixteen genes selected from various pathways were used for the validation of the Illumina RNA-seq by qRT-PCR analysis. The primer pairs were designed using primer3 program (http://bioinfo.ut.ee/primer3-0.4.0/) and details of the primers are shown in supplementary Table S11. After extraction, total RNA was reverse-transcribed using the PrimeScript RT Reagent Kit with gDNA Eraser (Takara, Dalian, China). Each qPCR reaction contains 10 μL 2× SYBR Green Master Mix Reagent (Applied Biosystems, CA, USA), 2.0 μL cDNA sample and 400 nM of gene-specific primer in a final volume of 20 μL.qRT-PCR was carried out using an ABI PRISM 7500 real-time PCR system (Applied Biosystems, CA, USA). PCR conditions were 2 min at 95 °C, followed by 40 cycles of heating at 95 °C for 10 s and annealing at 60 °C for 40 s. A template-free control for each primer pair was set for each cycle. The All PCR reactions were normalized using the Ct value corresponding to the Grapevine actin gene (XM_010659103). Three biological replications were used and three measurements were performed on each replicate.

4.7. Salt Stress and Recovery Assay

To screen out the marker genes following the oxidative stress severity caused by salt, 15 grapevine plants were treated with two different acute salt concentrations (1.5% and 3.0%) and then plants recovered by washing off the NaCl solution.

Everyday 3 potted grapevine seedlings were recovered from NaCl stress by washing off the salts with the distilled water; this step was repeated till the salinity content from the medium was reduced to the average level (around 0.1%), 1/2 strength of Hoagland nutrient solution with standard NaCl content was watered again. The salt treated plants were sampled and photographed every day during the treatment and recovered plants. All qRT-PCR reactions for the selected marker genes were the same as previously mentioned.

5. Conclusions

A comparative transcriptome analysis was explored on two libraries constructed from salt-treated and control grapevine leaf samples. Results revealed that 2472 genes were differentially expressed and were significantly involved in antioxidant system, hormonal signaling, ion homeostasis and disease and pathogenesis-related pathways. Besides, many regulatory proteins encoding transcription factors were also identified that induce the function of other genes (e.g., HSPs) requisite for stress-adaptive responses and tolerance. The GO annotations assisted to screen out the series of molecular and physiological cues, which revealed their critical role in salt stress-tolerance mechanism. Moreover, salt stress significantly affected the photosynthetic efficiency and ions uptake and transport in V. vinifera. Though, antioxidant enzyme (CAT, POD and GST) activities were enriched to counter the lipid peroxidation. In this study, we have also screened out and validated the four candidate genes to predict salt severity in grapevine.

Taken together, current study provided a deep overview of enriched genomic information along with physiological validation that will be useful for understanding the salt stress regulatory mechanism in grapevine.

Author Contributions: Conceived and designed the experiments: L.G., M.S.H., J.F. Perform the experiment: M.S.H., N.K., M.F., S.J., X.Z. Analyzed the data: L.G., K.Z., M.N., Manuscript writing: L.G., J.F. All the authors approved the final draft of the manuscript.

Abbreviations

CAT	Catalase
Cl^-	Chloride
DEGs	Differentially expressed genes
GO	Gene Ontology
GST	Glutathione S transferase
HSPs	Heat shock proteins
K^+	Potassium
KEGG	Kyoto Encyclopedia of Genes and Genomes
Na^+	Sodium
POD	Peroxidase
PRs	Pathogenesis-related proteins
qRT-PCR	Quantitative reverse transcriptome-PCR
ROS	Reactive oxygen species
RNA-seq	RNA-sequencing
SOD	Superoxide dismutase
SS	Salt stress
TFs	Transcription factors

References

1. Haider, M.S.; Kurjogi, M.M.; Khalil-Ur-Rehman, M.; Fiaz, M.; Pervaiz, T.; Jiu, S.; Haifeng, J.; Chen, W.; Fang, J. Grapevine immune signaling network in response to drought stress as revealed by transcriptomic analysis. *Plant Physiol. Biochem.* **2017**, *121*, 187–195. [CrossRef]

2. Allakhverdiev, S.I.; Sakamoto, A.; Nishiyama, Y.; Murata, N. Ionic and Osmotic Effects of Nacl-Induced Inactivation of Photosystems I and Ii in Synechococcus sp. *Plant Physiol.* **2000**, *123*, 1047–1056. [CrossRef]

3. Mahajan, S.; Tuteja, N. Cold salinity and drought stresses: An overview. *Arch. Biochem. Biophys.* **2005**, *444*, 139–158. [CrossRef]

4. Munns, R.; Tester, M. Mechanisms of salinity tolerance. *Annu. Rev. Plant Biol.* **2008**, *59*, 651. [CrossRef] [PubMed]

5. Chinnusamy, V.; Jagendorf, A.; Zhu, J.K. Understanding and Improving Salt Tolerance in Plants. *Crop Sci.* **2005**, *45*, 437–448. [CrossRef]

6. Tester, M.; Davenport, R. Na^+ Tolerance and Na^+ Transport in Higher Plants. *Ann. Bot.* **2003**, *91*, 503–527. [CrossRef] [PubMed]

7. Postnikova, O.A.; Shao, J.; Nemchinov, L.G. Analysis of the alfalfa root transcriptome in response to salinity stress. *Plant Cell Physiol.* **2013**, *54*, 1041–1055. [CrossRef]

8. Fan, X.D.; Wang, J.Q.; Yang, N.; Dong, Y.Y.; Liu, L.; Wang, F.W.; Wang, N.; Chen, H.; Liu, W.C.; Sun, Y.P. Gene expression profiling of soybean leaves and roots under salt, saline-alkali and drought stress by high-throughput Illumina sequencing. *Gene* **2013**, *512*, 392–402. [CrossRef]

9. Wang, B.; Lv, X.Q.; He, L.; Zhao, Q.; Xu, M.S.; Zhang, L.; Jia, Y.; Zhang, F.; Liu, F.L.; Liu, Q.L. Whole-transcriptome sequence analysis of verbena bonariensis in response to drought stress. *Int. J. Mol. Sci.* **2018**, *19*, 1751. [CrossRef] [PubMed]

10. Jaffar, M.A.; Song, A.; Faheem, M.; Chen, S.; Jiang, J.; Chen, L.; Fan, Q.; Chen, F. Involvement of cmwrky10 in drought tolerance of chrysanthemum through the aba-signaling pathway. *Int. J. Mol. Sci.* **2016**, *17*, 693. [CrossRef]

11. Wang, C.; Chen, H.F.; Hao, Q.N.; Shan, Z.H.; Zhou, R.; Zhi, H.J.; Zhou, X.A. Transcript profile of the response of two soybean genotypes to potassium deficiency. *PLoS ONE* **2012**, *7*, e39856. [CrossRef] [PubMed]

12. Wang, Y.; Tao, X.; Tang, X.M.; Xiao, L.; Sun, J.L.; Yan, X.F.; Li, D.; Deng, H.Y.; Ma, X.R. Comparative transcriptome analysis of tomato (Solanum lycopersicum) in response to exogenous abscisic acid. *BMC Genom.* **2013**, *14*, 841. [CrossRef] [PubMed]

13. Hernández, J.A.; Corpas, F.J.; Gómez, M.; Río, L.A.D.; Sevilla, F. Salt-induced oxidative stress mediated by activated oxygen species in pea leaf mitochondria. *Physiol. Plant.* **2010**, *89*, 103–110. [CrossRef]

14. Sreenivasulu, N.; Grimm, B.; Wobus, U.; Weschke, W. Differential response of antioxidant compounds to salinity stress in salt-tolerant and salt-sensitive seedlings of foxtail millet (Setaria italica). *Physiol. Plant.* **2010**, *109*, 435–442. [CrossRef]

15. Chaparzadeh, N.; D'Amico, M.L.; Khavari-Nejad, R.A.; Izzo, R.; Navari-Izzo, F. Antioxidative responses of Calendula officinalis under salinity conditions. *Plant Physiol. Biochem.* **2004**, *42*, 695–701. [CrossRef] [PubMed]

16. Marrs, K.A. The functions and regulation of glutathione-S-transferses in plants. *Annu. Rev. Plant Physiol. Plant Mol. Biol.* **1996**, *47*, 127–158. [CrossRef]

17. Edwards, R.; Dixon, D.P.; Walbot, V. Plant glutathione S-transferases: Enzymes with multiple functions in sickness and in health. *Trends Plant Sci.* **2000**, *5*, 193–198. [CrossRef]

18. Jakab, G.; Ton, J.; Flors, V.; Zimmerli, L.; Métraux, J.P.; Mauchmani, B. Enhancing Arabidopsis salt and drought stress tolerance by chemical priming for its abscisic acid responses. *Plant Physiol.* **2005**, *139*, 267–274. [CrossRef] [PubMed]

19. Kim, S.; Son, T.; Park, S.; Lee, I.; Lee, B.; Kim, H.; Lee, S. Influences of gibberellin and auxin on endogenous plant hormone and starch mobilization during rice seed germination under salt stress. *J. Environ. Biol.* **2006**, *27*, 181.

20. Peleg, Z.; Blumwald, E. Hormone balance and abiotic stress tolerance in crop plants. *Curr. Opin. Plant Biol.* **2011**, *14*, 290–295. [CrossRef] [PubMed]

21. Parida, A.K.; Das, A.B. Salt tolerance and salinity effects on plants: A review. *Ecotoxicol. Environ. Saf.* **2005**, *60*, 324–349. [CrossRef] [PubMed]

22. Carillo, P.; Annunziata, M.G.; Pontecorvo, G.; Fuggi, A.; Woodrow, P. Salinity stress and salt tolerance. In *Abiotic Stress in Plants-Mechanisms and Adaptations*; InTech: Rijeka, Croatia, 2011.

23. Horie, T.; Karahara, I.; Katsuhara, M. Salinity tolerance mechanisms in glycophytes. An overview with the central focus on rice plants. *Rice* **2012**, *5*, 11. [CrossRef] [PubMed]

24. Zhang, L.T.; Zhang, Z.S.; Gao, H.Y.; Xue, Z.C.; Yang, C.; Meng, X.L.; Meng, Q.W. Mitochondrial alternative oxidase pathway protects plants against photoinhibition by alleviating inhibition of the repair of photodamaged PSII through preventing formation of reactive oxygen species in Rumex, K.-1 leaves. *Physiol. Plant.* **2011**, *143*, 396–407. [CrossRef] [PubMed]

25. Haider, M.S.; Kurjogi, M.M.; Khalil-ur-Rehman, M.; Pervez, T.; Songtao, J.; Fiaz, M.; Jogaiah, S.; Wang, C.; Fang, J. Drought stress revealed physiological, biochemical and gene-expressional variations in "Yoshihime"peach (Prunus Persica, L.) cultivar. *J. Plant Interact.* **2018**, *13*, 83–90. [CrossRef]

26. Wang, N.; Qian, Z.; Luo, M.; Fan, S.; Zhang, X.; Zhang, L. Identification of Salt Stress Responding Genes Using Transcriptome Analysis in Green Alga Chlamydomonas reinhardtii. *Int. J. Mol. Sci.* **2018**, *19*, 3359. [CrossRef] [PubMed]

27. Hiroaki, F.; Verslues, P.E.; Jian-Kang, Z. Arabidopsis decuple mutant reveals the importance of SnRK2 kinases in osmotic stress responses in vivo. *Proc. Natl. Acad. Sci. USA* **2011**, *108*, 1717–1722.

28. Aydi, S.; Sassi, S.; Abdelly, C. Growth, nitrogen fixation and ion distribution in Medicago truncatula subjected to salt stress. *Plant Soil* **2008**, *312*, 59. [CrossRef]

29. Teakle, N.; Flowers, T.; Real, D.; Colmer, T. Lotus tenuis tolerates the interactive effects of salinity and waterlogging by "excluding"Na$^+$ and Cl$^-$ from the xylem. *J. Exp. Bot.* **2007**, *58*, 2169–2180. [CrossRef]

30. Tanveer, M.; Shabala, S. Targeting Redox Regulatory Mechanisms for Salinity Stress Tolerance in Crops. In *Salinity Responses and Tolerance in Plants*; Springer: Basel, Switzerland, 2018; Volume 1, pp. 213–234.

31. Mandhania, S.; Madan, S.; Sawhney, V. Antioxidant defense mechanism under salt stress in wheat seedlings. *Biol. Plant.* **2006**, *50*, 227–231. [CrossRef]

32. Abogadallah, G.M. Insights into the significance of antioxidative defense under salt stress. *Plant Signal. Behav.* **2010**, *5*, 369–374. [CrossRef]

33. Liu, D.; Liu, Y.; Rao, J.; Wang, G.; Li, H.; Ge, F.; Chen, C. Overexpression of the glutathione S-transferase gene from Pyrus pyrifolia fruit improves tolerance to abiotic stress in transgenic tobacco plants. *Mol. Biol.* **2013**, *47*, 515–523. [CrossRef]

34. Wu, Q.; Bai, X.; Zhao, W.; Xiang, D.; Wan, Y.; Yan, J.; Zou, L.; Zhao, G. De novo assembly and analysis of tartary buckwheat (fagopyrum tataricum Garetn.) transcriptome discloses key regulators involved in salt-stress response. *Genes* **2017**, *8*, 255. [CrossRef] [PubMed]

35. Yan, Z.; Zhou, L.; Yan, P.; Xiaojuan, W.; Dandan, P.; Yaping, L.; Xiaoshuang, H.; Xinquan, Z.; Xiao, M.; Linkai, H. Clones of FeSOD, MDHAR, DHAR Genes from White Clover and Gene Expression Analysis of ROS-Scavenging Enzymes during Abiotic Stress and Hormone Treatments. *Molecules* **2015**, *20*, 20939–20954.

36. Baneh, H.D.; Attari, H.; Hassani, A.; Abdollahi, R. Salinity effects on the physiological parameters and oxidative enzymatic activities of four Iranian grapevines (*Vitis vinifera* L.) cultivar. *Int. J. Agric. Crop Sci.* **2013**, *5*, 1022.

37. Weisany, W.; Sohrabi, Y.; Heidari, G.; Siosemardeh, A.; Ghassemi-Golezani, K. Changes in antioxidant enzymes activity and plant performance by salinity stress and zinc application in soybean (*Glycine max* L.). *Plant Omics* **2012**, *5*, 60.

38. Nollen, E.A.; Morimoto, R.I. Chaperoning signaling pathways. molecular chaperones as stress-sensingheat shock'proteins. *J. Cell Sci.* **2002**, *115*, 2809–2816.

39. Li, J.; He, Q.; Sun, H.; Liu, X. Acclimation-dependent expression of heat shock protein 70 in Pacific abalone (Haliotis discus hannai Ino) and its acute response to thermal exposure. *Chin. J. Oceanol. Limnol.* **2012**, *30*, 146–151. [CrossRef]

40. Shao, F.; Zhang, L.; Wilson, I.; Qiu, D. Transcriptomic Analysis of Betula halophila in Response to Salt Stress. *Int. J. Mol. Sci.* **2018**, *19*, 3412. [CrossRef]

41. Martin, G.B.; Bogdanove, A.J.; Sessa, G. Understanding the functions of plant disease resistance proteins. *Annu. Rev. Plant Biol.* **2003**, *54*, 23–61. [CrossRef]

42. Christensen, A.B.; Cho, B.H.; Næsby, M.; Gregersen, P.L.; Brandt, J.; Madriz-Ordeñana, K.; Collinge, D.B.; Thordal-Christensen, H. The molecular characterization of two barley proteins establishes the novel PR-17 family of pathogenesis-related proteins. *Mol. Plant Pathol.* **2002**, *3*, 135–144. [CrossRef]

43. Adams, P.; Thomas, J.C.; Vernon, D.M.; Bohnert, H.J.; Jensen, R.G. Distinct cellular and organismic responses to salt stress. *Plant Cell Physiol.* **1992**, *33*, 1215–1223.

44. Brini, F.; Masmoudi, K. Ion transporters and abiotic stress tolerance in plants. *ISRN Mol. Biol.* **2012**, *2012*, 927436. [CrossRef] [PubMed]

45. Reddy, M.; Sanish, S.; Iyengar, E. Compartmentation of ions and organic compounds in Salicornia brachiata Roxb. *Biol. Plant* **1993**, *35*, 547. [CrossRef]

46. Zhu, J.-K. Regulation of ion homeostasis under salt stress. *Curr. Opin. Plant Biol.* **2003**, *6*, 441–445. [CrossRef]

47. Rajendran, K.; Tester, M.; Roy, S.J. Quantifying the three main components of salinity tolerance in cereals. *Plant Cell Environ.* **2009**, *32*, 237–249. [CrossRef] [PubMed]

48. Huazhong, S.; Quintero, F.J.; Pardo, J.M.; Jian-Kang, Z. The putative plasma membrane Na(+)/H(+) antiporter SOS1 controls long-distance Na(+) transport in plants. *Plant Cell* **2002**, *14*, 465–477.

49. Quan-Sheng, Q.; Yan, G.; Dietrich, M.A.; Schumaker, K.S.; Jian-Kang, Z. Regulation of SOS1, a plasma membrane Na+/H+ exchanger in Arabidopsis thaliana, by SOS2 and SOS3. *Proc. Natl. Acad. Sci. USA* **2002**, *99*, 8436–8441.

50. Sottosanto, J.B.; Saranga, Y.; Blumwald, E. Impact of AtNHX1, a vacuolar Na + /H. + antiporter, upon gene expression during short- and long-term salt stress in Arabidopsis thaliana. *Bmc Plant Biol.* **2007**, *7*, 18. [CrossRef]

51. Zhu, J.K.; Shi, J.; Singh, U.; Wyatt, S.E.; Bressan, R.A.; Hasegawa, P.M.; Carpita, N.C. Enrichment of vitronectin-and fibronectin-like proteins in NaCl-adapted plant cells and evidence for their involvement in plasma membrane-cell wall adhesion. *Plant J.* **1993**, *3*, 637–646. [CrossRef]

52. Tomoaki, H.; Jo, M.; Masahiro, K.; Hua, Y.; Kinya, Y.; Rie, H.; Wai-Yin, C.; Ho-Yin, L.; Kazumi, H.; Mami, K. Enhanced salt tolerance mediated by AtHKT1 transporter-induced Na unloading from xylem vessels to xylem parenchyma cells. *Plant J. Cell Mol. Biol.* **2010**, *44*, 928–938.

53. Li, H.; Tang, X.; Zhu, J.; Yang, X.; Zhang, H. De Novo Transcriptome Characterization, Gene Expression Profiling and Ionic Responses of Nitraria sibirica Pall. under Salt Stress. *Forests* **2017**, *8*, 211. [CrossRef]

54. Henderson, S.W.; Baumann, U.; Blackmore, D.H.; Walker, A.R.; Walker, R.R.; Gilliham, M. Shoot chloride exclusion and salt tolerance in grapevine is associated with differential ion transporter expression in roots. *BMC Plant Biol.* **2014**, *14*, 1–18. [CrossRef] [PubMed]

55. Wei, L.I.; Wang, L.; Cao, J.; Bingjun, Y.U. Bioinformatics analysis of CLC homologous genes family in soybean genome. *J. Nanjing Agric. Univ.* **2014**, *37*, 35–43.

56. Magnan, F.; Ranty, B.; Charpenteau, M.; Sotta, B.; Galaud, J.P.; Aldon, D. Mutations in AtCML9, a calmodulin-like protein from Arabidopsis thaliana, alter plant responses to abiotic stress and abscisic acid. *Plant J.* **2008**, *56*, 575–589. [CrossRef] [PubMed]

57. Wang, H.; Liang, X.; Wan, Q.; Wang, X.; Bi, Y. Ethylene and nitric oxide are involved in maintaining ion homeostasis in *Arabidopsis callus* under salt stress. *Planta* **2009**, *230*, 293–307. [CrossRef] [PubMed]

58. Yang, Z.; Lu, R.; Dai, Z.; Yan, A.; Tang, Q.; Cheng, C.; Xu, Y.; Yang, W.; Su, J. Salt-Stress Response Mechanisms Using de Novo Transcriptome Sequencing of Salt-Tolerant and Sensitive *Corchorus* spp. Genotypes. *Genes* **2017**, *8*, 226. [CrossRef]

59. Haider, M.S.; Zhang, C.; Kurjogi, M.M.; Pervaiz, T.; Zheng, T.; Zhang, C.; Lide, C.; Shangguan, L.; Fang, J. Insights into grapevine defense response against drought as revealed by biochemical, physiological and RNA-Seq analysis. *Sci. Rep.* **2017**, *7*, 13134. [CrossRef]

60. Pedranzani, H.; Racagni, G.; Alemano, S.; Miersch, O.; Ramírez, I.; Peña-Cortés, H.; Taleisnik, E.; Machado-Domenech, E.; Abdala, G. Salt tolerant tomato plants show increased levels of jasmonic acid. *Plant Growth Regul.* **2003**, *41*, 149–158. [CrossRef]

61. Nagaoka, S.; Takano, T. Salt tolerance-related protein STO binds to a Myb transcription factor homologue and confers salt tolerance in Arabidopsis. *J. Exp. Bot.* **2003**, *54*, 2231–2237. [CrossRef]

62. Yang, A.; Dai, X.; Zhang, W.-H. A R2R3-type MYB gene, OsMYB2, is involved in salt, cold and dehydration tolerance in rice. *J. Exp. Bot.* **2012**, *63*, 2541–2556. [CrossRef]

63. Rahaie, M.; Xue, G.-P.; Naghavi, M.R.; Alizadeh, H.; Schenk, P.M. A MYB gene from wheat (*Triticum aestivum* L.) is up-regulated during salt and drought stresses and differentially regulated between salt-tolerant and sensitive genotypes. *Plant Cell Rep.* **2010**, *29*, 835–844. [CrossRef] [PubMed]

64. Dai, X.; Xu, Y.; Ma, Q.; Xu, W.; Wang, T.; Xue, Y.; Chong, K. Overexpression of an R1R2R3 MYB gene, OsMYB3R-2, increases tolerance to freezing, drought and salt stress in transgenic Arabidopsis. *Plant Physiol.* **2007**, *143*, 1739–1751. [CrossRef] [PubMed]

65. Xiong, H.; Li, J.; Liu, P.; Duan, J.; Zhao, Y.; Guo, X.; Li, Y.; Zhang, H.; Ali, J.; Li, Z. Overexpression of OsMYB48-1, a novel MYB-related transcription factor, enhances drought and salinity tolerance in rice. *PLoS ONE* **2014**, *9*, e92913. [CrossRef] [PubMed]

66. Li, H.; Gao, Y.; Xu, H.; Dai, Y.; Deng, D.; Chen, J. ZmWRKY33: A WRKY maize transcription factor conferring enhanced salt stress tolerances in Arabidopsis. *Plant Growth Regul.* **2013**, *70*, 207–216. [CrossRef]

67. Shi, W.; Liu, D.; Hao, L.; Wu, C.A.; Guo, X.; Li, H. GhWRKY39, a member of the WRKY transcription factor family in cotton, has a positive role in disease resistance and salt stress tolerance. *Plant Cell Tissue Organ Cult.* **2014**, *118*, 17–32. [CrossRef]

68. Wei, W.; Huang, J.; Hao, Y.J.; Zou, H.F.; Wang, H.W.; Zhao, J.Y.; Liu, X.Y.; Zhang, W.K.; Ma, B.; Zhang, J.S. Soybean GmPHD-Type Transcription Regulators Improve Stress Tolerance in Transgenic Arabidopsis Plants. *PLoS ONE* **2009**, *4*, e7209. [CrossRef] [PubMed]

69. Singh, N.; Bressan, R.A.; Carpita, N.C. Cell Walls of Tobacco Cells and Changes in Composition Associated with Reduced Growth upon Adaptation to Water and Saline Stress. *Plant Physiol.* **1989**, *91*, 48–53.

70. Cameron, K.D.; Teece, M.A.; Smart, L.B. Increased accumulation of cuticular wax and expression of lipid transfer protein in response to periodic drying events in leaves of tree tobacco. *Plant Physiol.* **2006**, *140*, 176–183.

71. Gothandam, K.M.; Nalini, E.; Karthikeyan, S.; Jeongsheop, S. OsPRP3, a flower specific proline-rich protein of rice, determines extracellular matrix structure of floral organs and its overexpression confers cold-tolerance. *Plant Mol. Biol.* **2010**, *72*, 125–135. [CrossRef] [PubMed]

72. Kampkötter, A.; Volkmann, T.E.; de Castro, S.H.; Leiers, B.; Klotz, L.O.; Johnson, T.E.; Link, C.D.; Henkle-Dührsen, K. Functional analysis of the glutathione S-transferase 3 from Onchocerca volvulus (Ov-GST-3): A parasite GST confers increased resistance to oxidative stress in Caenorhabditis elegans. *J. Mol. Biol.* **2003**, *325*, 25–37. [CrossRef]

73. Pervaiz, T.; Haifeng, J.; Salman, H.M.; Cheng, Z.; Cui, M.; Wang, M.; Cui, L.; Wang, X.; Fang, J. Transcriptomic Analysis of Grapevine (cv. Summer Black) Leaf, Using the Illumina Platform. *PLoS ONE* **2016**, *11*, e0147369.

74. Haider, M.S.; Khan, I.A.; Naqvi, S.A.; Jaskani, M.J.; Khan, R.W.; Nafees, M.; Pasha, I. Fruit developmental stages effects on biochemical attributes in date palm. *Pak. J. Agric. Sci.* **2013**, *50*, 577–583.

75. Haider, M.S.; Khan, I.A.; Jaskani, M.J.; Naqvi, S.A.; Khan, M.M. Biochemical attributes of dates at three maturation stages. *Emir. J. Food Agric.* **2014**, *11*, 953–962. [CrossRef]

76. Ma, Y.; Wang, J.; Zhong, Y.; Geng, F.; Cramer, G.R.; Cheng, Z.M. Subfunctionalization of cation/proton antiporter 1 genes in grapevine in response to salt stress in different organs. *Hortic. Res.* **2015**, *2*, 15031. [CrossRef] [PubMed]

Permissions

All chapters in this book were first published in MDPI; hereby published with permission under the Creative Commons Attribution License or equivalent. Every chapter published in this book has been scrutinized by our experts. Their significance has been extensively debated. The topics covered herein carry significant findings which will fuel the growth of the discipline. They may even be implemented as practical applications or may be referred to as a beginning point for another development.

The contributors of this book come from diverse backgrounds, making this book a truly international effort. This book will bring forth new frontiers with its revolutionizing research information and detailed analysis of the nascent developments around the world.

We would like to thank all the contributing authors for lending their expertise to make the book truly unique. They have played a crucial role in the development of this book. Without their invaluable contributions this book wouldn't have been possible. They have made vital efforts to compile up to date information on the varied aspects of this subject to make this book a valuable addition to the collection of many professionals and students.

This book was conceptualized with the vision of imparting up-to-date information and advanced data in this field. To ensure the same, a matchless editorial board was set up. Every individual on the board went through rigorous rounds of assessment to prove their worth. After which they invested a large part of their time researching and compiling the most relevant data for our readers.

The editorial board has been involved in producing this book since its inception. They have spent rigorous hours researching and exploring the diverse topics which have resulted in the successful publishing of this book. They have passed on their knowledge of decades through this book. To expedite this challenging task, the publisher supported the team at every step. A small team of assistant editors was also appointed to further simplify the editing procedure and attain best results for the readers.

Apart from the editorial board, the designing team has also invested a significant amount of their time in understanding the subject and creating the most relevant covers. They scrutinized every image to scout for the most suitable representation of the subject and create an appropriate cover for the book.

The publishing team has been an ardent support to the editorial, designing and production team. Their endless efforts to recruit the best for this project, has resulted in the accomplishment of this book. They are a veteran in the field of academics and their pool of knowledge is as vast as their experience in printing. Their expertise and guidance has proved useful at every step. Their uncompromising quality standards have made this book an exceptional effort. Their encouragement from time to time has been an inspiration for everyone.

The publisher and the editorial board hope that this book will prove to be a valuable piece of knowledge for researchers, students, practitioners and scholars across the globe.

List of Contributors

Mohamed Mahmoud Rowezek
Biology Department, College of Science, Jouf University, Sakaka, Saudi Arabia

Marwa Radawy Marghany and Mohamed Gabr Sheded
Botany Department, Faculty of Science, Aswan University, Aswan 81528, Egypt

Ibrahim Bayoumi Abdel-Farid
Biology Department, College of Science, Jouf University, Sakaka, Saudi Arabia
Botany Department, Faculty of Science, Aswan University, Aswan 81528, Egypt

Ginés Otálora, María Carmen Piñero, Jacinta Collado-González, Josefa López-Marín and Francisco M. del Amor
Department of Crop Production and Agri-Technology, Murcia Institute of Agri-Food Research and Development (IMIDA), C/Mayor s/n, 30150 Murcia, Spain

Eloy Navarro-León, Santiago Atero-Calvo, Juan Manuel Ruiz and Begoña Blasco
Department of Plant Physiology, Faculty of Sciences, University of Granada, 18071 Granada, Spain

Francisco Javier López-Moreno
IFAPA, Institute of Research and Training in Agriculture and Fisheries, 18004 Granada, Spain

Alfonso Albacete
Department of Plant Nutrition, CEBAS-CSIC, Campus Universitario de Espinardo, 30100 Murcia, Spain
Department of Plant Production and Agrotechnology, Institute for Agri-Food Research and Development of Murcia (IMIDA), C/Mayor s/n, 30150 La Alberca, Murcia, Spain

Tussipkan Dilnur, Zhen Peng, Zhaoe Pan, Koffi Kibalou Palanga, Yinhua Jia, Wenfang Gong and Xiongming Du
State Key Laboratory of Cotton Biology, Institute of Cotton Research, Chinese Academy of Agricultural Sciences, Anyang 455000, China

Kasavajhala V. S. K. Prasad and Anireddy S. N. Reddy
Department of Biology and Cell and Molecular Biology Program, Colorado State University, Fort Collins, CO 80523, USA

Denghui Xing
Department of Biology and Cell and Molecular Biology Program, Colorado State University, Fort Collins, CO 80523, USA
Genomics Core Lab, Division of Biological Sciences, University of Montana, Missoula, MT 59812, USA

Katarzyna Patrycja Szymańska, Lidia Polkowska-Kowalczyk, Małgorzata Lichocka, Justyna Maszkowska and Grażyna Dobrowolska
Institute of Biochemistry and Biophysics, Polish Academy of Sciences, Pawińskiego 5a, 02-106 Warsaw, Poland

Sajid Hussain, Chunquan Zhu, Zhigang Bai, Jie Huang, Lianfeng Zhu, Xiaochuang Cao, Satyabrata Nanda, Aamir Riaz, Qingduo Liang, Liping Wang, Yefeng Li, Qianyu Jin and Junhua Zhang
State Key Laboratory of Rice Biology, China National Rice Research Institute, Hangzhou 310006, Zhejiang, China

Saddam Hussain
Department of Agronomy, University of Agriculture Faisalabad, Punjab 38000, Pakistan

Chuthamas Boonchai
Center of Excellence in Environment and Plant Physiology, Department of Botany, Faculty of Science, Chulalongkorn University, Bangkok 10330, Thailand

Thanikarn Udomchalothorn
Center of Excellence in Environment and Plant Physiology, Department of Botany, Faculty of Science, Chulalongkorn University, Bangkok 10330, Thailand
Surawiwat School, Suranaree University of Technology, Nakhon Ratchasima 30000, Thailand

Siriporn Sripinyowanich
Faculty of Liberal Arts and Science, Kasetsart University, Kamphaeng Saen Campus, Nakhon Pathom 73140, Thailand

Luca Comai
Department of Plant Biology and Genome Center, University of California Davis, Davis, CA 95616, USA

Teerapong Buaboocha
Department of Biochemistry, Faculty of Science, Chulalongkorn University, Bangkok 10330, Thailand
Omics Science Center, Faculty of Science, Chulalongkorn University, Bangkok 10330, Thailand

Supachitra Chadchawan
Center of Excellence in Environment and Plant Physiology, Department of Botany, Faculty of Science, Chulalongkorn University, Bangkok 10330, Thailand
Omics Science Center, Faculty of Science, Chulalongkorn University, Bangkok 10330, Thailand

Haoshuang Zhan, Xiaojun Nie, Ting Zhang, Shuang Li, Xiaoyu Wang, Xianghong Du and Wei Tong
State Key Laboratory of Crop Stress Biology in Arid Areas, College of Agronomy and Yangling Branch of China Wheat Improvement Center, Northwest A&F University, Yangling 712100, China

Weining Song
State Key Laboratory of Crop Stress Biology in Arid Areas, College of Agronomy and Yangling Branch of China Wheat Improvement Center, Northwest A&F University, Yangling 712100, China
ICARDA-NWSUAF Joint Research Center for Agriculture Research in Arid Areas, Yangling 712100, China

Chunli Ji, Xue Mao, Jingyun Hao, Xiaodan Wang, Jinai Xue, Hongli Cui and Runzhi Li
Institute of Molecular Agriculture and Bioenergy, Shanxi Agricultural University, Taigu 030801, China

Cuihua Chen, Chengcheng Wang, Zixiu Liu, Lisi Zou, Jingjing Shi, Shuyu Chen, Jiali Chen and Mengxia Tan
College of Pharmacy, Nanjing University of Chinese Medicine, Nanjing 210023, China

Xunhong Liu
College of Pharmacy, Nanjing University of Chinese Medicine, Nanjing 210023, China
Collaborative Innovation Center of Chinese Medicinal Resources Industrialization, Nanjing 210023, China
National and Local Collaborative Engineering Center of Chinese Medicinal Resources Industrialization and Formulae Innovative Medicine, Nanjing 210023, China

Min Zhong, Yu Wang, Yuemei Zhang and Sheng Shu
Key Laboratory of Southern Vegetable Crop Genetic Improvement, Ministry of Agriculture, College of Horticulture, Nanjing Agricultural University, Nanjing 210095, China

Jin Sun and Shirong Guo
Key Laboratory of Southern Vegetable Crop Genetic Improvement, Ministry of Agriculture, College of Horticulture, Nanjing Agricultural University, Nanjing 210095, China
Suqian Academy of Protected Horticulture, Nanjing Agricultural University, Suqian 223800, China

Le Guan, Muhammad Salman Haider, Nadeem Khan, Maazullah Nasim, Muhammad Fiaz, Xudong Zhu, Kekun Zhang and Jinggui Fang
College of Horticulture, Nanjing Agricultural University, Nanjing 210095, China

Songtao Jiu
Department of Plant Science, School of Agriculture and Biology, Shanghai Jiao Tong University, Shanghai 200240, China

Index

Printed in the USA
CPSIA information can be obtained
at www.ICGtesting.com
JSHW062020261223
54313JS00023B/35